A PRACTICAL GUIDE TO GAS ANALYSIS BY GAS CHROMATOGRAPHY

A PRACTICAL GUIDE TO GAS ANALYSIS BY GAS CHROMATOGRAPHY

Piet de Coning

John Swinley

ELSEVIER

Elsevier
Radarweg 29, PO Box 211, 1000 AE Amsterdam, Netherlands
The Boulevard, Langford Lane, Kidlington, Oxford OX5 1GB, United Kingdom
50 Hampshire Street, 5th Floor, Cambridge, MA 02139, United States

Library of Congress Cataloging-in-Publication Data
A catalog record for this book is available from the Library of Congress

British Library Cataloguing-in-Publication Data
A catalogue record for this book is available from the British Library

ISBN: 978-0-12-818888-0

For information on all Elsevier publications visit our website at https://www.elsevier.com/books-and-journals

Publisher: Susan Dennis
Acquisition Editor: Kathryn Eryilmaz
Editorial Project Manager: Redding Morse
Production Project Manager: Nilesh Kumar Shah
Cover Designer: Mark Rogers

Typeset by TNQ Technologies

Contents

Visual glossary

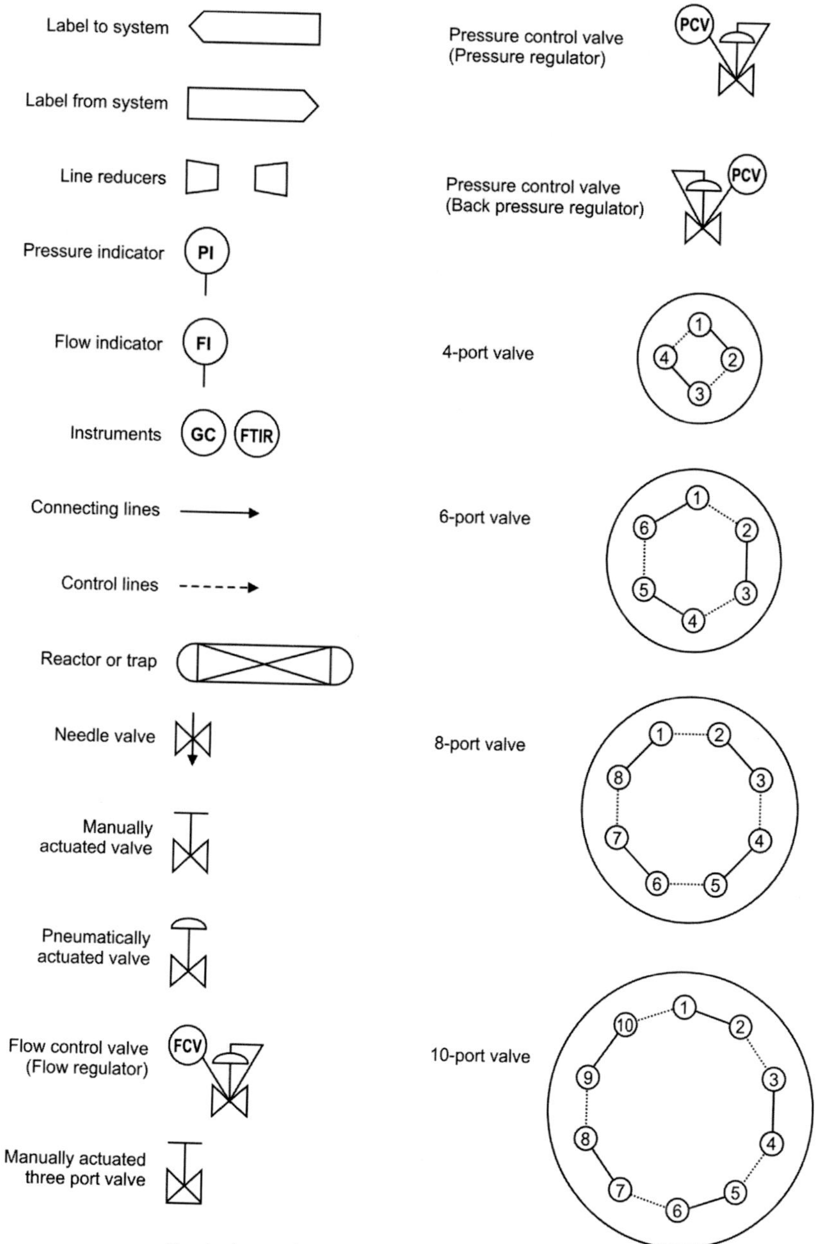

Symbols used in plumbing diagrams in this book.

Introduction

We have presented many courses in capillary gas chromatography, gas chromatography mass spectrometry and gas analysis by gas chromatography for over 20 years. While there are many good books and even more excellent references on gas chromatography on the Internet, the readily available information of gas analysis using gas chromatography is often difficult to locate. What is available is usually fragmented and spread over many sites and blogs. For this reason many people encouraged us to write this book, encompassing our hands-on experience and based on the content of our gas analysis course, allowing the essentials to be summarised in a single source.

When faced with this task, the most difficult challenges are what to include, what to leave out and how much detail is necessary to explain the subject. Starting off with GC theory we based this on course material that was given to us many years ago, with permission to use it for our own courses, by Ben Baars, then of Chrompack and later Varian. Over time we expanded on this and added more of what is specific to gas analysis. However, in the course of two lifetimes involved with gas chromatography the original sources of much of the material is simply not known any more. There are many excellent references and on-line resources on chromatography theory, gas laws and specific applications that can be consulted if the summary we provide is not sufficient.

Defining gas analysis can be rather complicated, after all the word 'gas' comes from the Dutch word, 'gaos', meaning chaos. One could be tempted to define gases as those compounds that are gases at atmospheric temperature and pressure, but then samples of these often contain small amounts of compounds that are liquids under those conditions. Examples would be water vapour and VOC's in air and also higher molecular weight hydrocarbons in natural gas. On the other hand there are gases that liquefy under their own vapour pressure such as butane and liquid petroleum gas. One could define gas analysis as the analysis of compounds having boiling points of less than a certain temperature but even then some compounds will be excluded that need to be analysed. In the end we rather loosely limited our discussion to the GC analysis of compounds and matrices that are best injected directly using valves rather than an auto-sampler using syringes to inject liquid samples.

In the text we gave preference to SI units but in practise, some imperial sizes are so common we stuck with them rather than converting them to SI – for example the tubing diameters and fittings for them. We used the double quote character " as an abbreviation for 'inch' or 'inches'. Where trade names and specific suppliers or manufacturers are mentioned or referenced it does not necessarily mean that we endorse those – it is simply the specific product we used. There may be many other similar products that are equally suitable.

The purpose is to provide an introduction and overview for the novice gas analyst and some, we hope, useful tips for the experienced gas analysis expert. We trust that by sharing our experience in gas analysis it will help some analysts to avoid a few of the pitfalls that we have encountered.

It is not possible to give a comprehensive list of things NOT to do in the analysis of gases although we have tried to point out potential hazards but we cannot be held responsible for any damage to equipment or personnel using any of the information in this book.

We gratefully acknowledge the permission given for the use of figures from various publications as referenced in the text. We also thank Sasol for permission to publish and use photographs taken in one of their laboratories. We are deeply indebted to our wives Eleanor Swinley and Freda de Coning for their support and assistance without which we would not have been able to complete this book. All the graphics is the book were painstakingly drawn by Freda to whom we owe special thanks.

We also acknowledge the cooperation we have had from Chemetrix in presenting courses and from Scientific Supply Services in supplying lots of support information. To everyone else who has helped to make this project a success we extend our sincere thanks.

CHAPTER 1

Overview and theory

Contents

A Practical Guide to Gas Analysis by Gas Chromatography
ISBN 978-0-12-818888-0
https://doi.org/10.1016/B978-0-12-818888-0.00001-2

This chapter has three sections all of which is covered extensively in various books, articles and on-line sites on the Internet. There are many excellent books on the history of chromatography and a few are listed as references to this chapter. Similarly, much more detail on the theory of chromatography and gas laws is available in textbooks as well as the Internet. Again we mention a few books but there are many more.

1.1 Historical overview

It is always prudent to look back at the beginning and the developments that led to what we have today. Although there is always much to learn from history, it falls outside the scope of this book. However a very brief and selective overview would probably be acceptable.

1.1.1 Early days

Probably the earliest separations resembling chromatography were performed by dye chemists who separated mixtures of dyes by dipping the ends of string, filter paper or cloth into the dye and allowing the dye to separate into bands as it moves up by capillary action. The principle of gas-solid adsorption was used as early as 1905 by W. Ramsey to separate gases on activated charcoal and can be seen as a forerunner of gas-solid chromatography [1]. It is, however, generally accepted that the first real chromatography was performed by the 30 year old Russian specialist in botany, biochemistry, and physiology, Mikhail Semenovich Tswett. He started experimenting with packing solid powders such as calcium carbonate into a vertical glass column, applying a small amount of plant pigments to the top and then washing it through the column with a solvent. This caused the compounds to separate into coloured bands and he gave the process the name chromatography, probably from the Greek words *chroma* and *graphein* meaning colour writing. Column chromatography is still practiced in much the same way today although colour is not used to detect the separated compounds. He presented a lecture entitled 'On a New Category of Adsorption Phenomena and its Application in Biochemical Analysis' at a meeting of the Warsaw Society of Natural Scientists on 21 March 1903.

This is a special date in the history of chromatography as it is generally accepted as the birth date of a technique that would become the leading separation technique of the 20th century.

Following on the work of Tswett was the use of chromatography by Leroy Sheldon Palmer who used the technique to isolate pigments, mainly carotenoids, in dairy products and biological tissues. His PhD thesis was published in 1913 and he published a book in 1922 on the separation of carotenoids. In 1931 Richard Kuhn from Germany and his student Edgar Lederer, from France, used the same technique to separate biological molecules. It took another ten years for the next major development. While studying the amino acid composition of wool using counter current liquid-liquid separation two British chemists, Archer J.P. Martin and Richard L.M. Synge, hit on the idea to coat one of the phases on a solid support. Using silica gel as support and tightly bonded water as the one phase held stationary by the support and percolating chloroform, the mobile phase, through it they could obtain remarkable separations. This innovative idea of partitioning the compounds to be separated between two liquid phases became known as partition chromatography. They also indicated that the moving phase might as well be a gas but did not pursue this any further at the time. They received the Nobel Prize for their work in 1956.

They experienced problems with reproducibility using silica gel and continued their work using paper chromatography which was much more repeatable and had become very popular during the 1940's for the separation of amino acids and other biological compounds. Nikolay A. Izmaylov and Maria S. Shrayber, two Soviet scientists, coated the stationary phase on glass plates and processed it similar to paper chromatography to obtain excellent separations.

This work was reported in 1938 but the technique of thin-layer chromatography only became popular after 1956 when the German scientist Egon Stahl started working on it.

1.1.2 Gas chromatography

The first gas-solid chromatography was performed by the Austrian scientist, Erika Cremer in 1944 using solid stationary phases for the separation of acetylene hydrogenation products. One of her students, Fritz Prior, did his PhD on gas chromatography, presented its qualitative abilities in 1947 and they published the work in 1951. Another student, Roland Müller, continued this work by investigating the qualitative and quantitative aspects and used the technique of determining the peak area by using the width of the peak at half the peak height that is described in Chapter 9 [2].

During the same time period A.J.P. Martin and another British chemist, Anthony T. James coated particles with non-volatile liquids, packed them into thin tubes and used compressed gas to move the compounds to be separated through the columns and they are therefore credited with the invention of gas-liquid chromatography. Due to much higher diffusivities of solutes in gases compared with liquids, the equilibrium processes involved in the chromatographic process would be much faster so that the columns would be much more efficient and separation times much shorter. The merits of this technique were rapidly realised by petrochemical companies and the laboratories of British Petroleum and Shell Oil as well as many instrument companies actively pursued this technology.

The first gas chromatographs were rather clumsy instruments that had large diameter packed columns and insensitive detectors. No commercial instruments were available and the pioneers of this technique designed and built their own instruments. Chromatograms were plotted on graph paper from retention time measurements made with a stopwatch and detector responses read from galvanometers. Later chromatograms were plotted on chart recorders and peak areas measured by cutting out the paper and weighing the piece of chart. A flurry of fundamental research and instrument development followed and established the gas chromatography technique firmly as one of the most versatile tools available to analytical scientists.

The Swiss mathematician and physicist Marcel J. E. Golay who at the time worked as a consultant for the Perkin-Elmer Corporation did a theoretical study of gas chromatography columns. He came to the conclusion that very long columns with small diameters of which the inner surface is coated with a thin layer of non-volatile liquid would be capable of better separations than could be obtained with the packed columns. Most of the columns used today are so-called open tubular columns also known as capillary columns and were earlier known as Golay columns. These columns are so efficient that it is not uncommon to separate several hundred compounds in a suitable sample within an hour or less. Not surprisingly, these columns revolutionised gas chromatography.

One of the major limitations in introducing capillary columns was that there were no detectors sensitive enough to detect the extremely small amounts of compounds exiting the column. Fortunately, at about this same time, detectors with extremely low limits of detection became available, which were compatible with the small sample sizes required by these new columns. The most popular of these is the flame ionisation detector or FID.

The Flame Ionisation detector was independently developed at the Central Research Laboratories of the Imperial Chemical Industries of Australia and New Zealand (ICIANZ) and the department of Physical Chemistry of the University of Pretoria in South Africa in 1957. The ICIANZ group of I. G MacWilliam and R.A. Dewar followed their work to completion and are therefore generally accepted as the inventors of the FID since the Pretoria group of J. Harley, W. Nel and V. Pretorius did not pursue their original work at that time. The ease of use, high sensitivity, large linear range and predictable response of the FID has made it into the most widely used GC detector today.

When discussing the theory of chromatography it is impossible not to mention the so-called Van Deemter equation. This famous equation describes the parameters that affect column efficiency and provides the optimum flow rate at which optimum efficiency can be attained from a specific column. It was the result of work done by Dr J.J. van Deemter and his colleagues A. Klinkenberg and F.J. Zuiderweg at Shell Research, Amsterdam, Netherlands in offering a theory that described the dispersion in packed chromatography columns. It was validated and presented by A.I.M. Keulemans and A. Kwantes in 1956 at the First International Symposium on Gas Chromatography. The Van Deemter equation applies equally well to packed and capillary GC columns as well as to liquid chromatography and remains the cornerstone of chromatographic theory. Chromatography is one of the few sciences where practical developments have been predicted by theoretical considerations.

1.1.3 Other forms of chromatography

Although this book deals with gas chromatography it is probably prudent to briefly acknowledge the pioneers of other forms of chromatography as well.

Per Flodin and Jerker Parker developed cellulose based polymers with pores of a consistent size in 1959. These proved ideal for separating very large molecules such as proteins which would diffuse into and out of the pores with the different sizes of molecules spending more or less time in the pores thus allowing them to be separated based on size. This was known as gel permeation chromatography and is most often called size exclusion chromatography. Size exclusion chromatography using Molecular sieves is quite prevalent in the separation of gases by gas chromatography and will be discussed in some detail later in this book.

An American scientist, J. Calvin Giddings used the gas chromatography theory to propose the conditions that would give similar separations using liquid chromatography. In 1964 he demonstrated this using small diameter columns packed with adsorptive solids and thus the technique of high performance liquid chromatography (HPLC) was born. Obviously pumps that could deliver a steady high pressure solvent flow and detectors capable of detecting the small amounts of compounds exiting the column had to be developed. Liquid phases were initially not possible as they were washed off by the solvents used as mobile phase. The German István Halász succeeded in bonding liquids as a thin film to the stationary phase and this turned HPLC into an exceptionally versatile analytical tool.

Ion exchange chromatography using zeolites as adsorbents, which are fancy types of rocks, was first reported in 1938 by T.I. Taylor and Harold C. Urey. During the Manhattan project this method was used to separate the fission products of uranium and other elements produced by the explosions. It has since evolved and is widely used in chemistry, biochemistry, clinical and water analysis. Modern ion exchange resins have made ion exchange chromatography into a very useful tool for the separation of both organic and inorganic ionic species.

The gap between GC and HPLC was bridged in 1962 when Ernst Klesper and co-workers separated porphyrins using a dense gas or supercritical fluid. The British scientist James Lovelock had predicted in 1958 that gases at high pressure could be used as mobile phases since they are capable of dissolving some compounds. Supercritical fluid chromatography uses supercritical fluids such as carbon dioxide at pressures of typically 40 MPa (400 atm) although Giddings and others have used pressures as high as 200 MPa (2000 atm) to separate polymers, sugars, carotenoids and many more species. Pressure programming is used in SFC similar to the way temperature programming is used in GC.

Pedro Cautrecasas and his colleagues described affinity chromatography in 1968. This technique provides the ultimate selectivity as it uses a substrate bound to the solid phase onto which biomolecules such as specific enzymes will attach themselves while other compounds are washed out of the column. The captured enzymes are then released by changing the mobile phase composition.

This type of 'digital chromatography' in which compounds either move or do not move as opposed to the continuous elution in normal chromatography is also used to 'extract' specific classes of compounds from complex matrices in solid phase extraction.

There are many others who made great contributions to chromatography and who probably deserve to have their names mentioned but since this book is not about the history of chromatography, we will stop here. It is also interesting to see that major contributions were made from a vast variety of countries. There are many excellent references on the history of chromatography and it is quite fascinating to read. An internet search will reveal many of these.

It is no exaggeration that the contribution of chromatography to the successful development of chemistry, as well as biosciences, just cannot be overestimated.

1.1.4 Why use chromatography?

The answer lies in the need for separation. To selectively determine the components of a sample that contains a mixture of chemical compounds it is vital to separate the compounds in order to determine them qualitatively or quantitatively. Selectivity can sometimes be obtained by using a method of detection that will measure a physical or chemical property of the compounds of interest to the exclusion of all other components that may be present in the sample. The other way is to physically separate the sample components prior to detection with a less selective detector and this is what is often done in chromatography.

It is clear that samples such as plant extracts, crude oil and blood plasma are extremely complex and contain a variety of compounds that simply cannot be individually determined by any single technique. These samples therefore have to be simplified to make them amenable to the different methods of analysis, each selected for its inherent ability to determine certain classes of compounds. Following identification and quantitation of the sample components the combined analyses enable the analyst to characterise the original sample. In many cases the analytes cannot be recovered after analysis as they are destroyed in the process but there are non-destructive methods that allow the analyst to look at different classes of compounds successively for example, when a thermal conductivity detector is used before a flame ionisation detector to selectively determine the permanent and hydrocarbon gases respectively. Chromatography is fundamentally a separation technique and the identification of the peaks obtained can only be done by comparison with suitable standards. The quantitative information obtained from the GC detector does not provide a result in useable units as in the case of a balance or pH meter. Instead the signal obtained must be quantified relative to suitable standards.

It is therefore imperative, as in any chemical investigation, to have an idea of what compounds, or compound classes, must be determined before selecting an analysis technique.

1.2 Chromatography fundamentals

There are a number of important fundamental principles in the understanding of chromatography but even more so in the practice of it. These concepts become second nature to chromatographers and they develop a kind of 'gut feel' for the systems they work with. This is what probably gave birth to the saying that chromatographers use the science of chromatography to practice the art of chromatography.

1.2.1 Chemical separation

Chemical separation requires phase separation. To separate any two chemical compounds they have to be in different phases. For example, we have two compounds, one of which is soluble and the other not, the first is dissolved and separated from the insoluble one by filtration. All chemical separations utilise phase separation in one form or another such as precipitating proteins from a solution or using immiscible solvents to separate polar from non-polar compounds by liquid-liquid extraction. In the latter case the two liquids are shaken together for some time to get intimate contact between the phases and to establish equilibrium between the phases prior to letting it stand to separate.

The mechanism of phase contact, equilibrium and separation is also used in chromatography albeit in a slightly different way. In chromatography one phase is held stationary either as a solid or by coating it onto particles or on the inside surface of the column. The other phase is the gas phase which transports the sample to be separated and is allowed to move over the stationary phase. The interaction between the phases will depend primarily on the type of stationary and mobile phases. Obviously the phases must be in contact for long enough to ensure that some interaction with the stationary phase can take place as shown in Fig. 1.1. Therefore the flow rate of the mobile phase cannot be too fast. On the other hand, diffusion will cause the area of maximum concentration to spread out causing broader peaks which means that the flow rate of the mobile phase cannot be too slow either, but a lot more about this later.

There are many different types of chromatography and they are normally characterised by the phases involved. If both phases are liquid, it is

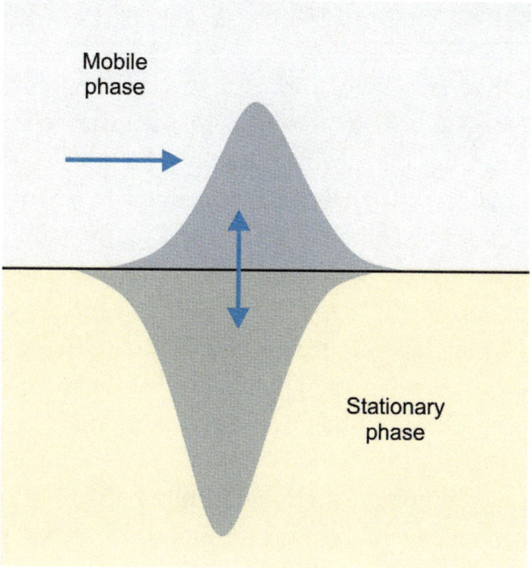

Fig. 1.1 A graphic illustration of the interaction of the sample component with the stationary phase.

called liquid-liquid chromatography and this is the common phase combination used in HPLC and TLC while liquid-solid chromatography, in which the stationary phase is a solid, is used mostly in gel-filtration chromatography for the separation of proteins, peptides and polymers. The molecular interaction mechanism is often added to further characterise the type of chromatography as in Reversed Phase chromatography and Ion Exchange chromatography. Normal Phase chromatography is defined as separation with an inert or non-polar mobile phase such as gas chromatography. However, in Reversed Phase chromatography as is typically used in HPLC, the mobile phase is more polar than the stationary phase and therefore plays a very significant role in the separation. In gas chromatography the stationary phase may be a solid, as is often used in gas analysis, or a high boiling liquid that may be either coated on a powdered support or coated on the inside of the column wall. Gas-liquid chromatography (GLC) is used in arguably more than 90% of all GC applications but it has little application in gas analysis. In gas analysis, gas-solid chromatography (GSC) is more often used although GLC columns are used when higher boiling compounds are entrained in a gaseous stream for example, higher hydrocarbons in liquefied petroleum gas. In some cases it happens that more than one separation mechanism acts simultaneously during the separation.

Today one tends to use the name GC in general and not to differentiate between GLC and GSC.

Two terms that are often confused are adsorption and absorption. Adsorption is temporary and reversible while absorption is permanent and irreversible. In gas liquid chromatography the mechanism of selective adsorption is solvation and analytes are retained by dissolving in the stationary phase. Desorption increases as the oven temperature is increased and since there are no by-ways or turn-offs, what goes in one side must eventually come out on the other side unless it is hijacked or absorbed by active sites with which it reacts irreversibly. In the solvation process the rule 'like dissolve like' applies. This simply means that polar compounds are best separated on polar columns while non-polar compounds are best separated on non-polar columns.

In gas chromatography, the mobile phase is always a gas, usually hydrogen, helium, nitrogen or argon. Since these gases are all relatively inert, the composition of the mobile phase does not contribute significantly to the separation process.

Many compounds can be analysed on two different types of chromatography but for the analysis of gaseous and very volatile samples, gas chromatography is the only suitable technique. For less volatile samples GC is not always suitable and other chromatographic techniques must be used. Prior to embarking on an analysis it is therefore imperative to understand that the sample dictates the technique.

1.2.2 The mobile phase

The gaseous mobile phase in GC moves through the column at a constant average linear velocity. The first principle to lock down is that nothing can move through the column faster than the carrier gas. It is the vehicle that carries the compounds from the injector through the column to the detector. Passengers on a train cannot move faster or slower than the train. The second part of this principle is that everything that moves through the column, when it moves, moves at the same velocity. In other words, all the passengers on the train move at the same average speed.

This might seem to contradict what is observed in the GC results. Compounds are separated and seem to take different times to move through the column. The answer lies not in the mobile phase but in the stationary phase. Different compounds have different affinities for the stationary phase and will therefore spend longer or shorter times in the stationary phase. The compounds are retained by the stationary phase for different lengths of time and this causes the separation. To stretch the train analogy – the only way a passenger can appear to travel slower than the train would be to jump off the front and then try to get back on at the rear end of the train.

The type of mobile phase, or in the case of GC, the carrier gas, will however affect the rate at which chromatography is done. Light gases like hydrogen and helium are typically used at much higher flow velocity than heavier gases such as nitrogen and argon. The reason for this will become clear later.

1.2.3 Linear velocity and volumetric flow

In the previous section we referred to the flow rate and velocity as if it was synonymous but that is not the case. The flow rate is measured in millilitre per minute and is clearly a function of the column volume. For example, a flow rate of 1 mL/min may be perfectly suitable for a column with a diameter of 0.25 mm but it is much too slow in a column with a diameter of 0.53 mm and too fast for a column with a diameter of 0.15 mm.

$$Fc = \frac{\pi r^2 L}{t_0} = \frac{\text{cm}^2 \times \text{cm}}{\text{min}} = \frac{\text{mL}}{\text{min}}$$

where Fc is the volumetric flow, r the radius of the column in cm, L the length of the column in cm and t_0 the time it takes in minutes for a compound to move through the column without any interaction with the stationary phase [3].

The flow in most capillary columns is so low that it cannot be accurately measured by most affordable flow meters. To calculate the flow, the time it takes for one column volume to be replaced must be determined. This is done by injecting a compound that will not interact with the stationary phase at all. As this compound cannot move through the column any faster or slower than the mobile phase; the time it takes for this compound to travel through the column is a very accurate measure of the time it takes to replace the entire column volume. Although it is not always easy to find a compound that has no interaction with the stationary phase, chromatographic conditions may be chosen to approximate zero retention, for example by elevating column temperature.

A more useful parameter than the volumetric flow is the average linear velocity, which is expressed in centimetre per second. Due to the compressibility of the carrier gas the velocity inside the column will vary along its length depending on the local pressure differential. We therefore use the concept 'average linear velocity' and this is defined as $\bar{u}$. The velocity will be slowest at the head of the column and increases towards the end of the column. The linear velocity is independent of the column volume and therefore a better measure of how fast the mobile phase is moving through the column. Since many of the factors that influence separation depend directly on the velocity of the mobile phase this is more significant than flow rate.

$$\bar{u} = \frac{L}{t_0} = \frac{\text{cm}}{\text{s}}$$

where $\bar{u}$ is the linear velocity, L the column length in cm and t_0 the unretained peak time in seconds [3].

Since linear velocity is absolute, it can be reproduced from one laboratory to another irrespective of laboratory temperature, ambient pressure or pressure units. Most of our work has been carried out at ambient pressures around 85 kPa and it could be quite confusing to publish a column head pressure figure as nobody at or near sea level, where the ambient pressure is closer to 100 kPa, would be able to repeat our work.

Devices that can measure the linear velocity inside a column do not exist and we again use the unretained peak time to calculate the average linear velocity. It is clear from the above equation that the linear velocity does not depend on the column diameter, only on the length.

Linear velocity can be converted to volumetric flow and vice versa since both are calculated using the column dimensions and the unretained peak time. In summary the volumetric flow is the time it takes for one column volume to be displaced while the linear velocity is the rate at which the column volume is replaced. For example, we measure an unretained peak time of 2 min and look at the column dimensions. The column volume is $V = \pi r^2 l$ which may give an answer of say 1500 mm^3 or 1.5 mL. This means that the volumetric flow is 1.5/2 = 0.75 mL/min.

Another factor that will influence the velocity of the mobile phase or carrier gas is the temperature as it affects the density of the gas. If we heat a gas in a closed container, it expands; the pressure increases and the container may explode. However if we leave the container open, the pressure will decrease. That is why a hot air balloon goes up as long as the gas inside is kept at a temperature higher than ambient. In a GC column at constant pressure an increase in temperature will cause the resistance to increase and therefore the flow will decrease. Mass flow controllers were used in packed column GC to control a constant flow by causing the column head pressure to increase with temperature increase. These controllers respond rather slowly and are not always able to control the low column flows needed for capillary GC. In modern gas chromatographs this is compensated for by electronic pneumatic controls which ensure a constant linear velocity, even when the column oven temperature is increased from ambient to more than 300 °C.

1.2.4 The stationary phase

Some stationary phases act similar to the flypaper one used to see on farms. It consisted of a strip of sticky paper and if a fly sat on it, it was stuck. If the fly struggled enough it could set itself free. Other phases have surfaces that are not smooth and contain many little crevices and holes. Molecules that enter these pores get stuck there for some time before they can escape and are therefore held back or retarded. Most gas-solid chromatography phases utilise one of these two mechanisms to selectively retain compounds present in the mobile phase and this is the prominent separation mechanism for low boiling compounds such as gases. In gas-liquid chromatography the mechanism is somewhat different and depends on the solubility of the compound in the stationary and mobile phases. The compound to be separated dissolves in the liquid phase, where it can move around until the concentration in the mobile phase is low and then it will escape (evaporate) back into the mobile phase and move on. Regardless of the mechanism, the net effect is that the stationary phase will attempt to hold a compound back while the mobile phase will attempt to move the compound along. This is illustrated in Fig. 1.2.

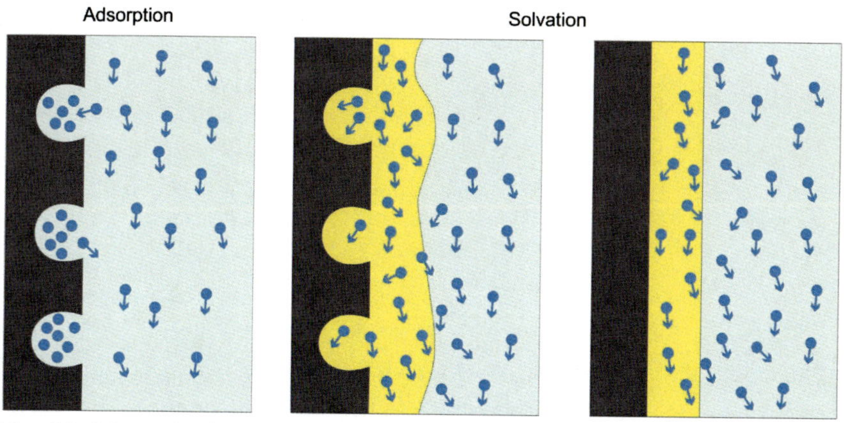

Fig. 1.2 Schematic showing the interaction of the gas phase compound molecules with the different types of stationary phases.

A more detailed discussion of the different stationary phases and their applications is described in Chapter 7 on columns.

1.2.5 Retention time and retention factor

Earlier we have established that it is quite easy to determine the time spent in the mobile phase by injecting something that does not interact with the stationary phase. We called this the unretained peak. The only way in which a compound can seem to move slower than the mobile phase is if it gets out of the mobile phase and spends time in the stationary phase, in other words it gets retained by the stationary phase. The difference in the retention time of that compound peak and the unretained peak, also known as the relative retention time, shown in Fig. 1.3 will tell us how long it spent in the stationary phase, remembering that everything that moves, when it moves, moves at the same velocity as the mobile phase.

The retention time is important for many other reasons as well. It marks the place of a peak in the chromatogram and can be used on its own or sometimes relative to the retention time of a standard, to identify peaks as the retention time will be the same for the same compound under the same conditions.

Retention times are most often measured in minutes or seconds and are influenced by the type of interaction between each component, the stationary phase and the mobile phase, the linear velocity of the mobile phase, the column length and the temperature of the column.

When performing extractions using two immiscible phases the distribution of a solute between the phases is described using the distribution

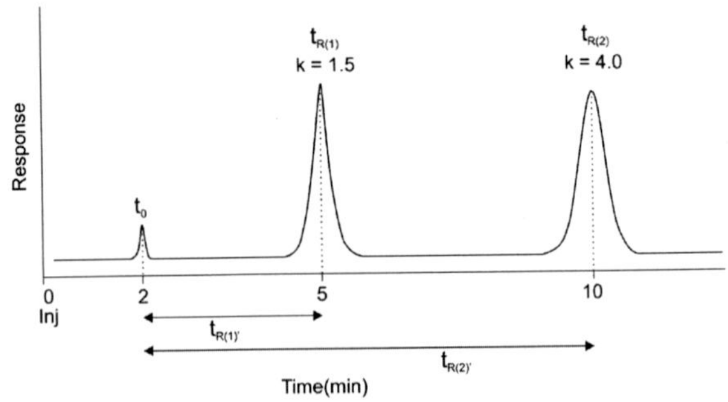

Fig. 1.3 Chromatogram showing the unretained time, t_0, the retention times of the two peaks, $t_{R(1)}$ and $t_{R(2)}$ with their relative retention times $t_{R(1)'}$ and $t_{R(2)'}$. Also shown is the determination of k′ using the unretained peak time and the retention times of the two peaks [3].

constant which is just the concentration in one phase divided by the concentration in the other phase. In chromatography the distribution constant can be defined similarly, but seeing as it is rather difficult, if not impossible, to determine the concentrations in the mobile and stationary phases, we use the time spent in the stationary phase divided by the time spent in the mobile phase to arrive at the so called retention factor, also known as the capacity factor and has the symbol k'.

It is worthwhile to make sure that you fully understand the concept of k'. If you fully comprehend it and how different parameters affect it, understanding and troubleshooting chromatography problems are made much easier. A few examples may help to illustrate the principle referring to Fig. 1.3. If a compound has a t_R of 2 min and the unretained peak time is 1 min, the k' will be 1 which means that the compound spends as much time in the stationary phase as in the mobile phase or half the total retention time in the stationary phase and the rest in the mobile phase. Similarly, a compound with a k' of 4 in a system with a t_0 of 2 min means that the compound spends four times as long in the stationary phase than in the mobile phase. One may be tempted to strive to get k' as large as possible, however if you analyse a sample with only one target compound and it has a k' of say 10 then you are wasting a lot of time. By adjusting the parameters this analysis can be optimised to get the k' to less than 4, which will result in more analyses per time and therefore more productivity, profit (or time off).

Because k' is the ratio of the time spent in the stationary phase divided by the time spent in the mobile phase it has no units and is independent of column length. The value of k' is influenced by the type of interaction between the component, the stationary phase and the mobile phase, the temperature of the column and the amount of stationary phase in the column. It is not dependent on the average linear velocity but is highly temperature dependent and typically could halve or double for a change of column temperature of only 20 degrees.

1.2.6 Resolution

So far a number of concepts have been explained about the interaction between two phases but we have not yet separated anything. Let us therefore return to the phase separation of immiscible liquids — at least we can see it! It is not too difficult to imagine a solvent system where one compound is more water-soluble and another more soluble in an organic phase. If we do successive extractions on both layers, by adding fresh solvent to the water and fresh water to the used solvent, we can effectively

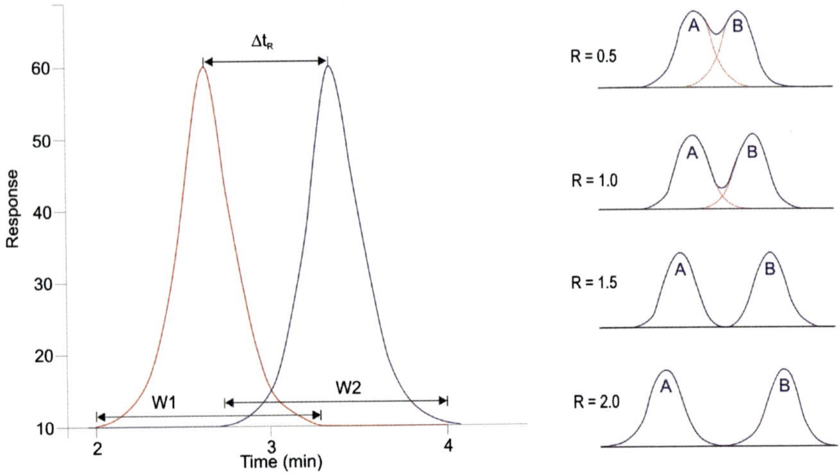

Fig. 1.4 Schematic showing the separation of the peaks at their highest points but with some overlap at their bases (left). The separation of peaks at different resolutions is shown on the right.

separate the two components if we do enough extraction steps. There is actually an apparatus, the so-called Craig counter-current extraction apparatus that can do this, although it is hardly used any more. The concentration distribution of the two compounds, after many steps, may look as in Fig. 1.4.

Clearly we need a way to determine, or calculate how well two compounds are separated and that is done by dividing the time difference between the apexes of the peaks by the average width of the peaks at their baseline level. This quantity is known as the resolution, abbreviated R.

$$R = \frac{2 \times \Delta t_R}{W_1 + W_2}$$

where Δt_R is the difference in retention time and W_1 and W_2 the widths of the two peaks at base [3].

The resolution depends on two factors, namely the distance between the regions of maximum concentration and the width of the region containing each compound. In chromatography terms it is the difference between peak maxima and the average of the peak widths. All parameters are measured in terms of the retention time. As in the discussion of k′ too much resolution is a waste of time and clearly too little resolution means that the compounds are not adequately separated. A resolution of less than

0.5 means that a significant amount of compound B will be present in the first peak and resolution is inadequate while a resolution of 1.5 means that 99.87% of compound A will be in the first peak and only 0.13% of compound B. This is normally considered adequate. A resolution of more than 2 means that the peaks are fully separated and that another peak of equal width can be fitted in between the two peaks. This is known as the Trennzahl number, calculated as $R - 1$, but this concept is not really used any more.

1.2.7 Selectivity, the alpha value

The term selectivity refers to the ability of a chromatographic system to distinguish between two different analytes in the sample being analysed. It is simply the ratio between the retention factors, or k's of the two compounds [3] and can be seen as the difference between the peak maxima of the two analytes in a chromatogram, similar to the Δt_R in Fig. 1.4.

$$\alpha = \frac{k'_2}{k'_1} = \frac{(t_{R_2} - t_0)}{(t_{R_1} - t_0)}$$

Clearly a high alpha value will show that the two analytes are well separated, however since the width of the peaks are not taken into account in calculating the alpha value, it is not the same as resolution. Per definition the alpha value will always be greater than 1 and if alpha is equal to one it means the two peaks are co-eluting. A high alpha value but with poor resolution often means that it is possible to separate the two compounds on the system but that the parameters are not optimised.

1.2.8 The plate theory

In the early years of chromatography, the process was best explained in terms of distillation theory which was well known at the time. As a refresher, distillation is the separation of compounds based on their boiling points and is used in many industries, for example in refineries to separate petrol, diesel and other products from crude oil. In the distillation column vapour travels upwards and condenses on the way, running down the sides of the column in a thin liquid film. Interaction between the upward moving vapour and the downward moving liquid effectively causes continuous counter-current distribution between the two phases. If two compounds with different boiling points are distilled, the mixture with the starting liquid composition will be in equilibrium with the vapour but the relative amounts in the vapour will not be the same as in the liquid. The vapour will contain more of the compound with the lower boiling point. If this vapour is condensed we get a liquid with the same composition as the vapour. However, when we heat this liquid again, the vapour that is formed will again contain more of the lower boiling compound as it evaporates easier than the higher boiling compound. When this vapour is condensed, the liquid will have the same composition as this second vapour. Using this principle the number of steps required for complete separation of the two compounds can be calculated. Each evaporation and re-condensation step can be seen as a theoretical plate and if we know the height of each plate the length of the column that will give complete separation can be calculated. It therefore follows that the lower the plate height, the shorter the column will be and the height of each plate will give us an indication of the column efficiency. These plates are not real physical plates, but virtual or theoretical plates.

In chromatography the number of plates determines the separating power of the column and the height equivalent of each theoretical plate, or HETP, is the measure of column efficiency. The HETP can be calculated from the retention time and the width of the peak under isothermal conditions. Attempting to measure HETP on any peak that has been subjected to temperature changes during analysis will give totally erroneous results. It is important to understand that column efficiency must be measured on a symmetrical peak which has been well retained on the column for example, having a k′ of 5 or more. Calculating HETP on any early eluting peak will only measure how well the system plumbing has been assembled.

Before the HETP can be calculated the number of plates is calculated using the retention time and the width at half height of a late eluting peak measured as in Fig. 1.5.

$$n = 5.54\left(\frac{t_R}{W^{1/2h}}\right)^2$$

Since no interaction with the stationary phase takes place while the solutes are in the mobile phase the relative retention time provides a more accurate value for the number of effective theoretical plates [3].

$$n_{eff} = 5.54\left(\frac{t'_R}{W^{1/2h}}\right)^2$$

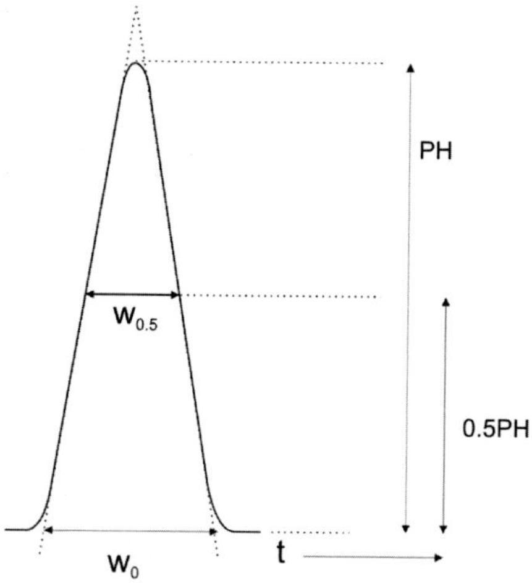

Fig. 1.5 Chromatographic peak showing where the measurement for the calculation of HETP is taken.

Knowing the number of plates and the column length, the number of plates in the column can be calculated although it is usually expressed as the number of plates per metre. The higher the number of plates per metre, the more efficient the column will be [3].

$$\frac{Plates}{m} = \frac{n}{L}$$

where n is the number of effective theoretical plates and L the column length in metre.

The column efficiency is however expressed as HETP which is the inverse of the number of plates per unit length and therefore the lower the HETP, the more efficient the column will be.

$$HETP = \frac{L}{n}$$

where L is the column length in millimetre and n the number of effective theoretical plates [3].

The HETP is influenced by many factors and it is important to state the conditions under which it was measured. Most column manufacturers will include a test report with their columns in which the HETP is given as well as the conditions under which it was determined. Generally the HETP should be less than the internal diameter of the column for all good quality columns. For example a column of 0.25 mm internal diameter should have an HETP of 0.25 mm or less and therefore this column should have at least 4000 plates per metre.

1.2.9 The rate theory

The plate theory is useful to explain the origin and effect of HETP but could not explain the chromatographic process adequately because of several shortcomings. The main problem was that it did not consider the dynamics inside the column; for example, it assumes that equilibrium between the stationary and mobile phases is instantaneous and that no lateral or longitudinal diffusion takes place. These problems were addressed in the rate theory that culminated in the Van Deemter equation.

The easiest way to understand the rate theory is to look at the factors that cause band broadening individually. All these factors act together and it is not really possible to isolate any one of them in practice. Looking at how each factor would affect the column efficiency, or HETP, provides a good basis for understanding how the combination of these factors affect separation and peak shape.

When the sample is injected, it occupies a very narrow zone inside the column but as it moves along the column it will tend to spread and the sharp edges that were there at the beginning will tend to become blurred. Eventually this zone or peak will take on a Gaussian distribution and the width of the broadening peak can be described as a standard deviation. By relating each of the factors that cause band broadening to a standard deviation, the different effects, say x, y and z can be added using statistical principles and the formula $(\sigma_{sum})^2 = \sigma_x^2 + \sigma_y^2 + \sigma_z^2$. The sum of the factors will therefore be larger than any of the individual factors but considerably less than the algebraic sum of their effects.

Earlier we established that both the resolution and the HETP or column efficiency depend on the peak width. In order to get optimal column efficiency and resolution therefore means that we have to endeavour to keep the peaks as narrow as possible. The peak width value is squared in these equations so a 10% improvement in peak width gives a 21% improvement in efficiency. Narrow peaks are also higher and more easily distinguished from baseline noise therefore they are easier to detect at low levels and hence provide better detection limits. More about this will follow later.

1.2.9.1 Longitudinal diffusion

As the name implies broadening of the peaks occur because of diffusion along the axis of the column or in other words along the length of the column. This is similar to the inter-diffusion of two molecular species in a gas stream flowing in an ordinary tube. There is also some diffusion in the stationary phases used in partition chromatography but because the stationary phase layers in modern columns are liquids and very thin, this effect is negligible compared to the diffusion in the mobile phase. Since longitudinal diffusion happens mostly in the mobile phase, it follows that the longer the compound spends in the mobile phase, the more diffusion occurs [4]. We know that all compounds spend the same amount of time in the mobile phase; the same time as the unretained peak (t_0), but different times in the stationary phase. It therefore follows that the slower the linear velocity the more diffusion will occur and conversely that to minimise longitudinal diffusion the linear velocity should be as fast as possible.

As the linear velocity is increased, the peak width reduces so that HETP becomes smaller which is highly desirable. However, if the linear velocity is decreased the increase in HETP is somewhat moderate up to a point where it will deteriorate rapidly as can be seen in Fig. 1.6. From this it should be clear that it is preferable to err on the fast side rather than the slow side when setting the flow rate.

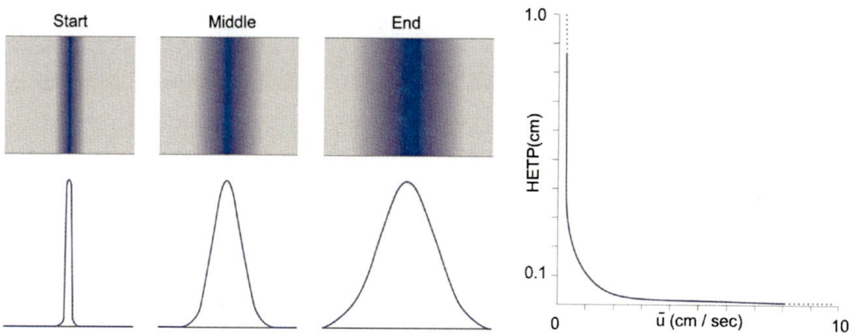

Fig. 1.6 Schematic of the distribution of analytes in a column showing the broadening that occurs due to longitudinal diffusion. The effect of longitudinal diffusion only is clear in the graph of HETP versus linear velocity showing that HETP improves with increased linear velocity due to the decrease in the effect of longitudinal diffusion.

1.2.9.2 Slow equilibration

Ideally the compounds in the mobile phase will cross the boundary between the mobile and stationary phases instantaneously. Unfortunately this does not happen because there is no kinetic control over the rate at which the analyte molecule will arrive at the interface between the two phases and furthermore there is no kinetic control over the rate at which the analyte molecules will cross the interface between the phases. These effects are called 'diffusion-controlled kinetics' and 'sorption-desorption kinetics' respectively.

Because of these two effects the analyte molecules in the stationary phase will lag behind those that are in the mobile phase as shown in Fig. 1.7. In a way they are trying to catch up with the rest but are being prevented from doing so because of their retention by the stationary phase. This happens continuously until the analytes exit the column and results in considerable broadening of the analyte peak. The mobile phase velocity is not the only culprit here; the thickness of the stationary phase will also contribute. If the stationary phase is very thin an analyte molecule at the bottom of the layer can reach the surface quickly and the lag time will be relatively short. The lag time will get significantly longer in thicker stationary phases which will require using a slower mobile phase velocity to allow for the slower equilibration. In adsorption chromatography there is no broadening due to diffusion in the stationary phase but the sorption-desorption effects are much larger in these phases, effectively cancelling this advantage. Similarly, columns with large internal diameters will require a slower mobile phase velocity to allow sufficient time for analyte molecules to diffuse from the centre of the column to the column walls where they can interact with the stationary phase.

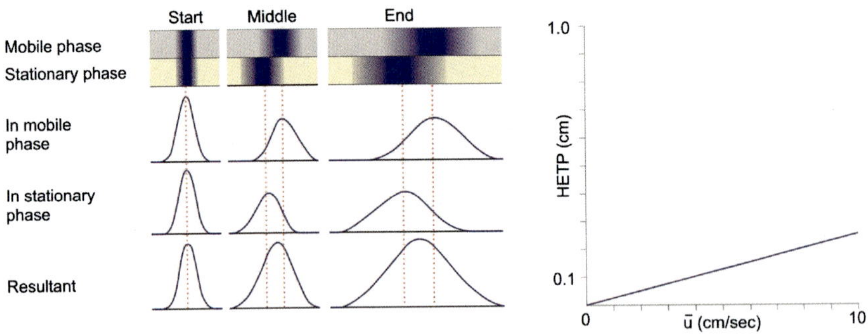

Fig. 1.7 Schematic of the distribution of analytes in a column showing the broadening that occurs due to slow equilibration between the mobile and stationary phases. The graph of HETP versus linear velocity for slow equilibration only shows that HETP improves with decreased linear velocity due to the increased time available for equilibration between the stationary and mobile phases.

In order to minimise the effect of slow equilibration the linear velocity should clearly be as slow as possible which is exactly opposite to the fast linear velocity that is required to minimise longitudinal diffusion. Obviously, the actual optimum velocity will be a compromise between fast enough to minimise the longitudinal diffusion but slow enough to minimise the effect of slow equilibration [4].

1.2.9.3 Flow paths

It is not difficult to imagine that there are countless pathways through the maze that is formed by the particles packed in a column. Similarly, in a liquid stationary phase there are also different paths that can be followed by the analyte molecules in it and in the mobile phase all analyte molecules also do not necessarily follow the straightest and shortest route. Some molecules will travel a slightly shorter distance than others depending on the route they take as illustrated in Fig. 1.8. This will add to the broadening of the narrow zone that was injected. The different paths that analytes can take in the stationary phase are not dependent on the flow of the mobile phase since it is not in contact with the mobile phase. However, the different paths in the mobile phase will be influenced by the velocity of the mobile phase by a rather complex mechanism called the 'coupling theory of eddy diffusion.' What it comes down to is that the faster the mobile phase moves, the bigger the difference will be between molecules that follow the straightest path and those that follow longer paths. Therefore a faster mobile

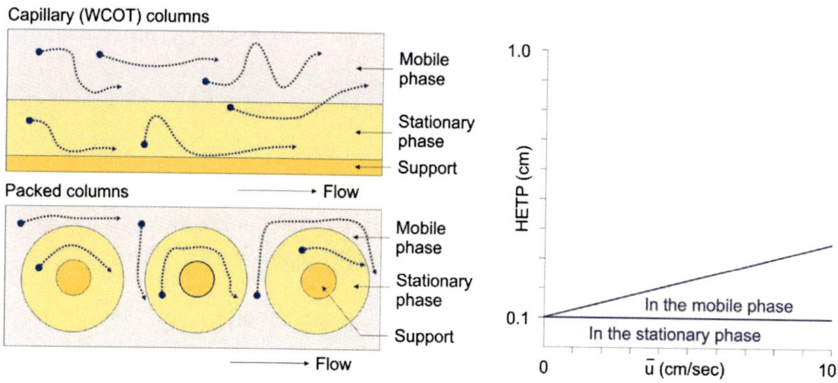

Fig. 1.8 Schematic of the distribution of analytes in capillary (top) and packed (bottom) columns showing the broadening that occurs due to different flow paths in the mobile phase. The graph of HETP versus linear velocity for different flow paths only shows that HETP improves with decreased linear velocity due to the different flow paths in the mobile phase, while the same effect in the stationary phase is constant and not dependent on the mobile phase velocity.

phase velocity will increase the broadening due to different flow paths in the mobile phase while the effect of the different flow paths in the stationary phase will remain constant, regardless of the mobile phase velocity. Once again better column efficiency will be obtained at lower linear velocities.

1.2.9.4 The Van Deemter equation

From the preceding discussions it is clear that the optimum column efficiency will be a compromise between the velocity required to minimise longitudinal diffusion, slow equilibration and the different flow patterns. This is exactly what Van Deemter et al. studied and the result is the famous Van Deemter plot. Fig. 1.9 shows the combination of the plots of column efficiency versus linear velocity with the contribution of the three main effects indicated.

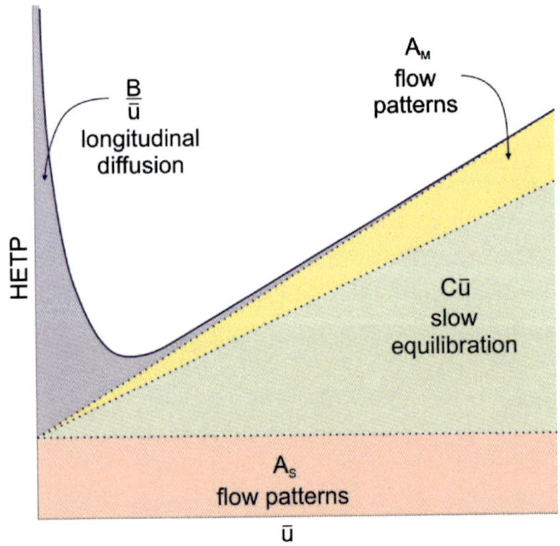

Fig. 1.9 Graph of HETP versus linear velocity showing the combined effect of longitudinal diffusion, slow equilibration and different flow paths to arrive at the Van Deemter curve. It clearly shows that HETP is the lowest and therefore column efficiency at its maximum at a specific narrow range of linear velocities.

The Van Deemter equation relates the variance per unit length of a separation column to the linear mobile phase velocity by considering the physical, kinetic and thermodynamic properties of a separation and looks rather formidable in its expanded form [5].

$$HETP = 2\lambda d_p + \frac{2\gamma D_{gas}}{\overline{u}} + \frac{8k'Df^2}{\pi^2(1+k')^2 D_{liq}}\overline{u}$$

where *HETP* is the column efficiency, λ a constant which is a measure of the packing irregularities, γ a correction factor accounting for the tortuosity of the gas channels in the column, d_p the average particle diameter of the solid support, D_{gas} the diffusivity of the solute in the gas phase, $\bar{u}$ the linear gas velocity, k' the capacity factor given by the ratio of the time spent in the stationary phase divided by the time spent in the mobile phase, Df the effective thickness of the film that is coated on the particles of the support and D_{liq} the diffusivity of the solute in the liquid phase [4]. The effective thickness of the film is squared and hence film thickness is of prime importance when selecting chromatography column dimensions.

Fortunately the equation becomes as simple as A,B,C in practice.

$$HETP = A + \frac{B}{\bar{u}} + C \times \bar{u}$$

where *HETP* is the column efficiency expressed as the height equivalent to a theoretical plate and therefore a measure of the resolving power of the column in cm, *A* the Eddy-diffusion parameter, related to channelling through a non-ideal packing in cm, *B* the diffusion coefficient of the eluting particles in the longitudinal direction, resulting in dispersion in $cm^2\ s^{-1}$, *C* the resistance to mass transfer coefficient of the analyte between mobile and stationary phase in seconds and $\bar{u}$ the linear velocity in cm/s [3].

It is clear from the sum of these effects, as shown in Fig. 1.9, that there is a fairly narrow range in which the combined effects are minimised and where the optimum column efficiency, or lowest HETP, will be obtained. Furthermore it is clear that deviating from the optimum linear velocity has little benefit. Lastly, the curve also shows that if non-optimum flows must be used, a faster linear velocity would be less detrimental to the column efficiency than a linear velocity that is too slow. In practice, deviations from the optimum are purposely used in specific applications for example in process GC's where getting results quickly becomes more important than getting optimum efficiency of separation. At the other extreme, slower than optimum linear velocities are sometimes tolerated in specific gas analysis applications to provide more time for switching between columns and detectors with valves. That will be discussed in much more detail in a later chapter.

Optimum average linear velocity will vary considerably for columns of different physical dimensions. As a starting point one would always set the initial velocity somewhat higher than the expected optimum so as to save analysis time. Much more significant is the variation from the use of different carrier gases that will be discussed in later chapters.

1.2.10 Dead volumes and unswept volumes

Apart from the broadening influenced by the mobile and stationary phases there are a few other factors that can contribute to band, or peak, broadening. Dead volumes refer to areas in the flow path where the sample can expand and become diluted. If such a volume is swept with additional gas added just before the expansion volume, its effect can be reduced. One of the challenges in setting up systems for gas analysis is to ensure that there are no unswept dead volumes in the system that will cause broadening or peak distortion and thereby adversely affect the separation.

1.2.10.1 Feed volume

In discussing the broadening effects due to the mobile phase we assumed that the sample is applied as a very narrow zone to the column. This is not really possible as the sample must be introduced into the carrier gas and will therefore mix with it and become diluted and broader. The major contributing factor to this is the design of the sample inlet which will be discussed in detail later. As shown in Fig. 1.10, the larger the injector volume the broader the sample application will become. This is more serious when performing liquid injections, for example if 3 μL methanol is vaporised at 250 °C and 200 kPa pressure it will occupy 1.6 mL (1600 μL), a relatively enormous volume, probably larger than the volume of the

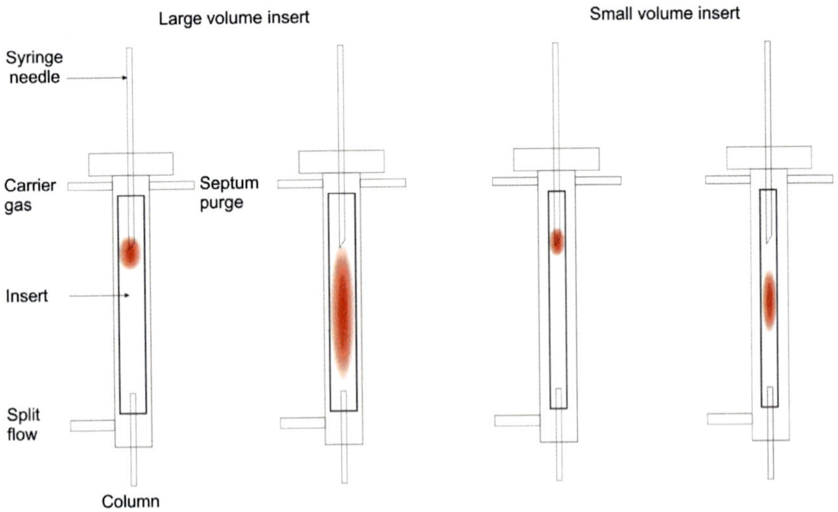

Fig. 1.10 Schematic of two injectors, one with a large internal volume that causes significant expansion of the sample upon vaporisation and therefore broadening the injection band and the other with a smaller internal volume will give much less broadening of the sample prior to injection.

whole column. In the design of modern split-splitless injectors the feed volume is optimised in such a way that it can be used to either dilute concentrated samples or concentrate dilute samples, although the latter is seldom applied in gas analysis. For now it is probably sufficient to realise that having a large feed volume (injector volume) will have an effect similar to injecting a very large volume of a dilute sample.

1.2.10.2 Detector volume

A potential dead volume is the internal volume of the detector or of the detector interface. If the volume of the detector exceeds the volume of mobile phase that contains the analyte to be determined, the analyte peak will be broadened, the concentration will be reduced and become more difficult to detect. It will also act as a mixing chamber that can allow perfectly separated compounds to become mixed again simply because they are allowed to expand into a large volume. In practice, this can occur when the column is not installed the correct distance into the detector body. In most well-designed detectors the laminar flow of the mobile phase is preserved, often by adding a make-up gas and the contribution to peak broadening minimised as in Fig. 1.11. There are however some detectors,

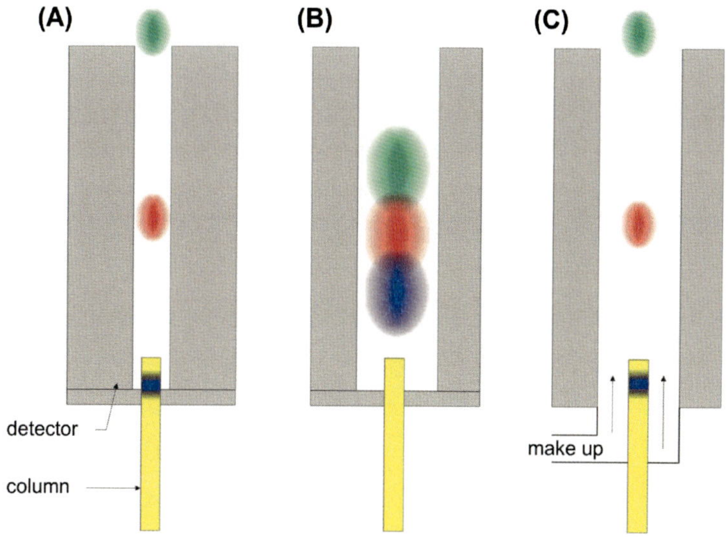

Fig. 1.11 Schematic of detectors with different internal volumes. With a small internal volume detector the compounds remain separated (A) but if the volume is too large the separated components mix with one another after exiting the column as it expands into the detector volume (B). With addition of make-up gas the separation is preserved despite the relatively large detector volume (C).

such as FTIR where a long optical path length is needed for detection, leading to relatively large detector volumes. In such cases large bore capillary or even packed columns must be used to ensure the analyte peaks are both wide enough and concentrated enough to allow for their detection.

1.2.11 Peak symmetry

Ideally all peaks in a chromatogram should be sharp and symmetrical with a perfect Gaussian shape on a flat baseline. This is seldom achieved although it is possible to get close with the correct choice of system components and operating parameters. Despite best efforts all peaks show some tailing which is sometimes only noticeable close to the baseline. To get a numerical value for the lack of symmetry, two methods are commonly used, the asymmetry factor and the tailing factor although there are differences in the way the software from different vendors define and use these terms.

For the calculation of both factors the peak is divided in two by dropping a vertical line from the peak apex to the baseline. For the asymmetry factor a horizontal line is drawn at 10% of the peak height above the baseline and for the tailing factor the line is drawn at 5% of the peak height. The front and back sections of the peak formed by the vertical line can be labelled a and b as in Fig. 1.12.

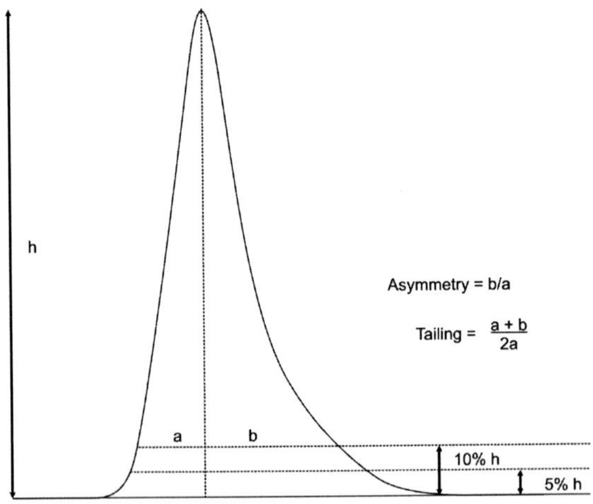

Fig. 1.12 Calculation of the asymmetry and tailing factors.

The asymmetry factor (*AF*) is calculated using the values at 10% of the peak height and is the ratio between the front and back parts of the chromatogram [6].

$$AF_{10\%} = \frac{b}{a}$$

The tailing factor (*TF*) is calculated using the values at 5% peak height and is the ratio between the peaks width and two times the front part of the chromatogram [6].

$$TF_{5\%} = \frac{(a+b)}{2a}$$

Because the two values are calculated at different heights above the baseline there is no direct relationship between them and one cannot be readily converted to the other. However, the values obtained are usually very similar with the asymmetry factor often a little higher than the tailing factor.

1.2.12 Overloading

When a capillary column is overloaded, the stationary phase becomes saturated with the analyte while there are still free analyte molecules that can move ahead. As the free molecules get in touch with unsaturated stationary phase they will be adsorbed but are retained slightly less than the bulk of the analyte. This creates a high concentration zone preceded by areas of lower concentration in the column and when these zones elute, the peak is asymmetrical with the front end sloping more than the back end of the peak. This effect is known as 'front tailing' – a nice contradiction in terms.

When a packed column is overloaded, saturation of the stationary phase also occurs and theoretically the same effect should be seen as in capillary columns. However, there are many more complex flow paths in which the analytes can be retained and therefore equilibrium is never really reached. The net effect of this is that some analyte molecules are left behind the high concentration front and when the peak elutes, it is also asymmetrical, but the tailing is at the back – no contradictions in this case.

The difference in tailing observed in capillary and packed columns is illustrated in Fig. 1.13. Gross overloading with a heavy sample gas matrix in a system using helium or hydrogen as a carrier gas, will temporarily

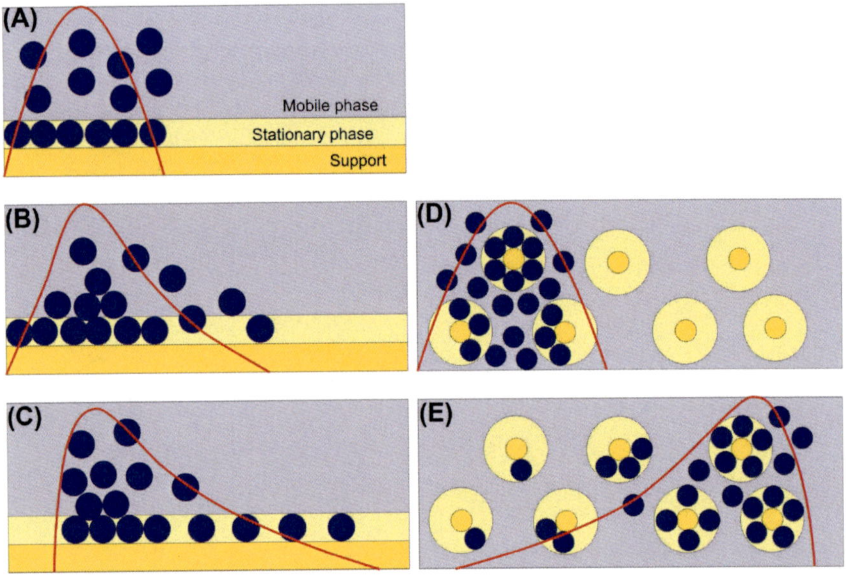

Fig. 1.13 Schematic showing that overloading on surface coated columns results in peak fronting while on porous layer and packed columns, overloading leads to peak tailing.

change the density of the carrier gas and will cause a reduction in retention times compared to standards mixed with helium or hydrogen as matrix. This can be frustrating in systems where gas standards are available in the same matrix as the carrier gas while the sample matrix has a substantially different density. Retention times that were determined with the standard will become shorter when the samples are analysed due to the density difference between the matrices of the standard blend and the sample.

In Fig. 1.13 the way in which tailing due to overload develops in capillary and packed columns is illustrated. The sample is injected (A) on a capillary column and the stationary phase becomes saturated. The sample excess now overtakes the part of the sample that was adsorbed (B) and adsorbs further down the column (C) where the concentration is now less than the bulk of the sample. This results in fronting or front-tailing. The opposite happens in a packed column where all the adsorption sites are initially saturated (D) but as the area of maximum concentration moves ahead a portion of the sample is slow to desorb (E) resulting in tailing.

1.2.13 Active sites

Sometimes peak tailing is observed, particularly with low concentration samples which seems to improve when injecting higher concentrations. In severe cases some analyte peaks can disappear altogether at trace levels while others, present at the same concentrations, remain visible. The culprit in this case is active sites; areas in the column or in the flow path of the sample where irreversible absorption takes place or where the sorption-desorption kinetics is very different from that which is deliberately sought in the column to get separation. At active sites, Fig. 1.14, analytes may be adsorbed for so long that the main analyte zone can pass by it and it only desorbs some time later, resulting in a tailing peak. There does not have to be many of these sites to make the effect noticeable since they act cumulatively. Common sources of active sites are non-deactivated glass or metal surfaces and areas where previous sample components decomposed, leaving traces of active carbon where adsorption can take place. More serious in trace analyses are absorption sites which may irreversibly absorb part or all of some low concentration compounds.

To minimise active sites, the glass surfaces such as the injector insert and the glass wool at the ends of packed columns are silanised to render them inactive. For the analysis of reactive gases all metal surfaces are similarly treated, for example using Silcosteel™ or by passivating the system with a very reactive gas. Passivation means that a compound such as acetic acid or dichloromethylsilane is allowed to absorb onto the active sites and forms such strong bonds that it is never released and therefore renders the active site, for all intents and purposes, inactive. Since deactivation of a column only covers the active sites and does not remove nor replace them with useable stationary phase, most analysts would rather replace the column than attempting to deactivate it.

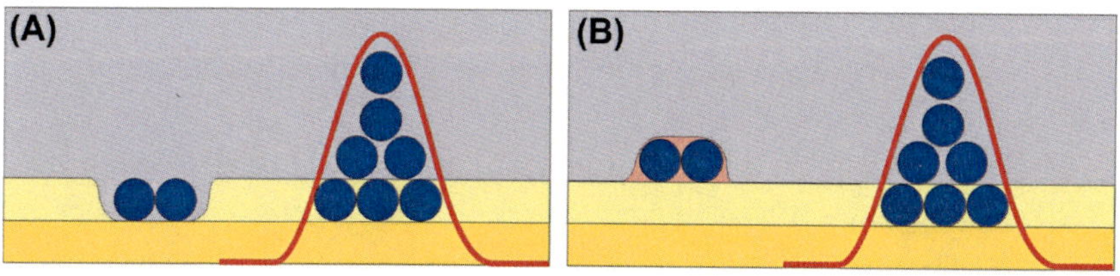

Fig. 1.14 Schematic showing how molecules can be permanently absorbed onto active sites in the column where the stationary phase has either been removed (A) or where carbon or other deposits created active spots (B).

1.2.14 Resolution revisited

Now that column efficiency and the other parameters have been defined, another look at resolution in terms of column parameters can provide valuable insight that will be useful for optimising the separations. Earlier in this chapter we introduced the concept of resolution and defined it in terms of the peak shapes of two closely eluting peaks. Resolution can also be described in terms of the chromatographic factors that affect it.

$$R = \frac{1}{4}\sqrt{\frac{L}{H}} \times \left(\frac{k}{k+1}\right) \times (\alpha - 1)$$

The first term is a measure of efficiency and is affected by the column length (L) and column efficiency (H) which is in turn dependent on the column diameter, carrier gas type and linear velocity. The second term is a measure of retention as defined by the retention factor (k) while the third term refers to the separation factor (α) [7].

What this means in practice is that the resolution will increase linearly with the square root of the number of theoretical plates and the resolution will also increase linearly with increasing k' values when these values are relatively small. At high values of k' the second term approaches 1 and its effect on resolution will become very small. Lastly, the resolution will always increase with increasing alpha values.

1.3 Gas laws

From as long ago as the time of Aristotle matter has been classified into one of three states namely solids, liquid and gases. If something has a fixed volume and shape it is considered a solid while a liquid will also have a fixed volume but will accept the shape of whatever container it is in. Gases have no fixed volume or shape and will accept both the volume and the shape of the container they are in. A detailed discussion of the gas laws and the associated thermodynamics and kinetics is beyond the scope of this book and is dealt with in much detail in subject specific books and articles.

When analysing gases the ultimate aim is generally to determine how much of a certain component gas is present in a sample. Gas samples are greatly affected by temperature and pressure and it is prudent to review the laws governing the behaviour of gases in the way it affects the analysis of gaseous samples and the interpretation of the results.

1.3.1 Measurement units

Preference should always be given to using the SI units. It is after all the international standard and is required by most scientific publications. Apart from that, using SI units for all measurements reduces the possibility of misinterpreting results because of site specific differences in definitions.

1.3.1.1 Units of pressure

Traditionally pressure was measured as the height of a liquid, such as mercury and was expressed as mm or inches of mercury. The modern SI definition of pressure is based on the force exerted per unit area expressed in fundamental SI units as $kg \cdot m \cdot s^{-2}/m^2$ or newton/m^2 or pascal (Pa). The pascal or its multiple the kilopascal (kPa) is the only SI unit of pressure and in practice, the kilopascal is more practical.

All other units of pressure have therefore been redefined in terms of kPa. The atmospheric pressure at sea level, 1 atmosphere (atm) is by definition exactly 101,325 Pa or 101.325 kPa. The torr, named after the inventor of the barometer, Torricelli, is 1/760 of an atmosphere or 13.3355 Pa. The almost identical unit of mmHg, is defined as 13.5951 Pa, by using a fixed density of mercury and a standard gravitational force. The term bar is used for 100 kPa, which is slightly below one standard atmosphere and from it the millibar, equal to one pascal, is derived. There are also some less common units for pressure such as kg/cm^2 which is approximately one atmosphere and in the USA the unit of pounds per square inch (psi) is often used. One atmosphere is approximately 14.696 psi. Since many GC's and pressure measuring equipment are either manufactured in the USA or intended for the US market, the psi is regrettably quite commonplace, but fortunately most modern GC's have the option to select the units of pressure that will be displayed by the instrument. We strongly recommend that users use the SI units and get used to it.

A conversion table for commonly used pressure units is given in Table 1.1. It is also useful to always quote pressure in absolute units and not in the so-called gauge pressure, for example, psig. Gauge pressure is the sum of atmospheric pressure and whatever the gauge displays. Since atmospheric pressure varies depending on the height above sea level and the current weather; it is not an exact measure and has no place in analytical science.

Table 1.1 Commonly used pressure units and their conversion factors.

	1 Pa	1 atm	1 at	1 bar	1 psi
1 Pa	1	0.987×10^{-5}	1.02×10^{-5}	1.0×10^{-5}	1.45×10^{-5}
1 atm	1.013×10^{5}	1	1.03	1.013	14.7
1 at	0.98×10^{5}	0.97	1	0.98	14.3
1 bar	1.0×10^{5}	0.987	1.02	1	14.5
1 psi	6.9×10^{3}	0.068	0.70	0.0689	1

1.3.1.2 Units of volume

The units of volume can conceivably be the cube of any convenient unit of length. The standard SI unit of volume is the cubic decimetre, dm^3, equal to 1 L. Other units are often more convenient in specific applications. For example, it is more convenient to indicate the size of the sample loop on a gas analysis GC in microlitre or in packed column systems in millilitre. In environmental studies the cubic metre, m^3, is the preferred unit of volume. A conversion table for commonly used volume units is given in Table 1.2.

1.3.1.3 Units of temperature

The temperature scale used by Charles was the Reaumur scale (water freezes at 0°Re and boils at 80°Re). Several other scales were also proposed of which the Fahrenheit scale (water freezes at 32 °F and boils at 212 °F) is probably best known and still used in many countries. After adoption of the Celsius scale (water freezes at 0 °C and boils at 100 °C) it was noted that when a graph of volume versus temperature was extrapolated it reaches zero volume at a temperature of -273.15 °C. Using this as the absolute zero the fundamental scale of temperature in the SI, called the absolute, thermodynamic or Kelvin scale was developed. On this scale the unit of temperature is called the Kelvin (not degree Kelvin), it has the same size as the degree Celsius and is abbreviated with the capital letter K. As an example 20 °C is the same as 293 K.

1.3.2 Pressure and Boyle's law

The English chemist Robert Boyle (1627—1691) was the first to perform quantitative gas measurements using a barometer and a manometer to study the pressures and volumes of different gases. This led to his law which is usually stated as: 'At any constant temperature, the product of the pressure and the volume of any size sample of any gas is a constant.' Mathematically this is expressed as $PV = k$ with k being a constant depending on a

Table 1.2 Conversion table for some commonly used volume units.

	Cubic m	Cubic dm (L)	Cubic cm (mL)	Cubic yard	Cubic foot	Cubic inch
Cubic metre	1	1000	1,000,000	1.308	35.315	61,024
Cubic decimetre (dm^3, L)	0.001	1	1000	0.0013	0.035	61.024
Cubic centimetre (mL)	1×10^{-6}	0.001	1	1.307×10^{-6}	3.531×10^{-5}	0.061
Cubic yard	0.765	764.56	764,555	1	27	46,656
Cubic foot	0.028	28.317	28,316.9	0.037	1	1728
Cubic inch	1.639×10^{-5}	0.0164	16.387	2.143×10^{-5}	0.0006	1

particular sample [8]. More useful is the derived equation $P_1V_1 = P_2V_2$ which shows that pressure and the volume vary inversely at fixed temperature. The pressure of the sample increases when the volume of the sample of gas is decreased for example as the plunger of a closed syringe is depressed, the volume becomes less and the pressure inside the syringe will increase.

Boyle's law is applied frequently in gas analysis and more examples will be described in the chapter on sampling. One application is in the sampling from a high pressure sample container where a pipette of known volume (it can be a section of tubing) is connected between the sample container and an expansion volume as in Fig. 1.15. Using valves between the sample container, the pipette and the expansion volume, the pipette is first filled with sample after which it is allowed into the expansion volume. In this way the pressure is reduced and it allows repeated sampling from the same sample container.

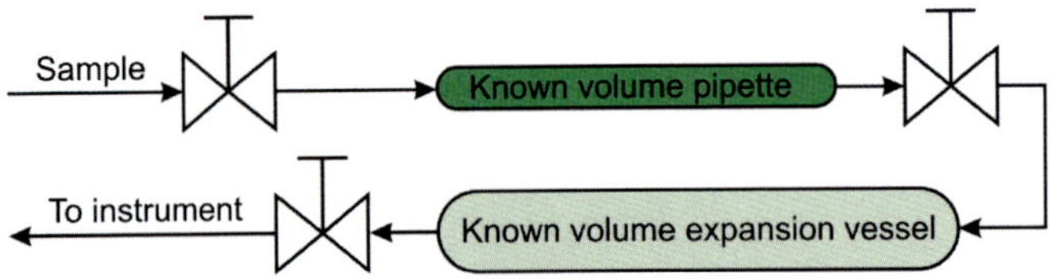

Fig. 1.15 Schematic of a sampling configuration with which a small volume of a high pressure sample is expanded to lower the pressure to a level that can be handled by the instrument.

1.3.3 Temperature and Charles' law

Although Boyle was aware that the volume of a gas changes when the temperature changes he could not establish a relationship between the two as there was no fixed temperature scale at that time. After the development of the temperature scale the French chemist Jacques Charles (1746–1823) developed what became known as Charles' law which states that: 'At any constant pressure, the volume of any sample of any gas is directly proportional to the temperature.' Mathematically this is expressed as $V = k't + k''$ where k' and k'' are constants and t the temperature on any scale [8]. After adoption of the Kelvin temperature scale and establishment of an absolute zero temperature, the k'' factor could be dropped and the equation became $V = kT$ or in its more useful form $V_1/T_1 = V_2/T_2$ where T denotes the absolute temperature in K.

Ignoring the effect of temperature can affect the repeatability of results. In the routine at-line monitoring of a gaseous analyte it was noted that the amount determined seemed to decrease during the day and increase during the night. Investigation showed that the actual amount injected decreased as the ambient temperature increased during the day and increased as the ambient temperature decreased simply because of the variation of volume with temperature. Mounting the injection valve in a temperature controlled oven solved the problem.

1.3.4 Volume and Avogadro's law

While studying chemical reactions that consumed and produced gases, the French chemist Joseph-Louis Gay-Lussac (1778–1850) postulated the law of combining volumes in 1809, which states: 'When measured under the same conditions of temperature and pressure, the volumes of gases which react together are in the ratio of small whole numbers.' For example one volume of chlorine will react with one volume of hydrogen to produce one volume of hydrochloric acid. This was used by the Italian chemist Amedeo Avogadro (1776–1856) in 1811 to postulate his hypothesis which states: 'Equal volumes of gases under the same conditions of temperature and pressure contain equal numbers of molecules.' This was not readily accepted at the time since there was considerable confusion regarding some gases, for example, most elemental gases such as nitrogen and oxygen were assumed to be monoatomic while we now know that they are actually diatomic. Eventually, after about 40 years this hypothesis became accepted and established that the volume of a gas is directly proportional to the number of atoms or molecules it contains. In mathematical terms this becomes $V = kn$ with k a constant and n the number of moles of the gas [8]. One mole contains 6.022×10^{23} molecules. This is an important law when it is necessary to calculate detector sensitivity in terms of response per number of molecules as is sometimes used in the manufacturer's specifications.

Most useful to realise from this relationship is that when a gas standard is certified as mole per mole of the components in a gaseous matrix this is the same as if it was certified as volume per volume. For example, if a component is present at 5 mol percent it is the same as 5 vol percent or similarly 5 ppm (mole) is the same as 5 ppm (volume).

1.3.5 Partial pressures and Dalton's law

Dalton formulated his law of partial pressures in 1803 following his study of the atomic-molecular theory of matter. This law states that: 'For a mixture of gases in any container, the total pressure exerted is the sum of the pressures that each gas would exert if it were alone.' In mathematical terms this is described as $p = p_1 + p_2 + p_3 + \ ...$ which means that the total pressure of air in a room would be made up of the sum of the partial pressures of nitrogen, oxygen, argon, carbon dioxide and the other gases present in the air [8].

This can be used to prepare calibration standards by filling a previously evacuated container to a certain pressure with a gas, then adding another gas to a second, higher pressure and so on until a specific partial pressure of each component is added and then finally topping it up to the desired pressure by adding the matrix gas. The amount of each component can be calculated using the partial pressures of each added component.

Another useful relationship stems from the fact that at ambient conditions, air behaves much like an ideal gas and the partial pressures of the component gases are directly proportional to the number of moles in a volume of air. This means that at a pressure of 100 kPa the partial pressures of the component gases of air are numerically equal to the mole percent of the component gases. Thus the partial pressures of the major components of dry air at 100 kPa are nitrogen, 78 kPa; oxygen, 21 kPa; argon, 0.9 kPa; and carbon dioxide, 0.03 kPa. The relative amounts of the 14 major component gases of air is given in Table 1.3.

Table 1.3 Composition of dry air at sea level in volume per volume.

	Gas	Volume
1	Nitrogen (N_2)	780,840 ppm (78.084%)
2	Oxygen (O_2)	209,478 ppm (20.9478%)
3	Argon (Ar)	9340 ppm (0.9340%)
4	Carbon dioxide (CO_2)	314 ppm (0.0314%)
5	Neon (Ne)	18.18 ppm (0.001818%)
6	Helium (He)	5.24 ppm (0.000524%)
7	Methane (CH_4)	1.79 ppm (0.000179%)
8	Krypton (Kr)	1.14 ppm (0.000114%)
9	Hydrogen (H_2)	0.55 ppm (0.000055%)
10	Nitrous oxide (N_2O)	0.325 ppm (0.0000325%)
11	Carbon monoxide (CO)	0.1 ppm (0.00001%)
12	Xenon (Xe)	0.09 ppm (9×10^{-6}%)
13	Ozone (O_3)	0.0–0.07 ppm ($0-7 \times 10^{-6}$%)
14	Nitrogen dioxide (NO_2)	0.02 ppm (2×10^{-6}%)

Data from various sources and different years. Actual amounts may differ slightly and are given as illustration and exclude variable amounts of water averaging about 1%.

1.3.6 The ideal gas law

The well-known ideal gas law combines the laws of Boyle and Charles as well as the hypothesis of Avogadro into one law which states that: 'The product of the pressure (p) and volume (V) is equal to the product of the number of moles (n), the universal gas constant (R) and the temperature (T), written as $pV = nRT$' [8]. The constant R is called the universal gas constant (Boltzmann constant) and is a combination of the constants in the equations from which the law was derived. The unit for R is $J \cdot mol^{-1} \cdot K^{-1}$ which is the same as $kPa \cdot L \cdot mol^{-1} \cdot K^{-1}$ or $kPa \cdot dm^3 \cdot mol^{-1} \cdot K^{-1}$ or $Pa \cdot m^3 \cdot mol^{-1} \cdot K^{-1}$ and R has the numerical value of 8.3143510 in any of the above units.

If different units of pressure are used, R will have a different value such as $0.08206\ L \cdot atm \cdot mol^{-1} \cdot K^{-1}$ which may be more useful in some cases. The difference is simply because one atmosphere is by definition exactly 101,325 Pa (101.325 kPa). One litre (L) is equal to 1000 mL or 1/1000 of a cubic metre which allows the values to be used for large and small gas volumes.

Rearranging the ideal gas law in the form $V/n = RT/p$ defines the molar volume of any gas as the ratio V/n which is always RT/p for any ideal gas. Using this relationship, the volume of one mole of gas can be determined at different temperatures and pressures, for example at 100 kPa pressure and 25 °C one mole of gas will occupy 24.789 L and at 0 °C the molar volume will be 22.7106 L. At one atmosphere pressure, the corresponding values at 25 °C and 0 °C will be 24.465 L/mol and 22.414 L/mol respectively.

Although all gases deviate from the ideal gas law, it is customary to consider all the gases normally encountered in gas analysis as ideal gases. This assumption, though technically incorrect, is acceptable since the temperatures and pressures involved are usually moderate, for example, at room temperature and atmospheric pressure most gases have a molar volume very close to that predicted by the ideal gas law. At increased pressure or decreased temperature, the molar volumes of real life gases start to deviate significantly from what the ideal gas law predicts.

1.3.7 Vapour pressure

One way to impress students is to boil some water in a round bottom flask until the steam has displaced all the air in the flask. The flask is then closed and the boiling ceases. The fun part then starts when you put your hands in ice water and then hold the flask above the water line — the water starts boiling again as if by magic. The rather less glamorous scientific explanation is that by cooling the vapour it condenses, reducing the pressure above the water to below the vapour pressure of the water which causes it to start bubbling as if boiling.

All condensed phases, liquids and solids, have a vapour pressure which is defined as the pressure the substance will establish at a specific temperature. The vapour pressures of a few substances are given in Table 1.4.

Table 1.4 Vapour pressures and densities of pure substances at 25 °C.

Substance	Vapour pressure (kPa)	Density (kg/m^3)
Water (H_2O)	3.1691	0.99702
Mercury (Hg)	0.2460	13.5340
Iodine (I_2)	0.1889	4.93
Methanol (CH_3OH)	16.8511	0.791
Ethanol (C_2H_5OH)	7.8279	0.785
Benzene (C_6H_6)	12.6893	0.899

The vapour pressure at a specific temperature is characteristic of any pure substance. An increase in temperature will, in many cases, cause a significant increase in vapour pressure. What is important to note is that substances will evaporate to generate a vapour pressure at temperatures much lower than the boiling point of the substance. This accounts for the formation of clouds from evaporating seawater despite the fact that the oceans are not boiling hot. Table 1.5 of vapour pressures of water at different temperatures illustrates this.

1.3.8 Vapour pressure and partial pressure

A common question when discussing injecting samples in a gas chromatograph is why the sample that has evaporated in the injector does not condense when the sample vapour reaches the column which is initially at a much lower temperature. The answer lies in the partial pressure and vapour pressure of the sample components. Substances will condense when the partial pressure exceeds the vapour pressure but since the sample is diluted substantially by the carrier gas in the GC injector its partial pressure is so low that it does not exceed the vapour pressure when it reaches the cooler column.

The same principle can be used when sampling streams with semi-volatile compounds that are gases at temperatures above that which the

Table 1.5 Vapour pressure and density of water at different temperatures.

Temperature (°C)	Vapour pressure (kPa)	Density (kg/m³)
0.01	0.61173	0.99978
1	0.65716	0.99985
4	0.81359	0.99995
5	0.87260	0.99994
10	1.2281	0.99969
15	1.7056	0.99909
20	2.3388	0.99819
25	3.1691	0.99702
30	4.2455	0.99561
35	5.6267	0.99399
40	7.3814	0.99217
45	9.5898	0.99017
50	12.344	0.98799
55	15.752	0.98565
60	19.932	0.98316
65	25.022	0.98053
70	31.176	0.97775
75	38.563	0.97484
80	47.373	0.97179
85	57.815	0.96991
90	70.117	0.96533
95	84.529	0.96192
100	101.32	0.95839
105	120.79	0.95475

GC can handle. If the stream is simply cooled down these compounds will condense which will of course change the composition of the stream and destroy the integrity of the sample. However, if the sample stream is diluted with a heated, inert gas the partial pressures can be lowered to the point where they will not exceed the vapour pressure at lower temperatures. The diluted stream can then be cooled down to a temperature that can be accommodated by the GC.

The effect of water vapour is often overlooked in samples, especially when analysed using a detector which does not respond to water or a column system that does not elute water vapour. If any liquid water is contained in a sample vessel in addition to the gas to be analysed, the vapour pressure of the water will add to the total pressure in the container, in mathematical terms $p_{(gas)} = p_{(Total)} - p_{(Water)}$. When the samples are injected at ambient temperature and pressure it should be clear that significant differences in sample composition can be expected between summer and winter due to the difference in the contribution of the partial pressure of the water vapour relative to the total pressure of the sample. A correction can be made for this using a table similar to the one given above or even better by quantifying the water in the sample gas using an appropriate technique such as FTIR.

1.3.9 Standard temperature and pressure (STP)

In order to compare results from different laboratories around the world where ambient temperature and pressure may be widely divergent, it is necessary to normalise results to some standard conditions. There are a number of different definitions of so-called standard temperature and pressure or STP. In Table 1.6 a few common definitions are given. There are more and therefore when reporting gas concentrations at STP it is important to know what is considered as standard conditions by the customer and the specific industry where the results will be applied. Converting from one set of conditions to the other is easy enough using the ideal gas law, but it remains important to clearly state what standard is being used for the report. This is important since one standard litre always contains the same number of moles, regardless of the gas.

It can happen that a sample must be injected at an elevated temperature to ensure that all compounds remain in the vapour phase or some samples may have to be injected at below ambient pressure because there is insufficient pressure in the sample container. Often the results are required at STP and the conversions are therefore necessary. Some examples of this will be discussed in the chapters on sampling and results processing.

Table 1.6 Common definitions of STP.

	Definition source	Temp. (°C)	$P_{(abs)}$ (kPa)
IUPAC	International Union of Pure and Applied Chemistry	0	100
EPA	U.S. Environmental Protection Agency	20	101.325
NIST	National Institute of Standards and Technology	20	101.325
ISO 13443	International Organization for Standardization	15	101.325
SPE	Society of Petroleum Engineers	15	100
OSHA	U.S. Occupational Safety and Health Administration	(60 °F)	(14.696 psia)

References

[1] S. Ahuja, Chapter 9. Gas chromatography, in: Chromatography and Separation Science, Academic Press, Elsevier Science (USA), San Diego, CA, USA, 2003, p. 134.
[2] L. Ettre, The beginnings of gas adsorption chromatography 60 years ago, LCGC North America 26 (1) (2008) 48–60.
[3] D. Rood, A Practical Guide to the Care, Maintenance, and Troubleshooting of Capillary Gas Chromatographic Systems, third revised ed., Wiley-VCH, New York, 1999.
[4] M. Lee, F. Yang, K. Bartle, Open Tubular Column Gas Chromatography, John Wiley & Sons, Inc., USA, 1984.
[5] H. Willard, L. Merritt, J. Dean, F. Settle, Instrumental Methods of Analysis, D. van Nostrand Company, New York, 1981.
[6] J. Dolan, Peak tailing and resolution, LCGC North America 20 (5) (2002) 430–436.
[7] J. de Zeeuw, ChromaBLOGraphy – Why Are There Different Resolution Equations in Chromatography? Restek.com, March 3, 2013. [Online]. Available: https://blog.restek.com/?p=7929.
[8] D. Tabor, Gases, Liquids and Solids, Penguin Books, Glasgow, UK, 1969.

Further reading

[1] L.S. Ettre, The centenary of "chromatography", LCGC North America (July 1, 2006) 680–692.
[2] L. Ettre, in: J. Hinshaw (Ed.), Chapters in the Evolution of Chromatography, Imperial College Press, 2008.
[3] L. Ettre, Chromatography: the separation technique of the twentieth century, in: A Century of Separation Science, Marcel Dekker, Inc., New York, 2002.
[4] C. Gehrke, R. Wixom, E. Bayer, Chromatography – A Century of Discovery 1900–2000. The Bridge to the Sciences/Technology, Elsevier, Amsterdam, 2001, pp. 7–8.
[5] R. Wixom, C. Gehrke, D. Chance, T. Mawhinney, Chromatography. A Science of Discovery, John Wiley & Sons, Inc., New Jersey, 2010.
[6] J. Cazes, R. Scott, Chromatography Theory, CRC Press, New York, USA, 2002.
[7] K. Dettmer-Wilde, W. Engewald, Practical Gas Chromatography. A Comprehensive Reference, Springer-Verlag, Berlin, Heidelberg, 2014.
[8] T. Griffith, Chapter 4. Gas laws: pressure, volume, and temperature, in: Basic Principles and Calculations in Process Technology, first ed., Prentice Hall, New Jersey, USA, 2015, pp. 93–119.
[9] R. Grob, E. Barry, Modern Practice of Gas Chromatography, fourth ed., John Wiley & Sons, Inc., New Jersey, USA, 2004.
[10] C. Holbrow, J. Lloyd, J. Amato, E. Galvez, M. Parks, Chapter 4. Gas laws, in: Modern Introductory Physics, second ed., Springer, New York, USA, 2010, pp. 83–108.
[11] W. Jennings, E. Mittlefehldt, P. Stremple, Analytical Gas Chromatography, second ed., Academic Press, 1997.
[12] H. McNair, J. Miller, Basic Gas Chromatography, second ed., Wiley & Sons Inc., New Jersey. USA, 2009.

CHAPTER 2

Gas analysis laboratory

Contents

2.1 Laboratory utilities

When planning a new gas analysis laboratory it is imperative that there is a clear understanding of exactly what is to be analysed and what matrices are to be disposed of. To be able to quantify impurities in nitrogen is a totally different analysis from doing the same impurities in ammonia for example. The GC is only a part of the whole system which will be as strong as the weakest link.

A practical guide to gas analysis by gas chromatography
ISBN 978-0-12-818888-0
https://doi.org/10.1016/B978-0-12-818888-0.00002-4

Many of the comments and suggestions in this section might seem to be common sense but it is uncanny to notice how many of these fundamentals are being ignored in many laboratories, despite being featured in the pre-installation instructions of most GC's [1].

2.1.1 Ambient conditions

Most instruments will work satisfactorily in a laboratory where the temperature ranges between 5 °C and 40 °C, with an optimum at around 22 °C. For a gas analysis laboratory it is very important to control the temperature within ±2 °C or better, given the effect that temperature has on gases as explained in the previous chapter. Keeping the temperature constant will ensure that the samples and standards that are kept in the laboratory are all at the same temperature. Direct sunlight and drafts from windows or air conditioners should be eliminated, not only because it affects the temperature but also because strong drafts can cause baseline disturbances on some detectors. All flow controllers have some sensitivity to temperature changes and it is not uncommon to find that retention times change at certain times of the day due to direct sunlight changing the temperature of the pneumatic compartment of the GC.

Since the GC oven is capable of rapid heating and must cool down after each analysis, it is recommended to add an additional 3000 Btu to the capacity of air conditioners for each GC and an additional 1000 Btu for major accessories such as permeation ovens and vacuum pumps. The relative humidity should be maintained at 20–80%, non-condensing. The unit of measurement for air conditioner capacity is the British Thermal Unit (BTU) and 1 Btu is approximately equal to 1.055 kJ.

2.1.2 Ventilation

There are several places where, if not considered, lack of adequate ventilation can result in hazardous vapours being released into the laboratory. Consideration should be given to the venting of excess sample, the outlet of the vacuum pump, the outlet of the split vent and septum purge lines as well as the exhaust from non-destructive detectors. When working with samples that contain poisonous gases consider placing the entire GC in a ventilated cabinet. Keep in mind that the standard laboratory personal protective clothing, such as gloves, safety glasses and overcoat offer no protection against gases and secondary containment is usually the safest way to go. Furthermore most gases, including many extremely toxic gases, are colourless and odourless making them particularly hazardous. If at all possible a ventilated cabinet, kept at the same temperature as the laboratory, should be used for the storage of samples and standards.

2.1.3 Electrical

At maximum demand, most GC's consume between 2000 and 3000 VA (2–3 kW) and require 20 A at 120 V or 15 A at 200–230 V with a frequency of 50–60 Hz (±2 Hz) depending on the country where they are used. It is therefore important to specify this when ordering a GC. All the plugs, sockets and wiring should match the instrument power rating and adhere to local electrical codes.

In addition, it is prudent to have the GC and its data system on a separate circuit that is not shared with vacuum pumps and any other equipment likely to introduce spikes in the electrical supply. It is also important to avoid running the power cables behind the GC as this is where the hot air is exhausted every time the GC oven cools down. This can result in overheating of the electrical wiring, increasing its resistance, and even melting of the electrical insulation.

2.1.4 Furniture

The bench on which the GC will reside should be able to carry about 120 kg–200 kg per linear meter. Allow about 2 m per instrument to have enough space for the data system and samples and leave at least 30 cm at the rear for the venting of the oven. Island benches are ideal since it is possible to easily access the rear of the instrument for testing, maintenance and changing connections. Some laboratories have shelving mounted above the instruments for keeping printed manuals and files for QC and other data. This is not a good idea since it firstly restricts the free venting of the hot air from the oven. Secondly the hot air can cause drying out of the shelving material, especially if it is made from compressed wood, causing it to dry out, become brittle and eventually break.

2.2 Getting the gases to the GC

The purpose of this section is to provide a better understanding of the hardware necessary to connect the gas supply to the GC system and how to reduce the pressure to the levels that can be handled by the GC system typically 5–700 kPa. Although we try to use only SI units, many GC's use Imperial sizes for the outside diameter of tubing and connections. This often leads to mixed units in specifications, for example, it is commonplace to specify a packed column as 2.4 m long and made from $^1/_8''$ tubing with an internal diameter of 2.2 mm. There are many unit converters available on the Internet and in manuals, similar to Table 2.1.

Table 2.1 Conversion of imperial length measures to metric units for commonly used tubing and fittings.

Imperial (inches)	Nominal metric (mm)	Exact metric (mm)
$^1/_{16}$	1.5	1.5875
$^1/_8$	3	3.175
$^1/_4$	6	6.35
$^3/_8$	10	9.525
$^7/_{16}$	11	11.11
$^1/_2$	13	12.7
$^3/_4$	20	19.05
1	25	25.4

There are three common sources of gas supplied to the GC laboratory, namely gas generators, plant gas and high pressure cylinders. Each has its own advantages and disadvantages.

2.2.1 Gas generators

Gas Generators are available for hydrogen, nitrogen and zero air and are gaining popularity in laboratories, mainly because they can supply high purity gas at sufficient pressure for most GC applications. The main advantage of these generators is that the gas is produced close to the point of use thereby avoiding the problems associated with plumbing high pressure gas to the laboratory from remote locations. There are several commercially available systems which all work on the same fundamental principles.

Hydrogen is produced from high purity distilled and deionised water by electrolysis. The hydrogen is collected and dried before leaving the system while the oxygen formed is vented. Apart from regular topping up of the water and periodically replacing seals these units require very little maintenance. The disadvantage is that the pressure and flow delivered is relatively low and therefore a unit of the correct size must be purchased depending on the number of GC's and detectors requiring hydrogen in the laboratory.

Zero air is very clean compressed air with the intake usually inside the laboratory. After compression, usually with an oil free pump, the air is dried and filtered to remove impurities, especially hydrocarbons, resulting in very clean air suitable for most GC instruments. In the nitrogen generator, the nitrogen is separated from the oxygen in zero air and also dried and filtered as needed to provide high purity nitrogen while the oxygen is vented.

From a safety perspective gas generators are a far better choice than high pressure gas supplied from production plants or cylinders. In laboratories that are in remote locations and where delivery of cylinders is unreliable or locations here the cylinder gas quality is questionable, gas generators offer the best way to keep the laboratory supplied with high quality gas. The quality of gas produced by any gas generator must be maintained by the regular and correct preventive maintenance.

2.2.2 Plant gas

In many production facilities, such as refineries, bulk hydrogen, nitrogen and air may be available at high pressure for use in various processes in the plant. These gases are usually not pure enough for use in instruments, but they can be purified using large capacity filters to a sufficient purity for instrument use. This is often the cheapest way of getting high quality gas but a continuous supply may be a problem as production plants routinely have to shut down for maintenance purposes. While this can be overcome by keeping a few cylinders in stock as backup a more serious problem is contamination of the plant gas. Reactors downstream of the laboratory may, for various reasons, spill some reagents or products in the gas lines at levels that cannot be removed by the filters and so result in contamination of the instrument and the lines leading to it. The danger of using plant gas is that every GC connected will be contaminated and may require very expensive module replacement. On one occasion of this nature the carrier gas contained oils that destroyed all the flow controllers and detectors of several GC's. The pressure of the gas may vary and high pressure plant lines suffer from the same disadvantages as gas cylinders as discussed below. However, in production plants where these gases are available there is normally sufficient in-house expertise to install and maintain these systems to ensure a constant supply to the laboratory.

2.2.3 Pressurised gas

Gas supplied in high pressure cylinders with gas pressures of around 20 MPa, depending on the gas, is still the most common source of gas for GC laboratories [2]. Gases, such as helium, that cannot be manufactured are necessarily supplied in high pressure containers. Should this gas pressure be released rapidly due to a breakage, the cylinder would become a missile such as those fired from submarines. Therefore, it is vital that safe laboratory practices are followed in the transportation, storage, and usage of these gases. As a bare minimum, adhere to the following rules:

- Always keep the cylinders capped and never remove the protective sleeve around the valve when not in use, even when empty.
- Never move a cylinder with a regulator installed and make sure the safety cap is in place over the gas valve when transporting the cylinder.
- Always keep the cylinder upright and chained securely to stationary objects in the laboratory, even when empty.
- Do not expose cylinders to temperatures above 40 °C.
- Never breathe any gas in; while it may be pure it is not certified for human use and may contain poisonous impurities.
- Always leave at least 500 kPa residual gas in a depleted cylinder.
- Ensure that flammable and oxidising gases are stored separately. It is useful to label the cylinders FULL, IN USE, or EMPTY and to date this condition.

2.2.4 Pressure reduction

The first step that is required is to reduce the high pressure using a regulator. The purpose of a pressure regulator is to maintain constant gas pressure to the GC. Regulators may be classified as two types: cylinder regulators and line regulators. Cylinder regulators attach directly to the cylinder valve. The cylinder regulator reduces the gas pressure from the cylinder pressure, usually 15–25 MPa for a new cylinder, down to a more usable pressure, around 600–700 kPa for gas chromatography. Cylinder regulators have two pressure gauges: an inlet, or high pressure gauge which reads the cylinder pressure; and a delivery, or outlet pressure gauge. This final delivery pressure is user adjustable by turning the large knob on the front of the regulator.

Pressure controllers use a diaphragm set under stress by a spring loaded control and the pressure is regulated as shown in Fig. 2.1. Since pressure is increased by increasing the mechanical stress on the diaphragm, this means that pressure increases with a clockwise turning of the shaft. The spring is very strong and cylinder valves should never be opened when the pressure regulator is set. Always back off the pressure setting by turning the control shaft to the fully counter clockwise position before opening the cylinder valve.

Cylinder regulators may be either single-stage or two-stage regulators, shown in Fig. 2.1. Two-stage regulators actually employ two regulators back-to-back in a single housing. The first stage reduces the cylinder pressure to 1.2–4.0 MPa, while the second stage performs the final pressure reduction. Two-stage regulators are less prone to 'creep' (a slow increase in

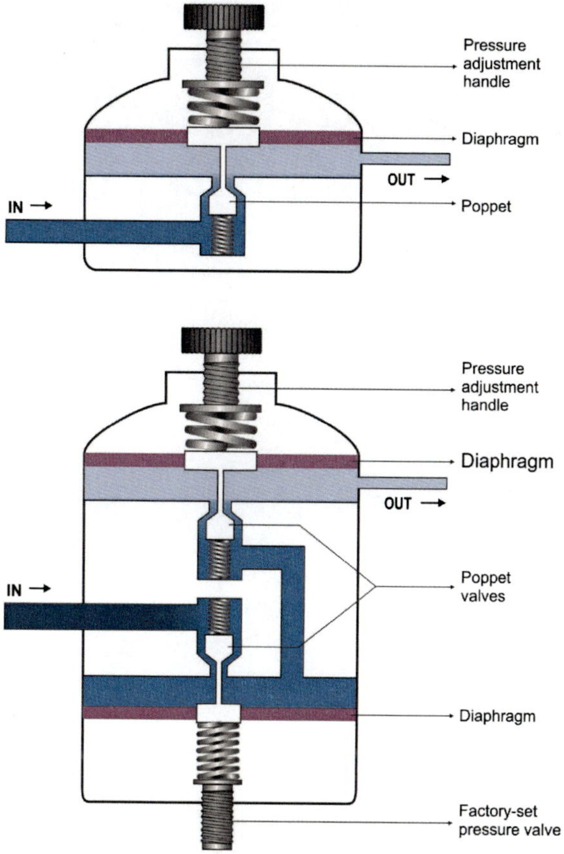

Fig. 2.1 Schematics showing the difference between a single (top) and a dual stage (bottom) pressure regulator.

delivery pressure as the cylinder empties) but have a lower flow capacity than single-stage regulators. Although more expensive, two-stage regulators should be used when a very constant delivery pressure is required, such as when controlling gas flows for a gas chromatograph. Line regulators have a lower allowable inlet pressure, typically 2 MPa, and must never be attached directly to a gas cylinder. Line regulators are used to further reduce the pressure of a gas from the supply line to that required at the point of use. Line regulators are always single-stage regulators and may be equipped with a single pressure gauge to indicate outlet pressure.

Pressure gauges are precision pieces of equipment and quite fragile despite their ability to handle high pressures. The functioning of the pressure gauge is deceptively simple. A flattened metal tube, usually copper,

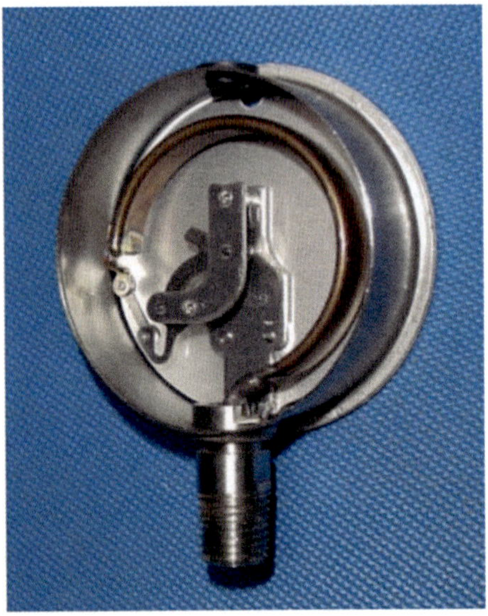

Fig. 2.2 Pressure gauge with the back removed showing the curved tube and lever mechanism by which the deformation of the tube is amplified and transferred to the needle on the front of the gauge.

is bent into a C-shape and sealed off on one end as can be seen in Fig. 2.2. The other side is connected to the gas via the screw-in fitting with which it is mounted onto the device in which the pressure must be measured. When the bent tube is pressurised, it tends to straighten much like a long balloon when one starts to blow it up. The movement of this C-shaped-tube is picked up by a set of interconnected levers that cause the needle to turn. Obviously the lower the pressures to be measured the thinner the walls of the C-tube has to be and conversely the higher the pressure to be measured, the thicker the tube will have to be. They are rated for specific ranges of pressures and when connected to higher than the rated maximum pressure will be damaged beyond repair and might even rupture, sending bits and pieces of glass and metal flying. As the regulator is often mounted on a cylinder and the gauges are then at eye level it is not difficult to see how dangerous a bursting gauge can be to an operator opening the valve on the cylinder. The deformation of the tube and the associated movement of the needle is precisely calibrated and opening a regulator to the point where the maximum reading on the gauge is exceeded will not only damage the gauge but might also cause it to burst. It is therefore very important to inspect the gauges regularly. One of the signs that the gauge is damaged is when the needle does not return to zero when there is no pressure on the gauge. This usually indicates that the C-tube has been deformed permanently and that readings displayed by the needle are simply not accurate any more. This is often caused when cylinders are opened too quickly and the regulator is not backed off, resulting in the gauge being exposed to a pressure shock. Many laboratories forbid the removal or replacement of these gauges by their own personnel. It is also essential to ensure that gauges that contain oils must not be used with certain gases such as oxygen since this will cause an explosive reaction. Replace the regulators when the gauges are faulty – it is not worth the risk to continue using faulty equipment.

When connecting a pressure reducer or regulator to cylinders, ensure that the threads are clean of dirt and of oil. Check the round surface of the bull-nose for scratching and inspect that the seat is clean. If the bull-nose is scratched, replace it as the cost of a replacement bull-nose is usually less than the cost of the contents of one cylinder. The seal at the bull-nose is metal to metal and thread sealing tape must not be used as it will flow under these very high pressures. High purity connections can use metal crush seals and it is advised to replace these if they have been loosened. The regulator with its gauges and bull-nose are filled with air before being used for the first time or having been disconnected for some time. This small volume of air is sufficient to contaminate the contents of a high purity cylinder or gas standard if allowed to mix together. To get rid of the air, the bull-nose and regulator should be purged, especially when connecting gas standards. With the outlet tubing connected to a suitable vent, connect the regulator with the regulator valve fully counter-clockwise, then open the cylinder valve and close immediately. Turn the control valve clockwise to expel the small amount of gas in the regulator as shown in Fig. 2.3. Turn the regulator fully counter clockwise and repeat this procedure at least five times. The outlet tubing can now be connected to where the gas is needed. The gas volume in the bullnose is a swept volume that is easily flushed but the gauge volume is a dead volume that requires several flushes to purge out the air.

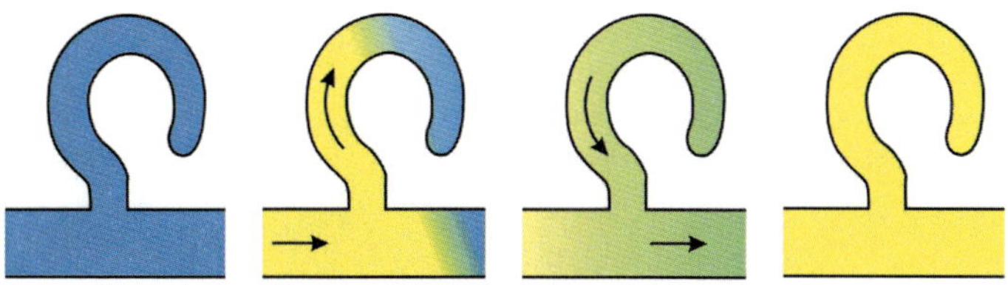

Fig. 2.3 Schematic showing the removal of air from a pressure gauge with repetitive flushing following a regulator change.

In some gas installations the regulator is connected to the cylinder with a flexible hose which increases the volume of air to be replaced. In such cases it is even more important to thoroughly flush the connection. It is advisable to have a changeover system for the gases if at all possible. These systems allow the connection of two or more cylinders at a time and this makes it possible to disconnect and reconnect cylinders without interruption of the gas flow to the laboratory. Most of these systems also allow for the independent purging of the connection between the cylinder and the regulator.

At this stage one would probably be tempted to use a tracer such as soapy water to check the bull-nose for leaks. The danger is that soap solution can be sucked into a fitting despite the positive pressure in the gas line by the Venturi effect. To avoid this possibility we suggest that soapy solutions be replaced with a 1:1 mixture of isopropanol and water. Should some enter the gas line it can be more easily baked out with a little heat than soap which contains a mixture of high boiling surfactants or saponified fatty acids. Obviously this technique cannot be used to check hot connections.

As we have previously said, for connecting cylinder gas to the GC the use of dual stage regulators, is highly recommended. Single stage low pressure regulators are found inside the GC and are discussed in more detail in Chapter 3.

For trace gas analysis and for PDHID systems it is mandatory to use electronic grade regulators. These do not use tapered threads with PTFE tape but are all metal. For cleanliness they are shipped under nitrogen gas so do not open the plastic container until ready for installation.

Electronic grade regulators use Swagelok® VCR metal gasket sealed unions as shown in Fig. 2.4. The seal is achieved by compressing a metal gasket between two highly polished beads. The most common gasket is silver plated for the best seal but un-plated gaskets of varying materials are also available for applications where the silver might react with the gas. It is very easy to assemble these unions on the bench by lying the gasket horizontally onto the union and tightening. Retainer assemblies are

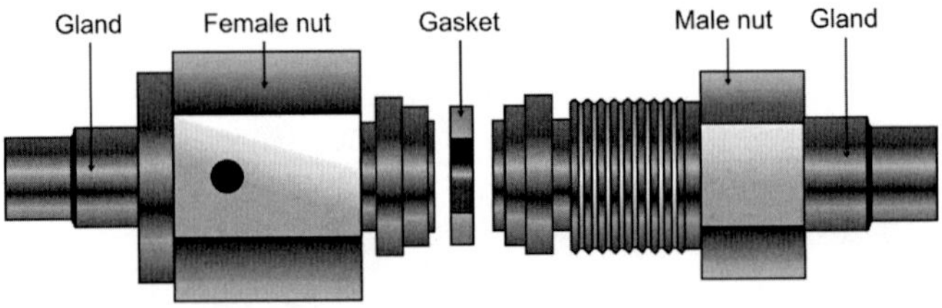

Fig. 2.4 VCR fittings for adapting to Swagelok or VICI fittings.

available to eliminate the challenges of fitting a gasket when the fitting is horizontal and the gasket vertical. Hand-tighten the fitting and then use two spanners to tighten to manufacturer's specification. For all fittings it is recommended to use two spanners, one as a counter-wrench to support the fixed side of the fitting and the other to tighten the nut as shown in Fig. 2.13. All-metal adapters for connecting VCR fittings to Swagelok or VICI fittings are available so that the regulator outlet can be connected to the analytical system with standard metal tubing.

2.2.5 Tubing

The next step in plumbing GC gases to the instrument is choosing the appropriate tubing size and material. All tubing has some degree of permeability that causes not only the escape from the contents inside the tubing but will also cause ingress of ambient air into the tubing, even against the pressure gradient as shown in Fig. 2.5. Metal tubing is far less permeable than polymeric tubing. Some columns have very low oxygen tolerance and will be rapidly destroyed if the carrier gas has a high concentration of oxygen.

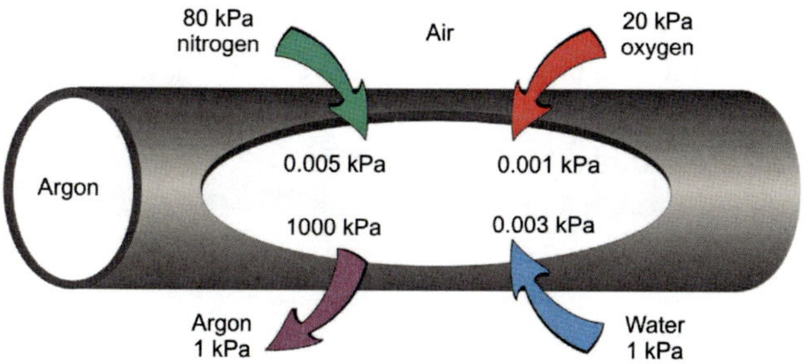

Fig. 2.5 Schematic showing the ingress of oxygen, nitrogen and water vapour into an argon line against a pressure gradient as well as the loss of argon through the side walls of the tubing.

Many GC's are supplied with $^1/_8''$ bulkhead fittings which means that $^1/_8''$ tubing should be the default size. The distance from the cylinder as well as the number of systems running off the same gas lines will determine whether larger diameter manifold tubing will be required. Copper and thin walled $^1/_8''$ tubing can be bent by hand whereas heavy walled $^1/_8''$ and $^1/_4''$ stainless steel tubing require a tubing bender. The bender incorporates lever arms that reduce the amount of force required to bend the tubing and prevents flattening of the tubing at the bend. Bends should be made with a smooth and uniform radius and should not kink or deform the tubing in any manner that obstructs flow. A few coils of tubing, also known as shock loops, should be made where the lines meet the instrument to allow the instrument to be moved for maintenance purposes and to prevent mechanical noise transmission to the gas lines that could cause compression fittings to loosen. These are shown in Fig. 2.6.

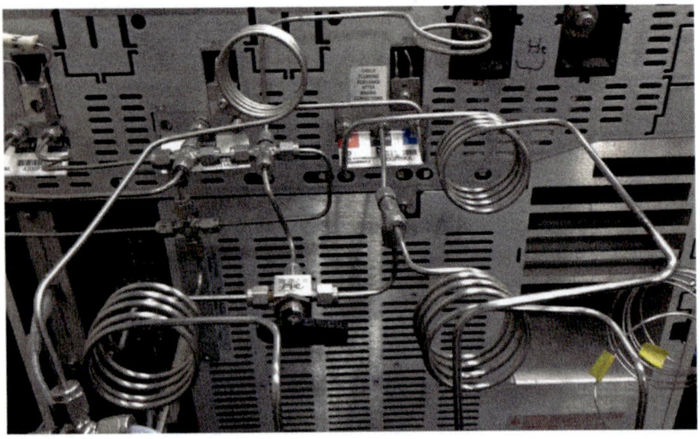

Fig. 2.6 Coiled stainless steel tubing to create shock loops behind the instrument.

It is important to realise that there will be a pressure drop across any length of tubing that will depend on the amount and nature of the gas flowing through the tube. An equation that could be used for this is the Darcy-Weisbach formula [3] for which there are many web sites that offer on-line calculation, such as the Engineering ToolBox [4].

$$\Delta p = \lambda \left(\frac{l}{d_h} \right) \left(\frac{\rho v^2}{2} \right)$$

where Δp is the pressure loss in Pa or N/m^2, λ the Darcy-Weisbach friction coefficient, l the length of tube in m, d_h the hydraulic diameter in m, v the velocity in ms^{-1} and ρ the density in kg/m^3.

In practice, we have found it possible to run a flow of 750 mL/min of helium through 1⁄16″ tubing having a 0.75 mm internal diameter but with a significant pressure drop. Care must be taken to ensure that the pressure shown on the regulator is close to the pressure delivered to the instrument. As a guideline 1⁄8″ tubing is adequate for up to two systems with Flame Ionisation Detectors, but 1⁄4″ tubing should be used from the gas supply when more instruments are connected on a common manifold.

The choice of gas line tubing must be made according to the application. Copper tubing is cheaper and easier to work with and if not pre-cleaned, must be cleaned internally before use. Cleaning of tubing requires removal of inorganics by rinsing with dilute HCl, followed by a polar solvent such as acetone and a non-polar solvent such as hexane to remove the hydrocarbon contamination from the drawing process. It must then be dried by flushing with dry nitrogen preferably at elevated temperature for which an existing packed column GC is ideal. The solvents and acid used should not interfere with the analysis, for example, tubing washed with hydrochloric acid or dichloromethane will retain some solvent and make it impossible to operate an electron capture detector within its specifications. Even without steam cleaning, the internal surface of any tubing will have a water vapour layer when exposed to room air and this layer will gradually be removed once a leak-tight system starts to operate. In our experience this will require at least one week of usage at room temperature and flows of 50—100 mL/min before the water vapour is reduced sufficiently for use with detectors that can detect water. If possible, heating of these tubes will help reduce this time considerably. This has no influence on the Flame Ionisation Detector (FID) but serious implications on water sensitive detectors such as the Pulsed Discharge Helium Ionisation Detector (PDHID) and Electron Capture Detector (ECD).

Sample lines that are used for samples or standards must be inert. Copper tubing and brass fittings cannot be used with any gas that contains reactive components. Thick walled stainless steel is preferred for trace analysis and 1/16″ stainless steel tubing with an internal diameter no more than 0.75 mm is necessary when very low levels of oxygen must be achieved such as with PDHID's. For such applications passivated metal tubing should be used and is imperative when exposed to elevated temperatures. These are available as Sulfinert™, Silcosteel™ and Ultimetal™ and can be handled like stainless steel tubing except that it must be bent with a smooth radius that has a minimum radius of 25 mm for 1/16″ tubing, 50 mm for 1/8″ tubing and 100 mm for 1/4″ tubing. Also this tubing cannot be used above 360 °C while acids and bases should not come in contact with the inner surface.

Inert thick walled polymeric tubing such as PEEK and TEFLON can be used for sample lines but will have significantly more contamination through permeation of ambient air. For trace analyses, all types of polymeric tubing should be avoided not only because of its permeability but also because it can adsorb organic components in the sample and later give them off resulting in increased baselines as well as ghost peaks.

All metal tubing can be cut using a standard metal tubing cutter. The cutting wheel is usually replaceable and must be kept sharp. When cutting or deburring tubing hold the tubing end pointing downwards so that any possible metal filings will drop away from the tubing and not end up inside it. As an extra precaution one can rinse the tubing with a suitable volatile solvent and blow some gas through the tubing before being connected.

As for metal tubing, polymeric tubing should also be clean, with ends cut square to the tube axis and free of external and internal burrs. Several tubing cutters are available and do a good job although a very sharp knife can work well.

In most gas analysis GC's, 1/16″ tubing constitutes an important part of the system and care must to be taken to ensure the integrity of the cut tubing. Particulates left over from improper cutting and cleaning of metal tubing eventually ends up in the rotary valves of the GC and is the major reason for premature valve failure. High performance systems require truly zero dead volume connections and uniform flows that can only be achieved using precision cut and finished tubing.

As the wall thickness of 1/16″ tubing increases so does the difficulty of achieving a clean cut. It is therefore strongly recommended to purchase pre-cut, pre-cleaned lengths of tubing that can simply be used without worrying over the formation of particulates and deformed inner and outer diameters as shown in Fig. 2.7. This tubing is highly polished, has flat ends and is free from contaminants.

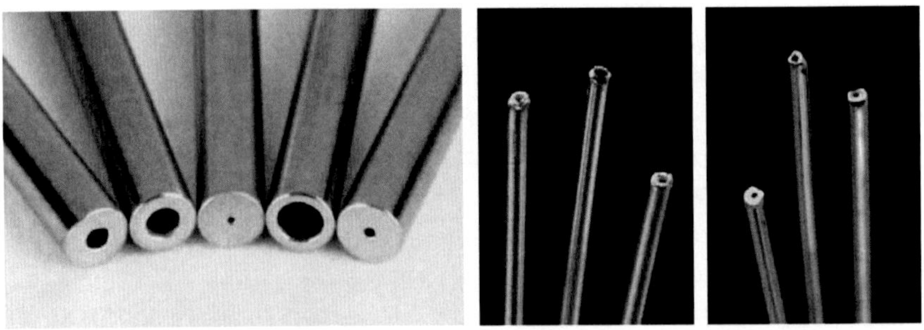

Fig. 2.7 These pictures clearly show the difference between clean cut tubing and tubing that was cut by filing or using pliers [5]. ©VICI AG International. *(Reproduced with permission, courtesy of VICI AG International.)*

When pre-cut tubing is not used, inspect the ends for a square cut and that the internal and external surfaces are deburred. Also check the ends of the tubing for any visible scratches along the length of the tubing where the ferrule will seal as in Fig. 2.8. If small scratches are visible the tubing can be polished using very fine emery paper and turning the tubing between

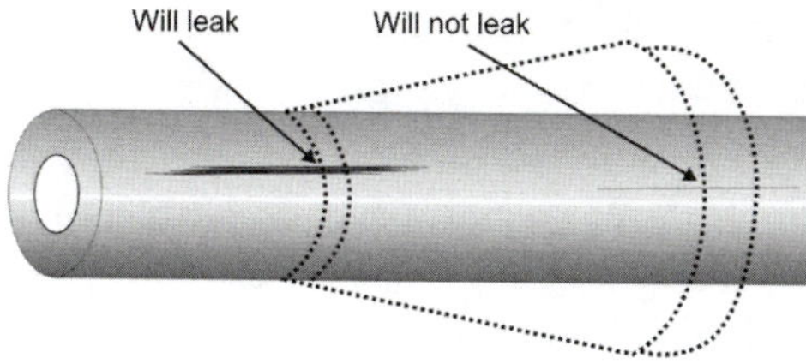

Fig. 2.8 Tubing showing longitudinal scratches at, and away from, the seal [6]. ©VICI AG International. *(Reproduced with permission, courtesy of VICI AG International.)*

the fingers. The small concentric scratches are not ideal, but will be less prone to leaking than longitudinal scratches. Remember to remove the sandpaper residue inside and out by rinsing with a suitable solvent and drying afterwards. Electro-polishing can be used to try and repair scratches but is not always successful as the edges of the scratch may be rounded off rather than repaired [6].

2.2.6 Fittings

The number of leaks is obviously proportional to the number of connections made although, in practice, this should realistically read 'exponentially proportional to the number of connections.' Remember that each well-made connection may add as much as 0.5 ppm ambient air to the carrier gas. Even worse is PTFE sealing tape where each tape sealed joint may add as much as 2 ppm ambient air to the gas in the tubing. There is little sense in buying high quality gas and contaminating it before it reaches the chromatograph. Gas line filters may help to reduce these levels but will have a limited lifetime. Clearly prevention is far better than cure.

Most major manufacturers have standardised on Swagelok®-type fittings, Fig. 2.9, a two-ferrule, mechanical grip-type fitting which use a nut, a back ferrule and a front ferrule to connect tubing via a union, elbow, tee or cross. These all-metal compression fittings provide leak-tight connections without the use of Teflon® tape or adhesives. When tightened the front ferrule is

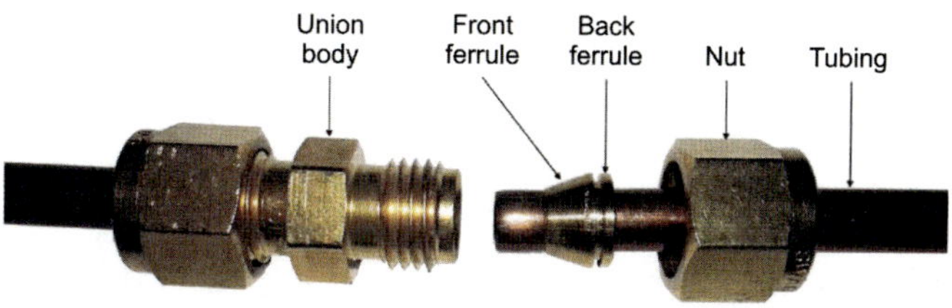

Fig. 2.9 Disconnected Swagelok fitting showing the ferrules gripping the tubing to provide a leak tight seal.

driven into the fitting body and the tubing to create primary seals, while the back ferrule hinges inward to create a strong grip on the tubing. Polymeric ferrules are available for use with glass or fused silica tubing.

To make a good connection slide the nut and ferrules onto clean, deburred tubing and insert the tubing into the fitting as far as possible. Hand tighten the nut, then withdraw the tubing about 1 mm. Failure to do this will result in the tubing being forced sideways when the ferrule grips the tubing and this makes it difficult to disconnect the fitting. This is especially important if the fitting is being installed on tubing that will probably be moved during its lifetime. Also it will be difficult to re-use the male connector in a different fitting. Where possible, use one spanner to support the fitting while using another open ended spanner to tighten the nut $^1/_2$ turn past finger-tight for $^1/_8''$ tubing as shown in Fig. 2.13. For $^1/_4''$ tubing, tighten the nut one turn past finger-tight. The manufacturers recommend more degrees of rotation than we do, which means that our figures will tend to slightly under-tighten the fitting, but we think it is advisable to leak check the fitting first and then tighten a little more if necessary. Never over-tighten the nut as the tubing and ferrules can become deformed and will not seal. A thickness gauge is available to measure the distance between the nut and the fitting and a well tightened fitting will typically show one thread between the nut and the fitting.

For 1⁄16″ fittings Swagelok® uses front and back ferrules that are small, quite difficult to handle and easily broken when tightened too much. Unlike the Swagelok system VICI AG® use a single ferrule that does not require the tubing to be slightly withdrawn before tightening the fitting. They are available in various metals for use with appropriate metal tubing and for trace gas analysis it is recommended to use gold plated ferrules as the malleable gold film gives an even better seal. We have had good success with these fittings.

The VICI zero dead volume fitting is made up of four parts: a female zero volume fitting detail, a male nut, a ferrule, and a length of tubing as shown in Fig. 2.10. This fitting is very leak-tight if correctly assembled and the tubing has been properly prepared. It is also important to note that the internal bore size in the fitting is matched in diameter by the tubing internal diameter so that turbulent flow is avoided.

Fittings are available in many different materials including stainless steel, Hastelloy, brass, PEEK, Teflon and more. The golden rule is to always match the material of the tubing and the fitting. Brass fittings should never be used to connect stainless steel tubing and vice versa as the differences in

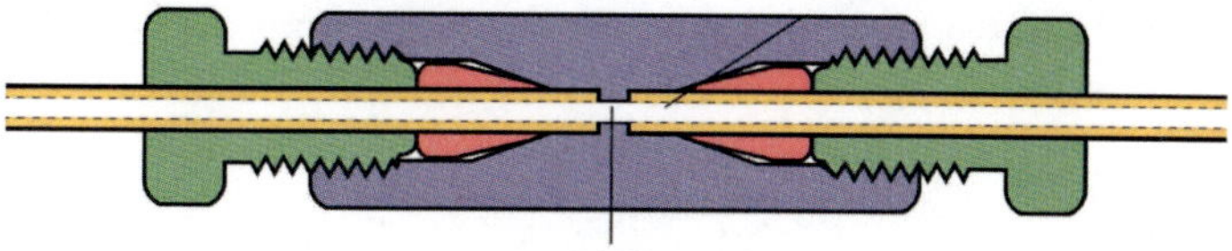

Fig. 2.10 Parts of a Valco fitting showing that sealing of the ferrule and tubing happens inside the connection [5]. ©VICI AG International. *(Reproduced with permission, courtesy of VICI AG International.)*

the expansion coefficients of the metals will result in leaks. The same applies to polymeric tubing and fittings in general but there are many cases where mixed fittings are used. Metal ferrules can be used with polymeric tubing but polymeric ferrules are not recommended for use with metal tubing. However, in some environmental sampling systems, this is done to allow the tightening of fittings by hand when in the field. Glass tubes are also connected using polymeric ferrules and Vespel, graphite or graphitised Vespel ferrules are used with metal fittings for the connection of fused silica and metal columns.

2.2.7 Finally

The old saying that prevention is always better than cure holds good so that when GC plumbing is done properly, there should be very little chance of contamination to the equipment from the plumbing. It is worth reiterating that it does not make sense to purchase very high quality gases, contaminate them with ambient air through poor plumbing and then try to remove those added impurities again.

2.3 Gases

It goes without saying that a gas chromatography laboratory for the analysis of gases requires gases. The gases that should be present in the laboratory will be determined by the specific application including which detectors will be used. The first that comes to mind is the carrier gas the purpose of which was already discussed in Chapter 1.

2.3.1 Carrier gas

The most important consideration when choosing carrier gas is compatibility with the detector to be used for the specific analysis. Four gases are commonly used as carrier gases; nitrogen, argon, helium and hydrogen. Theoretically the heavier carrier gases like nitrogen and argon gives optimum column efficiency but only over a very narrow range of relatively low velocity as can be seen in the Van Deemter plots of the different gases in Fig. 2.11.

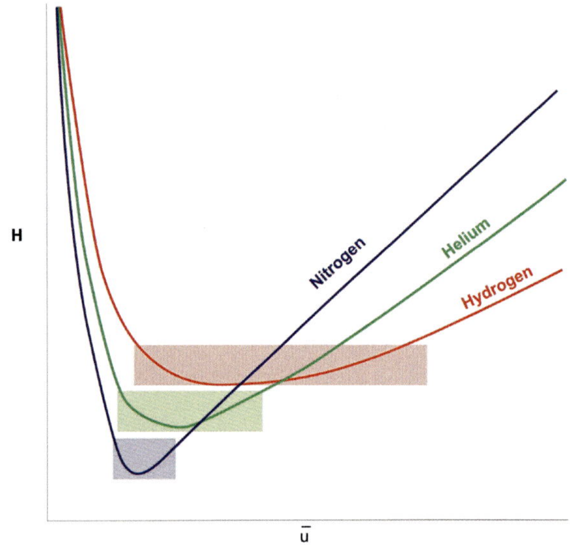

Fig. 2.11 Van Deemter curves showing the differences in efficiency and useable ranges of the three most common carrier gases, hydrogen, helium and nitrogen.

These low flow rates means that analysis times will be much longer than with the lighter carrier gases such as helium and hydrogen. In practice, especially in gas analysis, the gain in column efficiency by using nitrogen or argon is not great. However, when using a thermal conductivity detector for the analysis of hydrogen and helium, a heavy carrier must be used. Fortunately this is not a difficult separation. Nitrogen is popular because it is relatively inexpensive and can be supplied at high purity although in many instances argon is preferred because it is usually cleaner and more inert despite being more expensive. Using a very light carrier gas like hydrogen results in some loss of efficiency but it can be used over a much larger velocity range. This means much shorter analysis times and better productivity. As a starting point, linear velocities for argon carrier gas can be set at 10–15 cm/s, for nitrogen 12–18 cm/s, for helium 20–25 cm/s and for hydrogen 35–50 cm/s and further optimised when necessary. Although the optimum values will vary with column dimensions these values are above optimum and will give adequate performance for start-up conditions.

It is important that all carrier gases have a very low level of oxygen impurity as many of the column materials used in GC are susceptible to oxidation at elevated temperatures.

Hydrogen is reactive and may react with some compounds in the sample and in such cases helium would be preferred. A traditional fear of hydrogen originated in the days when large bore glass columns were used. These columns could break causing build-up of hydrogen inside the oven which could be ignited by sparks from the fan motor, causing an explosion. Modern systems use electronic flow controls that shut down automatically when the back pressure falls away as would happen when a column breaks at the inlet side. This makes it much safer to use hydrogen to the extent that the explosion risk is negligible. In gas analysis systems that use methanisers, it is preferable to use hydrogen as carrier gas.

Helium as carrier gas offers reasonable column efficiency and flow rate range and is inert. However it is quite expensive and since it is not a sustainable resource is becoming rather scarce. Some detectors, such as the PDHID, MS, and applications in which analytes may react with hydrogen make the use of helium mandatory but in most other cases it can be replaced by hydrogen.

2.3.2 Gas standards

A gas analysis laboratory will typically have gas standards and pure matrix and diluent gases as well as permeation, diffusion and dynamic dilution systems for the preparation of different concentration levels of standards. Standards are discussed in Chapter 4 while sample inlet systems for the GC will be discussed in detail in Chapter 6 on sampling.

When in use, gas standards in cylinders should be stored in a temperature controlled environment to avoid extremes in ambient temperature variation that could lead to condensation of some of the condensable compounds in the mixture.

2.3.3 Detector gases

Different detectors require certain fuel, oxidising and make-up gases in addition to the carrier gas and these requirements will be discussed in much more detail in Chapter 8. Some recommended carrier and make-up gases for the different detectors are given in Table 2.2. To analyse heavy gases on a TCD, hydrogen or helium is required as carrier while the same detector using nitrogen or argon is used for the analysis of light gases. The PDHID is best used with helium as carrier while nitrogen carrier is usually used with an electron capture detector.

Table 2.2 Recommended gas purity, regulators and filters for carrier and make-up gas [7].

Gas	Detector	Purity	Regulators[a]	Filters
Carrier	FID, TCD, PID FPD and PFPD[b]	He, H_2, N_2, or Ar 5.0	High purity	Hydrocarbon Moisture Oxygen
	PDHID[b] (carrier & ionisation gas)	He 6.0 or better	Ultra-high electronic grade	Zirconium getter based He purifier (from 5.0 to 6.0).
	ECD[b]	N_2 5.0 or 5% CH_4 + 95% Ar	Ultra-high	Hydrocarbon Moisture Oxygen
Make up	ECD[b]	N_2 5.0 or 5% CH_4 + 95% Ar	Ultra-high	Hydrocarbon Moisture Oxygen
	FID, TCD, PID, PFPD and FPD[b]	Same as carrier		

[a]Dual stage stainless steel regulators using stainless steel diaphragms are highly recommended.
[b]*Abbreviations:* FID – Flame ionisation detector, TCD – Thermal conductivity detector, PDHID – Pulsed Discharge Helium ionisation detector, ECD – Electron capture detector, PID – Photo-ionisation detector, PFPD – Pulsed flame photometric detector, FPD – Flame photometric detector, 5.0 = 99.999% and 6.0 = 99.9999%.

The combustion detectors such as the FID, NPD, FPD and PFPD all require hydrogen as a fuel gas as well as synthetic air as an oxidising gas. Although carrier gas can be used as a make-up gas for the FID, it is better to use a denser gas such as nitrogen when using a light carrier gas. Also check with the supplier of any specific detector to ensure that the recommended gases and filtration are in place for the detector they supply.

2.3.4 Other gases and gas lines

Strictly speaking the cryogenic gases such as liquid nitrogen and liquid carbon dioxide are not gases but liquids in the cylinders, but they also require high pressure lines in order to get the coolant to the instrument in liquid form. Obviously these are only required for sub-ambient operation of the gas chromatograph and may or may not be necessary depending on the types of analyses being done in the laboratory.

If a system requires liquid CO_2 for cryogenic cooling, the cylinder is connected without a regulator as the cryogenic control valve can handle full cylinder pressure. Note that a dip-tube cylinder must be used. Cylinders without dip tubes must be inverted to dispense liquid but this is not only clumsy but also an unsafe laboratory practice. The liquid CO_2 is often dirty and in-line particulate filters must be used since small diameter nozzles are used to cool the oven by the Joule-Thomson effect. These filters need not be sophisticated and a simple way of preventing clogging of the nozzle is to pack a small amount of glass wool into the compression fitting.

Liquid nitrogen is usually supplied in Dewars or so-called mini-tanks which are usually placed inside the laboratory. As pressure can build up in these containers a relief valve must be mounted in line such that excess pressure can be vented when necessary. It is best to vent the excess to the outside of the building or to a fume hood.

Both liquid nitrogen and carbon dioxide can replace the air in the laboratory and may pose a health risk, especially if the analysis times are short and cooling must be done frequently. As they are both odourless and invisible it is necessary to equip the laboratory with an oxygen depletion sensor mounted at bench level to warn the analysts when there is not sufficient oxygen in the laboratory.

It is strongly recommended to install a vent line in the laboratory used for gas analysis. This line may be polymeric tubing with an internal diameter of 15–20 mm leading from the instruments to a fume hood. Sample inlets, GC split injectors, vacuum pump exhausts and non-destructive detector outlets can all be connected to this line to ensure that sample excess, which may contain hazardous compounds, is safely removed from the laboratory. It is imperative when working with poisonous gases that the vent line exhaust should pass through a suitable deactivation trap. If compressed air or nitrogen is available, it is also prudent to connect that to the starting point of the vent line and allow a flow of a few millilitre per minute to continuously purge the vent line to prevent any build-up of chemicals in it which may cause problems through back diffusion into the sample inlets.

2.3.5 Removing gaseous impurities

All gases do contain some impurities that are not desirable to the equipment, especially water vapour, oxygen and hydrocarbons. In Fig. 2.12 a chromatogram of helium 5.0 before passing through a purifier illustrates this. Also these gases often flow continuously so that the actual amount of impurity becomes cumulatively significant. Various types of in-line gas purifiers are available to clean the gases. Getter systems remove oxygen and water vapour effectively but will not remove noble gas impurities. Some operate at high temperature while others operate at ambient temperature. Getters vary considerably in purity and capacity. Care must be taken in disposing of spent

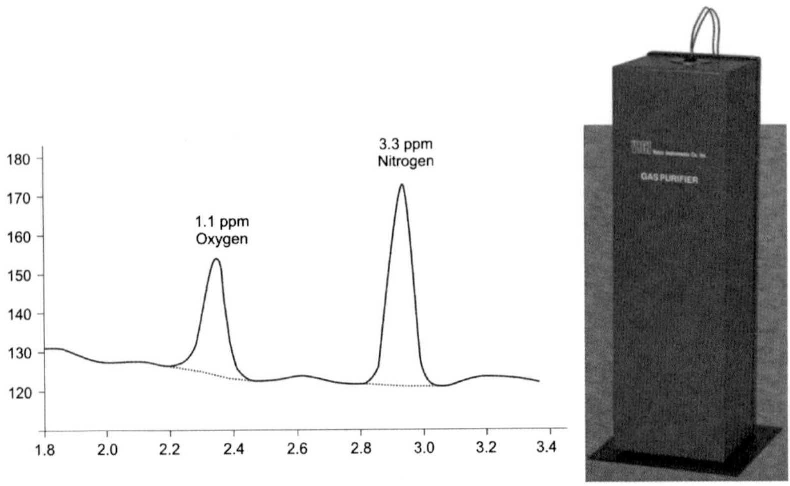

Fig. 2.12 A chromatogram of commercial helium 5.0 analysed on a custom built gas chromatograph. The same helium is used as carrier and PDHID ionisation gas after purification using a VICI getter system pictured on the right [5]. ©VICI AG International. *(Reproduced with Permission, Courtesy of VICI AG International.)*

getters as they may well be highly reactive when exposed to air. Although getters usually will also remove water vapour, Molecular sieve traps exhibit excellent capacity for removing trace levels of moisture from carrier gas. When saturated, these traps can be regenerated by baking them out in a GC oven at 300 °C with a flow of dry carrier gas.

Hydrocarbon impurities are rarely found in any significant concentration in commercial gases but their presence in the carrier gas and air lines can result in high detector background, instability and ghost peaks. Hydrocarbon and solvent contamination can be removed using activated charcoal. When using air supplied from a compressor it is essential that water and oil filters be used and monitored regularly. Generally compressors are not recommended for air generation because, apart from these potential contamination problems, there is a possibility of the pressure fluctuating at the speed of the diaphragm unless a large header tank is used. When analysing samples for low concentrations of hydrocarbons it is recommended to use 'Zero' air.

Some manufacturers supply traps that show a colour change when the traps are partially or fully saturated. These are made from glass or from a plastic material. Glass is rather fragile and must be mounted well out of the way of any area where metal objects such as sample containers are handled. Plastic materials may be permeable to oxygen and not advisable on a carrier gas line. Non-indicating traps are generally contained in a metal housing for strength and ruggedness. Indicating traps have the advantage that it is possible to visually determine when a trap must be exchanged. Non-indicating traps are marked with the date when they are installed and the expected exchange date is diarised, usually on the control PC. Some installations use a high capacity non-indicating trap in series with an indicating trap The non-indicating trap is installed upstream of the indicating trap so that a colour change on the indicating trap shows that both traps need replacement.

The order in which the traps are placed in the carrier gas line and their proximity to the GC is very important. The Molecular sieve trap should be placed closest to the carrier gas cylinder to remove moisture, and prevent condensation in the carrier gas line. If a hydrocarbon trap is used it should be placed next to prevent hydrocarbons contaminating the oxygen trap. The oxygen trap must be placed closest to the GC and as close as possible to the GC bulkhead. If the detector used requires a separate make-up gas ensure that this line also contains traps to reduce background. For example, the make-up line for an Electron Capture Detector must have water vapour and an oxygen trap installed. In many cases the make-up gas flow is considerably higher than the carrier gas flow and therefore the purity of the make-up needs to be as good as or even better than the carrier gas purity.

Traps are installed vertically to prevent channelling due to the settling of the trap material. A horizontal trap could allow gas to pass through a channel having little interaction with the packing material. One of the most important decisions regarding gas purification will be the sensitivity of the detector towards certain impurities. Care must be taken to ensure that the correct gas is coupled to the correct bulkhead on the back of the equipment.

Since each gas line purifier has a finite capacity, it is recommended to fit one purifier per instrument rather than having one filter per line supplying a manifold. The danger of any manifold system is that any contamination from the feed gas will simultaneously contaminate all instruments that are pulling gas from the manifold. If any reactor vessels are using the same gas from any manifold it is essential that non-return valves be fitted on the reactors. However, these non-return valves have been known to fail and cause serious contamination in the GC lines. Clearly it is important to exchange saturated filters as rapidly as possible, especially those that are sensitive to oxygen and water vapour. Any exposure to air will reduce the capacity of that filter cartridge. Some filter bases are fitted with valves that automatically isolate the gas lines while the filter is exchanged. If the flow of gas is to be interrupted by changing a gas line filter ensure that the GC system is cooled beforehand.

Some form of trap may be necessary to prevent toxic gases being vented into the laboratory. When the inlet uses a split vent this would require a trap. Also any system such as backflush-to-vent requires attention as to where and how this effluent is handled. These traps will typically have an activated charcoal stationary phase and must not have high back pressure otherwise the electronic flow controllers could have difficulty in regulating correctly. The capacity of this trap will determine how many injections can be done before the trap needs replacing. Sometimes it is possible to vent the sample onto a trap that is mounted inside the GC oven and this is regenerated each time a temperature program is run.

If the detector is sensitive to air, such as the PDHID, a long thin length of tubing must be installed at the detector outlet to prevent back diffusion of the ambient air. Even if a total flow of 40 mL/min is used, a length at least 5 m of 1⁄16″ × 0.75 mm i.d. metal tubing is required.

2.3.6 Leak checking

While plumbing any system it is important to leak check the system periodically during assembly and before any heating is done. Contamination of GC gases by the ingress of air will reduce trap lifetime and can cause damage to columns as well as some detectors that are particularly oxygen sensitive. Leak checking the instrument before the installation and conditioning of columns prevents column degradation which may produce high bleed and short lifetimes. Irreversible damage can be done by exposing the column to oxygen at elevated temperatures. Leak checking must be performed from the cylinders to the GC, as well as all the fittings inside the GC.

The first test done is to monitor the loss of pressure on the primary gauge of the cylinder regulator when the cylinder valve is closed. Depending on the type of flow controller used it may be easy to reduce flows or it may be necessary to blank off some connections to reduce flows to normal operating levels. Although the system will still be using gas, a single leak tight system should not deplete the internal volume of the regulator within 20 min. Any significant pressure loss within say 5 min represents a major leak that must be repaired before continuing with leak checking. Remember that each sampling and switching valve has two pathways and must be tested in both positions.

As discussed earlier in this chapter, if the gas cannot be detected by a leak detector, leak check the system using a 50:50 water-isopropanol mixture.

A portable electronic leak detector that operates on the same principle as a thermal conductivity detector is the best way of leak detection. Since these detectors use air as a reference only gases that have a thermal conductivity different from air will give adequate sensitivity. Helium and hydrogen give a signal that is regarded as positive while argon will give a so-called negative signal. For example the hydrogen analysis channel of a GC may use nitrogen as carrier and can therefore not be leak checked with an electronic leak detector. The same GC has a Flame ionisation detector whose hydrogen line can be interchanged during leak detection and the channel changed back to nitrogen once the leak has been corrected. Clearly the thermal conductivity detector will require some time to stabilise its sensitivity after switching carrier gas.

When leak checking inside the GC oven the oven fan must be turned off for several reasons. Firstly the oven temperature controller cannot regulate. Next there will be hot air blowing in your face and most importantly the leaking gases are mixed by the fan to mask the source of leakage and could even generate an explosive mixture inside the oven.

If one is searching for leaks on a hydrogen or helium line you must start from the lowest level and work systematically upwards to the highest level fittings. Conversely when looking for argon leaks start from the highest level fittings and work systematically downwards.

2.4 Tools, accessories and spares

Apart from the electronic leak detector mentioned above there are a number of tools that should be kept in the gas analysis laboratory. Many of these are available in kit form from some suppliers but these kits are often for general purpose GC applications which makes them incomplete as far as a gas analysis laboratory is concerned and need to be augmented by collecting the right tools for the job.

2.4.1 Gas handling

A reliable flow meter is essential. Several electronic flow meters are available and when choosing one, go for a meter that is pre-calibrated for hydrogen, helium, nitrogen and air. While these flow meters are very good for measuring relatively large flows such as the detector gas flows and column flow through a packed column, it cannot be used to measure the flow through a capillary column. To confirm that there is flow through the column the detector end of the column can be placed in a vial of suitable solvent such as methanol and checked for the presence of a stream of bubbles. To measure the column flow, use the unretained peak time as explained in Chapter 1. The bubble flow meter is arguably the most accurate way of measuring gas flows, but it should be kept in mind that the soap bubble is permeable to light gases such as hydrogen and helium. For really accurate flow measurement bubble flow meters should be corrected for changes in ambient pressure especially when not working at sea level. Electronic flow meters need to have a regular calibration schedule if accurate flows must be measured or reported. In practice, it is not always the absolute accuracy of a flow measurement that is important as very fine optimisation adjustment is done empirically for detector sensitivity or selectivity.

Another essential accessory is a good vacuum pump. When connecting sample containers to the sampling system on the GC, the line will be full of air and it must be removed to ensure the integrity of the sample being injected. While the air can be purged out using the sample if sufficient sample is available, it can create problems with discrimination as will be explained in the chapter on sampling. For small samples and samples close to or below ambient pressure, the sample line and sample inlet must be evacuated. For the same reasons the lines carrying the gas standards to the GC must be evacuated whenever a standard is changed.

2.4.2 GC tools

A few tools that no chromatography laboratory should be without include a ceramic column cutter for cutting fused silica capillary columns as well as a cutter for metal columns. While triangular needle files can be used to cut metal columns it is very difficult to obtain a clean, straight cut. The smaller the internal diameter the more easily the cut will block the tubing. Improperly cut tubing and especially columns, will affect the chromatography and it is therefore much better to use the correct tool for the job. A magnifying glass is necessary to inspect the ends of columns after cutting to ensure that the cut is clean and straight. Burring and cutting may also be responsible for fine particulates that damage valve cores and stators. Some deactivated glass wool may be necessary for packing into the injector inserts if equipped with an injector and it is useful as a filter whenever samples which may contain particulates are sampled. A pair of gloves must be used when handling glass wool and the hands should be washed afterwards to ensure that small glass particles are kept away from one's eyes.

A pair of tweezers to handle clean parts such as new injector inserts and septa is necessary and a ferrule removal tool that looks like a miniature corkscrew will make removing stubborn ferrules much easier. On some GC's the injector nut requires a special spanner and it should therefore form part of the toolkit.

2.4.3 General tools

It was already mentioned that in gas chromatography imperial and metric sizes are often mixed. Another anomaly is the spanner sizes necessary to correctly assemble fittings. $^1/_4''$ fittings require $^9/_{16}''$ spanners and not 14 mm, $^1/_8''$ fittings require $^7/_{16}''$ spanners and not 11 mm and $^1/_{16}''$ fittings require $^1/_4''$ spanners. The use of the correct sizes certainly prolongs the lifetime of any nut without rounding of the hexagon. While adjustable wrenches are certainly handy at home, their use in the laboratory should be avoided for the same reason. So a set of open-ended imperial sized spanners are needed and two of each of $^9/_{16}$, $^1/_2$, $^7/_{16}$, $^5/_{16}$ and $^1/_4$ inch spanners are required to ensure that the fitting can be held while tightening as in Fig. 2.13. The fitting or union must be supported while tightening each nut and the two nuts should not be tightened against each other. Some suppliers offer ring spanners with a cut-out to allow it to pass over the appropriate sized tubing but these spanners can be distorted when used too vigorously. Depending on the gases used there is also a need for large open ended spanners for tightening and loosening the bullnose nuts.

As was discussed earlier, good quality, sharp tubing cutters are needed for $^1/_4''$ and $^1/_8''$ tubing. Some cutters allow cutting both of these diameters but a special cutter is required for $^1/_{16}''$ tubing. A set of fine files including flat, triangular and round files are useful to smooth edges. When installing

Fig. 2.13 Two spanners are used with one supporting the fitting while the other is used to tighten the nut.

tubing, a tube bender is required for thick walled $^1/_8''$ tubing as well as for $^1/_4''$ and thicker tubing as described earlier.

One way of regulating the flow at a fixed pressure without the risk of ingress of ambient air, is by using flow restrictors in the $^1/_{16}''$ tubing used for the connection of columns and valves. These flow restrictors can be very precise and the authors have built entire GC's and standard dilutions systems using flow restrictors made by flattening the tubing to the point where a specific flow is obtained at a certain pressure as shown in Fig. 2.14.

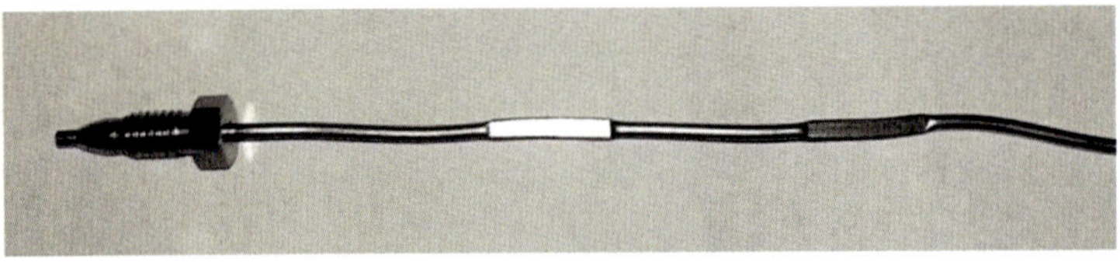

Fig. 2.14 Two series fixed restrictors in $^1/_{16}''$ tubing made by flattening the tubing.

Restrictors can be made by connecting the tubing to a source of gas at a set pressure at the one end and connecting a flow meter to the other end. The tubing is clamped between two pieces of square bar in a bench vice and flattened until the desired flow is achieved as measured with the flow meter. A useful tool for making these restrictors can be made by replacing the cutting edges of a bolt cutter with pieces of square bar as in Fig. 2.15. It is used in the same way as the vice but is more portable and allows restrictors to be made in situ.

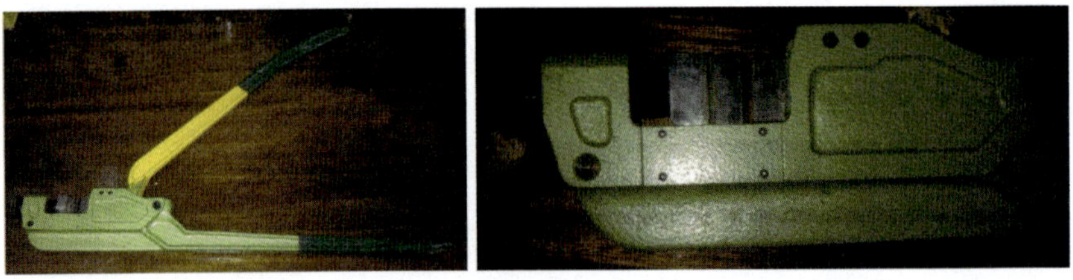

Fig. 2.15 A bolt cutter that was modified to press restrictors into $^1/_{16}''$ tubing.

Typically we use $^1/_{16}''$ tubing with a 0.75 mm internal diameter and first lightly crimp the tube. Then the gas supply is connected to the tubing at the operating pressure anticipated in the application and the tubing further crimped until the desired flow is achieved. The tubing does tend to spring back a little when the pressure of the crimping jaws is released. One quickly gets the feeling and crimps to a flow of say 15 mL/min which will give a nominal flow of 20 mL/min when released. The tubing must be kept straight after crimping and a mechanical support improves the robustness of the restrictor. For very low flows it is often easier to make two restrictors in the same tubing a few centimetres apart. The first will reduce the flow to within the range of flow required while the second fine tunes the flow to the desired flow. Each restrictor can be calibrated at a certain pressure for a particular gas and at a certain temperature. For example a restrictor can be made to deliver a flow of 24 mL/min helium at room temperature and at a pressure of 400 kPa.

A set of small screwdrivers with flat and Philips heads as well as two sets of Allen keys, imperial and metric, should be available. Many instruments are now supplied with screws having Torx™ heads and therefore appropriate Torx™ keys are required although these are often supplied with the original purchase. There are many screws of different sizes in a GC and it is useful to have these tools at hand. Another useful tool is a magnet on a stick – one corollary of Murphy's Law states that a small screw, when dropped will roll into the most inaccessible corner. With the magnet this can usually be retrieved without dismantling the entire GC. A small pair of side-cutting pliers as well as a pair of long nose pliers will also not be out of place in the toolbox. Also have a look at the list of tools at the end of Chapter 12 on troubleshooting.

A small flashlight is always handy since GC ovens, unlike their counterparts in the kitchen do not often come with oven lights. An essential tool is a multi-meter as it will help a lot when doing troubleshooting provided the person using it is trained to work with electricity. While one might argue that many of the tasks for which these tools are needed fall within the duties of the service engineers of the instrument suppliers, some GC's for specialised gas analysis can be so customised that the operator often knows the system better than the supplier. Furthermore, being able to trace the problem can save a lot on service calls and we cannot even begin to count how many problems were resolved by combining a skilled operator and remote telephonic support.

2.4.4 Spares

It is very frustrating when you need to make a connection, say between the sample container and the gas inlet but you do not have the correct fittings to make a gas tight connection. It is therefore advisable to keep some spares in a drawer for this purpose. Also, once a fitting was made for connecting, for example a container with a $^{1}/_{4}''$ NPT fitting to the $^{1}/_{16}''$ gas inlet on the GC, keep it as it might be needed again. The list in Table 2.3 is by no means comprehensive but may serve as a useful starting point for your collection.

Table 2.3 A list of fittings that could conceivably be kept in the laboratory. Depending on the type of analyses and sample containers received the list may be modified to satisfy the requirements of the laboratory.

Item	Number
$^{1}/_{4}''$ stainless steel or brass if copper tubing is used	2× unions 2× T-pieces 2× Crosses
$^{1}/_{16}''$ stainless steel	4× unions 4× T-pieces 2× crosses
$^{1}/_{16}''$ stainless steel pre-cut and cleaned tubing	2× 10 m roll, 5× 500 mm, 5× 300 mm, 5× 100 mm lengths
$^{1}/_{4}''$ to $^{1}/_{8}''$ reducers	4×
Ferrules — stainless steel or brass if copper tubing is used	1× set of 10 — $^{1}/_{4}''$ 1× set of 10 — $^{1}/_{8}''$ 1× set of 10 — $^{1}/_{16}''$
Reducing ferrules — graphite Vespel	5× $^{1}/_{4}''$ with $^{1}/_{16}''$ hole 5× $^{1}/_{8}''$ with $^{1}/_{16}''$ hole
$^{1}/_{4}''$ graphite Vespel ferrules for injector insert	5×
$^{1}/_{8}''$ stainless steel or brass if copper tubing is used	4× unions 4× T-pieces 2× crosses
$^{1}/_{16}''$ stainless steel VICI fittings for trace gas analysis	4× unions 4× T-pieces 2× crosses
Polymeric tubing (Teflon or PEEK)	1× 5 m × $^{1}/_{8}''$ and 1× 5 m × $^{1}/_{16}''$
$^{1}/_{8}''$ to $^{1}/_{16}''$ reducers	4×
Ferrules — stainless steel VICI for trace gas analysis	10× stainless steel 10× gold plated stainless steel
Graphite Vespel ferrules for column connection	10× for 0.53 mm id 10× for 0.32 mm id 10× for 0.25 mm id
Manual valves ($^{1}/_{4}''$, $^{1}/_{8}''$ or $^{1}/_{16}''$ depending on the tubing used)	2× on-off valves 2× three way valves

If the system contains an injector, at least 50 low bleed septa for each different injector in the laboratory as well as at least 10 clean injector inserts should be kept in stock and it is also a very good idea to keep one additional multi-port valve for each type (4, 6, 8, 10 port) of valve used in the GC system. One additional micro-electric actuator is usually sufficient as they can be used on any of the valves but if pneumatic actuators are used then a

90 degrees, 60 degrees, 45 degrees and 36 degrees actuator will be needed respectively for the types of valves (4, 6, 8, 10 port) used in the laboratory.

Additional bits and pieces to keep in a drawer would typically include thread sealing tape, cable ties and insulation tape. A small amount of Presstik® on the end of a screwdriver can also be a great help in picking up small parts in inaccessible areas.: Presstik®, a reusable adhesive made from papyrus plant material was invented in South Africa and is a licensed trademark of Bostik Ltd. England. It is also known as 'Blu Tack' in USA, 'White Tac' in Germany and 'Elefantsnot' in Denmark. Apart from the isopropanol for leak checking, solvents like acetone and methanol are good to have available for cleaning and it is also a good idea to keep vacuum pump oil in stock.

While on the topic of oil, under no circumstances should any form of thread lubricant be used on fittings and tubing. There is always a danger that some of the lubricant could enter the system which will cause some reaction between this and the sample or carrier gas. However, high temperature thread lubricant is required on some screw threads that hold heated components in place for example a detector tower.

References

[1] J. Hinshaw, Preparing the laboratory for a gas chromatograph, LCGC Europe 21 (2) (2003) 130–135.

[2] E. Guthrie, D. Vassilaros, Frequently Asked Questions About Chromatographic Gases – A Practical Primer, Air Products and Chemicals, Inc., 2003 (40604), 2003. [Online]. Available: http://www.airproducts.com/~/media/Files/PDF/industries/analytical-laboratories/analytical-labs-faqs-chromatographic-gases.pdf.

[3] F. Miller, A. Vandome, J. McBrewster, Darcy-Weisbach Equation, VDM Publishing, 2010.

[4] Darcy-Weisbach Pressure and Head Loss Equation, The Engineering Toolbox [Online]. Available: http://www.engineeringtoolbox.com/darcy-weisbach-equation-d_646.html.

[5] Valco Instruments Company Inc, Valco VICI Cheminert Catalog 65 INT, VICI AG International, Switzerland, 2012.

[6] Valco Instruments Co Inc, Fitting Instructions. Technical Note 503 Rev 11/06, Switzerland: VICI AG International.

[7] J. Hinshaw, Optimizing detector set-up and operating conditions, LCGC Europe 22 (3) (2009) 138–142.

CHAPTER 3

The gas chromatograph

Contents

In Chapter 2 we considered the correct way of getting our gases to the GC hardware and reducing the cylinder pressure from the MPa range to a value of several hundreds of kPa. Here we will discuss the precision pressure and flow controllers used inside our gas analysis system for our carrier and fuel gases. We will also look at the building blocks that make up a modern gas chromatograph and finally try to offer advice on the choice the 'ideal' gas analysis system.

The modern gas chromatograph consists of four interdependent chromatography modules, controlled by several temperature controllers that may communicate with one or more microprocessor systems. A fifth module that is vital to the overall performance is the chromatography data system which will probably be stand alone.

3.1 Gas supply

The first module is the gas supply (pneumatic) unit which provides all the necessary gas supplies and may involve a number of different gases, depending on the type of detector used. Each unit must be calibrated for

A practical guide to gas analysis by gas chromatography
ISBN 978-0-12-818888-0
https://doi.org/10.1016/B978-0-12-818888-0.00003-6

the specific gas that will be used and will probably compensate for ambient temperature and pressure changes as well as compressibility of that gas. For example, a thermal conductivity detector may operate with only carrier gas while a flame ionization detector will require hydrogen, air or oxygen to support combustion as well as carrier gas. Thus, for the FID, a minimum of three different gases would be required which will also involve the use of three flow controllers, three flow monitors and possibly a flow programmer for the carrier gas. These gas supply units could be controlled by a microprocessor to set the required flow rates, to monitor the actual flow rates, and adjust when and if necessary. The pneumatic system must be designed and installed in such a way to ensure that no degradation to the quality of gas in the cylinders occurs before it enters the analytical system. The quality of gas in the cylinders can quickly be destroyed by using a sub-standard regulator or the wrong tubing or inferior fittings and connectors. Once assembled, the system has to be checked for leaks as described previously. Only when this is all correct should we consider how to do pressure and flow regulation for the GC.

Cylinders pressure regulators and their pressure gauges have been covered in the previous chapter so only line pressure regulation will be discussed. The simplest form of control is the pressure controller which responds rapidly to changes in setting and the pressure is commonly increased by a clockwise rotation of the control knob. Although this gives a precision control of pressure it does not compensate for any resistance changes in the column system. The column resistance is temperature dependent and constant carrier gas velocity is important for optimum separation.

3.1.1 Manual pressure and flow control

We have seen that pressure controllers use a diaphragm set under stress by a spring loaded control and the pressure is regulated. We also saw that pressure was increased by clockwise turning of the shaft.

A manual flow controller is similar to a pressure regulator with a feedback across the membrane that controls the differential pressure across the diaphragm, Fig. 3.1. This means that it will compensate for changes in resistance but takes some time to stabilise after changing the setting of the flow controller. The flow is commonly increased by turning the control knob in an anticlockwise direction. This is analogous to opening a water tap which is also a flow controller. However some controllers use counting dials which mechanically make it appear as though the flow is increased by turning the knob clockwise.

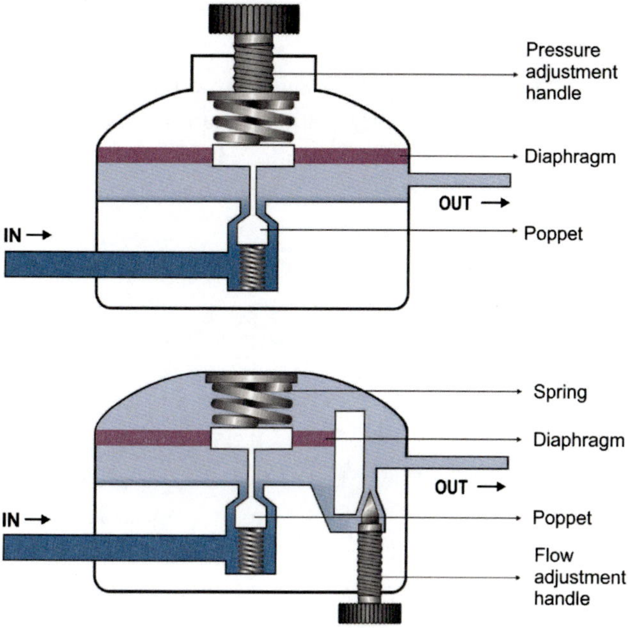

Fig. 3.1 Schematic of a pressure regulator (top) and manual flow regulator (bottom) showing the difference in operation.

Flow controllers are set for a particular flow and the head pressure is regulated to achieve this flow and will vary according to the resistance to flow. They are therefore rather slow to respond to changes in resistance or pressure. While this was acceptable in packed column applications it is not good enough for modern applications. Another drawback of these flow controllers is that they do not have the sensitivity to deliver low flow rates. These controllers require that the inlet pressure be at least 80 kPa higher than the outlet pressure before they are able to control the flow effectively. Higher pressure differentials improve their performance. There are two versions available; one is referenced to upstream pressure while the other is referenced to downstream pressure.

Even with packed columns the slowness of the flow controller has always been frustrating with valve systems, especially those using column switching. For sampling as the valve is actuated we go from a normal and stable flow, through a short period of zero flow as the valve core rotates, to a situation where a significant sample volume is added that is most likely at ambient pressure. The result is that the mass flow controller will drop its flow by a large amount and the flow will gradually increase back to the set value over a period that may be a minute or more. Despite the shock absorbing effect of the column system, the detector flow will also reduce and slowly recover. On any flow sensitive detector this will be seen as a sloping baseline. This means that any early eluting component will be very difficult to quantify especially at low concentration. Hydrogen in transformer oil is just such a situation and is a vitally important diagnostic in the health of the transformer.

This problem can be partially overcome using a pressure controller in parallel with the mass flow controller as in Fig. 3.2. In this configuration, the pressure controller is backed off while the flow controller is set to the required flow at the temperature of injection and the column head pressure is allowed to stabilise then read. For example, the head pressure is 220 kPa gauge at a flow of 20 mL/min. Now back off the flow controller and the pressure will drop for the lower flow. Increase the pressure regulator to 220 kPa on the pressure gauge and reset the flow controller to 20 mL/min. At the point of injection when the mass flow controller drops its pressure, the pressure controller will assume control and maintain flow. Flow control is thus being done by the mass flow controller but when there is a valve switch the pressure controller will take instantaneous control of flow. The mass flow controller will take some considerable time to recover but should take control of the flow before the temperature program begins. This minimises baseline fluctuations due to flow changes. Using modern electronic pneumatic controllers this principle can be similarly applied.

Needle valves are also flow controllers although they do not operate as mass flow controllers. They have limited flow ranges and the correct procedure is to operate the needle valve close to its midway position.

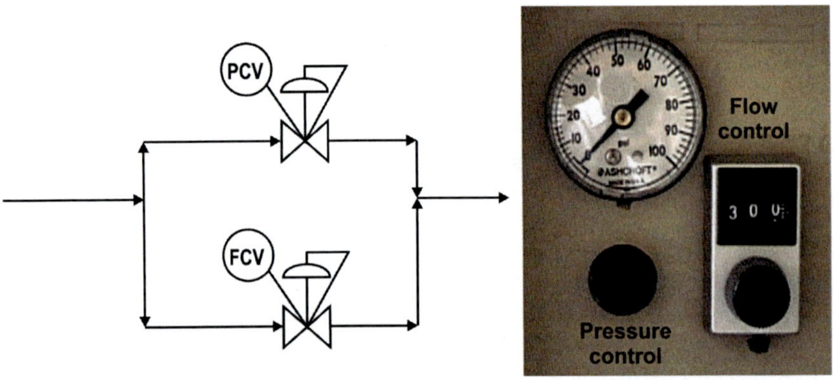

Fig. 3.2 A manual flow controller in parallel with a manual pressure controller.

They are used to balance system resistance after a valve switch as they are not stable enough for accurate column flow control, especially if there are resistance changes due to temperature programming.

Micro-metering needle valves, Fig. 3.3, with standard Viton® seals can handle temperatures up to 225 °C and for high temperature operation, valves with Kalrez® can be used continuously at 315 °C. Because of this, these needle valves can be mounted directly within a heated oven preventing the chance of cold spots in the analytical system. These micro-metering valves can handle gas pressures up to 7 MPa. For stringent applications such as pulsed discharge detectors, un-lubricated valves are available and must be fitted prior to any in-line gas purification system.

Fig. 3.3 A selection of micro-metering valves [1]. ©VICI AG International. *(Reproduced with permission, courtesy of VICI AG International.)*

Needle valves should never be used as shut-off valves otherwise the valve seat will soon deteriorate. Needle valves are also useful to control the sample flow through the sample loop when doing dynamic sampling.

The so-called combo valves combine a needle valve with a shut-off valve and have typically been used for control of hydrogen and air flow to detectors, Fig. 3.4. These valves have a maximum temperature of 100 °C and a maximum pressure of 700 kPa.

In capillary GC the use of back pressure regulation (BPR) on the split vent became customary as this responds as quickly as a normal pressure regulator when large flow changes are made by valve switching such as splitless injection. Although this injection technique is not applicable to gas analysis, back pressure regulators are used for split injection as discussed in

Fig. 3.4 A selection of combo valves combining needle and shut-off valves with different coloured caps for use with different gases [1]. ©VICI AG International. *(Reproduced with permission, courtesy of VICI AG International.)*

the following section. Back pressure regulation can also be used on the vent line of a sampling valve to control the sample loop pressure even when the pressures in the sample lines are not constant. A schematic of back pressure regulator is shown in Fig. 3.5. The pressure on the BPR is increased with clockwise rotation of the knob in the same way as the pressure regulator discussed above.

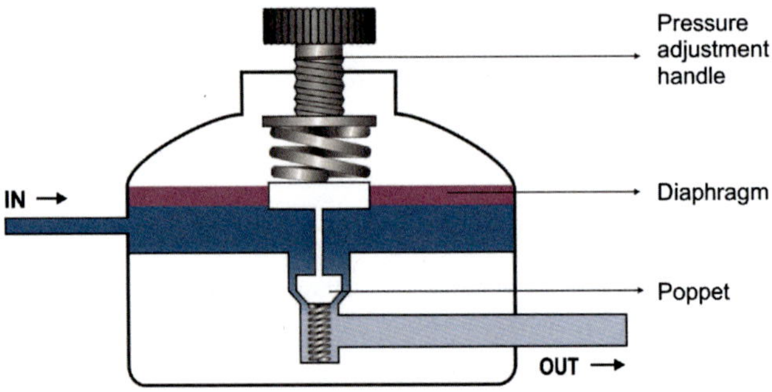

Fig. 3.5 A schematic diagram illustrating the operation of a back pressure regulator.

As described in Chapter 2, fixed restrictors can be used for flow control and can be made simply by crimping clean, thin walled, metal tubing between two jaws until a particular flow is reached through the tubing. Fixed restrictors can also be made by connecting a length of narrow bore tubing or even a piece of narrow bore column, into the line in which the flow must be controlled. It is however quite difficult to determine the length and diameter of this piece of tubing required to obtain the desired

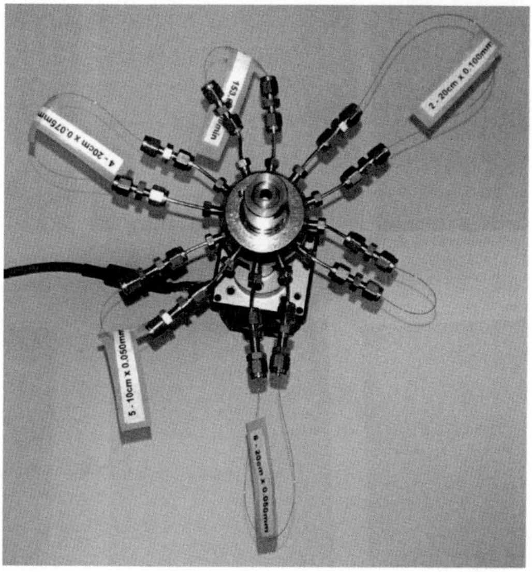

Fig. 3.6 Photograph of narrow bore fused silica tubing used as fixed restrictors in 1⁄16″ tubing on a stream selector valve.

flow at a specific pressure. The stream selector valve in Fig. 3.6 has a number of restrictors providing different flows and was used in an in-house constructed, inexpensive, gas dilution system. Fixed restrictors are also commercially available to provide a specific flow at a set inlet pressure and specific temperature.

Systems that are designed for the analysis of ultra-trace amounts of oxygen and nitrogen must be constructed using fixed in-line restrictors for all flow controls as the ingress of air through polymeric materials cannot be tolerated. Clearly each restrictor must be made individually as changing the resistance in any leg of a flow system will influence the flow in every other leg of the system.

3.1.2 Electronic flow controllers (EFCs)

The gas flow through any GC column will change with temperature, but unlike a liquid, a gas flows slower at elevated temperature. At constant pressure, the flow through a capillary column will almost halve if the temperature is increased from 50 to 300 °C. Because the efficiency of columns is dependent on the velocity of the carrier gas, the temperature dependence of gas flow prevents optimisation of the separation during the temperature program. To overcome this shortcoming a controller that effectively increases the column head pressure with the increasing temperature is needed such that a constant flow and therefore constant velocity is maintained throughout the analysis.

One of the most accurate ways to obtain constant flow rates is by using thermal mass flow controllers (MFCs). There are commercial units available that will do this on a rather large scale and are typically used to control the gas feed to reactors. However for gas chromatography we have special requirements and modern instruments are equipped with miniature electronic flow controllers, as shown in Fig. 3.7, that provide extremely high resolution and are software controlled to provide excellent flow control. As this technology has developed over the past years the resolution of these digital flow controllers has improved and extremely fine control of gas flows is a reality.

Fig. 3.7 Picture of a bank of electronic flow controllers that would be typical of a GC configured for multi-dimensional gas analysis.

The system configuration determines the type of carrier gas used and the system monitors ambient pressure and temperature and uses this information to correct for the desired flow. Since the typical GC used for capillary columns uses flow control for the column, flow control of split outlet and flow control of the septum purge, a three channel electronic flow module is used. For gas analysis, this may be a little over the top but is still very convenient to have. Some GC's also have auxiliary flow modules which can be configured for different gases and may be used to blend external gas mixtures or to do sample and standard dilution.

Most of the modern systems also use electronic flow controllers for their detector gases. Each is calibrated for the appropriate gas and cannot be exchanged without reconfiguring the flow controller for the correct gas. This gives excellent resolution of flow adjustments for detector optimisation and also long term stability. These flow controllers have built-in start up routines that program optimal flows for detector ignition and then control the flows at the set values for operation.

These modules can also be called electronic pneumatic controllers (EPCs) and control the gas flow by opening and closing proportional solenoid valves which must be calibrated for the specific gas used. The system must be configured in the software for control of the correct gas so that it will automatically apply the correct values of the physical properties of that gas in calculating the correct flow, pressure and velocity. If the system is configured for the incorrect gas, all calculated flows will be incorrect and the system might never go to a 'Ready' state. Even if the microprocessor does allow the system to go 'Ready' the chromatography will be compromised. Only the operator knows which gases are connected to which control modules and should this be changed, the software must be reconfigured. Detector fuel gas controllers are similarly calibrated for the specific gas chosen in the system configuration.

A significant advantage of EPC is that applications can be readily transferred from one instrument to another by simply transferring the method.

3.2 Sample introduction

The second module of the GC is the sample introduction unit, usually an injector that is situated inside a thermostatically controlled enclosure. For gases, a gas sampling valve is typically used for injection and can share the column oven although it is best situated in its own temperature controlled oven otherwise the column oven will be limited to the maximum temperature that the valve can tolerate. This could, for example, prevent column conditioning in that oven. For liquid samples there are many automated sample introduction systems that can carry out a very complex series of sample preparation procedures before injecting, but we will concentrate only on injectors that may be used in gas analysis. Generally gas samples are prepared off-line in the gas sampling system and are therefore regarded as ready for injection. Certain applications require headspace sampling and trapping and desorption of trace analytes; these are discussed in some more detail in Chapter 6. It is possible to have the gas analysis GC connected to one or more sample selection valves that can be controlled from the GC or externally allowing automated analysis of selected streams in a predetermined sequence.

A packed column injector may be used when only packed columns are used and a standard split-splitless (capillary) column injector for capillary columns, Fig. 3.8. There are several options available for the coupling of

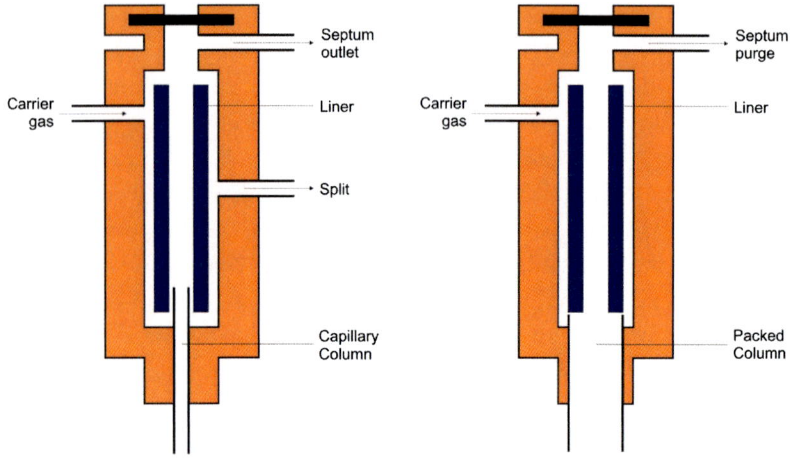

Fig. 3.8 Schematics of a split-splitless (left) and a packed injector (right).

wide bore, 0.53 mm i.d., capillary or micro packed columns and adaptors, for example, when using packed columns with a split-splitless injector.

In liquid analysis the injector is a vital system component to rapidly vaporise the sample and convert it to the gas state. As this transition takes place there is usually a large expansion as the gas volume may be several hundred times the liquid volume. The injection bandwidth must still be as narrow as possible for good peak shape. For this reason there are a variety of injectors and injection techniques that need to be selected for the liquid analysis to be done. In gas analysis only a few high pressure gas samples may liquefy under their own vapour pressure and require introduction using a liquid sampling valve. Since the majority of gas samples are already in the gas phase the injector is purely a mechanical connection between the sample valve and the column whose main requirement is to have zero dead volume. To minimise the chances of dead or unswept volumes it is preferable to mount the column directly on to the injection valve using appropriate fittings.

The internal volume of the injector is determined by the internal diameter of a deactivated glass liner. Typically a liner with the smallest internal volume would be chosen for gas injection using a sample loop. For gas injections less than 250 μL a liner with a 1 or 2 mm internal diameter is recommended while for volumes larger than 250 μL a 4 mm internal diameter liner is preferred, all with tapered inlets. When the sample is injected with a gas-tight syringe and split injection is done, the recommended liner is a large volume, 4 mm liner with no restrictions [2].

The different methods for introducing a gas sample into the GC are discussed in detail in Chapter 6.

3.2.1 Split injection

Because of the volume expansion of liquid samples upon vaporisation, it becomes nearly impossible to directly inject a small enough representative sample from a liquid. The most satisfactory way of reducing sample amount injected onto the column is to split a large portion of the evaporated sample to vent as shown in Fig. 3.9. This is done in the split injector. A very simple split injector can be made from a T-piece and the split leg controlled with a needle valve or restrictor but the reproducibility of split will not be acceptable for precision analysis. The commercial GC will probably be equipped with a so-called 'split/splitless' injector, Fig. 3.8, but the splitless injection technique with sample or solvent focussing cannot be applied in gas analysis since the initial column temperature could not easily be set below the boiling point of the matrix gas. Likewise pulsed splitless injection and other modern injection techniques such as programmed temperature vaporisation (PTV) are not used in gas analysis and therefore not discussed further. We do, however, briefly describe their application in the analysis of volatile organic compounds and introduce the concept of 'stacked injection' in Chapter 6.

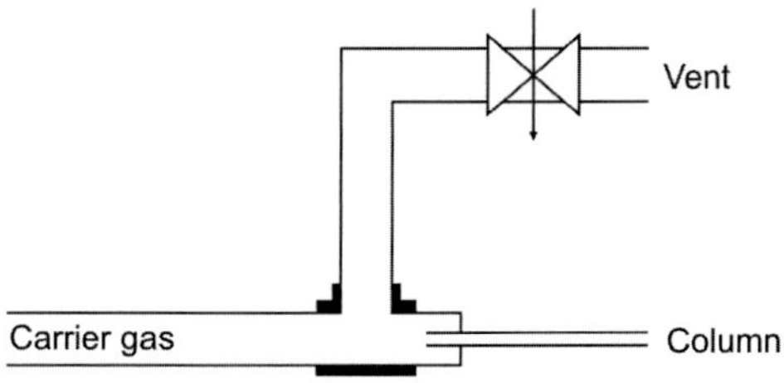

Fig. 3.9 Schematic of a simple splitter.

Split injectors are known to show discrimination towards high boiling compounds but this is not relevant in gas analysis. When performed properly, split injections are easy to automate, produce narrow peaks and yield consistent run to run peak areas.

When the split ratio on the split injector is set to 100:1 it means that only 1% of sample enters the column and 99% is vented as shown schematically in Fig. 3.10. By using the split, which has the effect of diluting the sample with carrier gas, high concentration samples can be successfully introduced without overloading the column. If a large sample volume must be injected

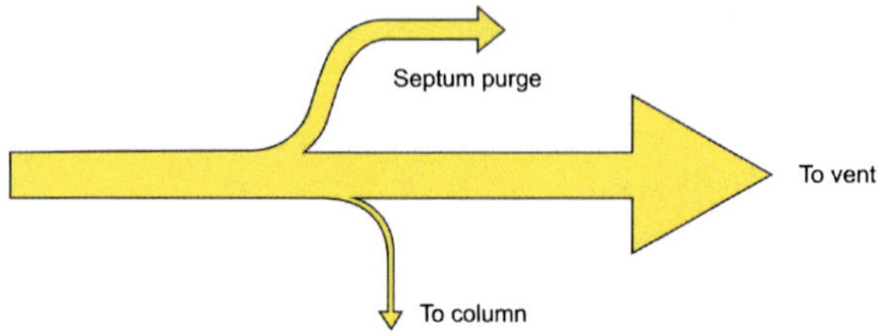

Fig. 3.10 The relative flows through the septum purge, vent and column.

into a low capacity column, the injected bandwidth will be broad. Consider that if one injects a gas sample of 1 mL into a column having a flow of 3 mL/min, the narrowest peak will be 20 s wide and will also be flat topped. By splitting the sample, the effective size of the sample loop can be adjusted for different sample concentrations and the injection bandwidth kept small.

3.2.2 Injector pneumatics

The split/splitless injector also has a septum purge which is a small flow of gas across the face of the septum to prevent volatile material from the septum or compounds that might have adsorbed onto it from entering the column. Although this is not really a problem in gas analysis the septum purge and its control is part of the injector controller. In gas analysis it is common to use only small split ratios which means that the split flow and septum purge flow are of similar magnitude and the influence of septum purge cannot be ignored. Closing off the septum purge would create an unswept volume that could contribute to peak broadening.

The split injector has an incoming flow that is divided into three streams, the column flow, septum purge flow and split flow. Obviously the sum of the three outlet flows is equal to the inlet flow. For example at a split ratio of 100:1 we would have 3 mL/min through the column, 297 mL/min out of the splitter and 5 mL/min through septum purge. The total flow will therefore be 305 mL/min which is controlled by the main proportioning valve. When the split is closed, the flows will be recalculated and adjusted to maintain the column flow as shown in Fig. 3.11. The large split flow must be properly disposed of especially if there is any possibility of toxic compounds. The electronic flow module has the ability to control flow or pressure. If any temperature programming is done flow control should be used. The controller then calculates the pressure necessary to maintain the set flow, split flow and septum purge flow at the current temperature. The split outlet is typically controlled using a back pressure regulator.

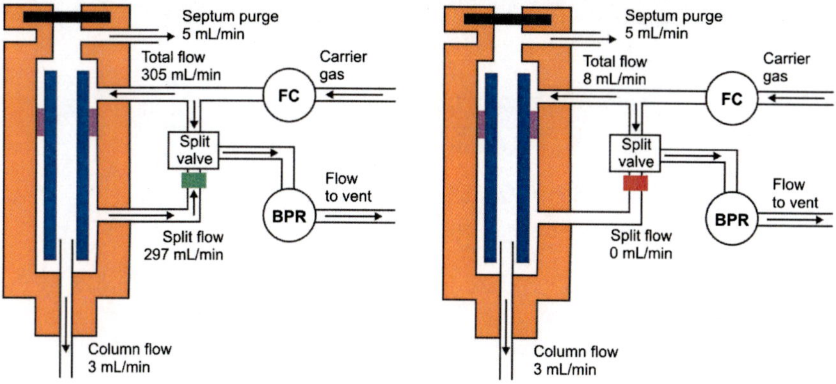

Fig. 3.11 The typical electronic injection flow controller flow paths in the split (left) and splitless (right) modes are shown where FC is the flow controller and BPR the back pressure regulator.

For most applications the column flow or velocity is set to a value slightly above optimum for the column and carrier gas used and then the split ratio is set according to the sample concentrations to be analysed. The column head pressure and the total flow are monitored but not controlled and are an important diagnostic which shows increasing head pressure as the column oven temperature increases so that constant flow is maintained.

Even when using electronic pneumatic control, there is merit in confirming injector performance from first principles particularly when doing troubleshooting. First measure the flows of septum purge and split outlets using a bubble flow meter or an electronic flow meter. As discussed in Chapters 1 and 2, the flow through capillary columns is difficult to measure accurately using bubble flow meters or electronic flow meters. To accurately determine the flow through a capillary column an unretained compound is injected and the linear velocity (and flow) is calculated from the retention time of the unretained compound. In gas analysis, this is usually done using the retention time of the least retained component and elevating the column temperature to minimise retention. Cleary there may be some error in this value but at least the calculated velocity will err on the slow side which means that the actual velocity will be marginally higher than calculated.

Once the carrier gas velocity has been set, inject a sample that contains an unretained component. For example, a 30 m column with an internal diameter of 0.53 mm column, operated at a linear velocity of 20 cm/s with helium carrier gas may elute hydrogen at 2.5 min. The split ratio was set to 100:1 and the flow meter gives a value of 3 mL/min for the septum purge and a split vent flow of 250 mL/min. The volumetric flow through the column can be calculated:

$$Flow = \frac{(\pi)(column\ radius\ in\ cm)^2(column\ length\ in\ cm)}{Unretained\ peak\ time\ in\ minutes}$$

$$Flow = \frac{(3.14159)(0.0265\ \text{cm})^2(3000\ \text{cm})}{2.50\ \text{min}} = 2.65\ \text{cm}^3/\text{min}$$

Since the septum purge flow has little influence in the splitting of the sample we can calculate the split ratio as [3]:

$$Slit\ \ ratio = \frac{Column\ flow + Split\ vent\ flow}{Column\ flow}$$

$$Split\ ratio = \frac{2.65 + 250\ \text{mL/min}}{2.65\ \text{mL/min}} = 95.34$$

This result of 95:1 is totally acceptable bearing in mind that there will always be some retention for every so-called unretained component.

The approximate amount that is injected on the column when using a split can be calculated with C_{inj} being the concentration injected, C_{Sample} the concentration of the sample in μg/μL and V_{Sample} the sample volume injected in μL.

$$C_{inj} = \frac{C_{Sample} \times V_{Sample}}{Split\ ratio}$$

3.2.3 Splitting and quantitation

Despite the use of splitting to accommodate a wide range of sample and standard concentrations, differing split ratios should not be used in quantitation. Although it might be temptingly convenient to use changes in split ratio to establish multipoint calibration curves, this should be avoided as it will not be valid. If the calibration curve has been established for a certain concentration range it is not good analytical practice to change the split ratio and assume that the concentrations can be calculated. If the standard is run at a particular split ratio and a calibration curve set up, the sample must be run under identical conditions. It is not acceptable to calibrate at a split ratio of 10:1 but to run the sample at 100:1 and extrapolate the result by multiplying by the ratio of the ratios as the detector linear dynamic range is being ignored.

3.3 The column and column oven

The heart of the separation takes place in the column that is mounted in the oven having very precise and stable temperature control. The actual separation occurs in a length of tubing packed with a stationary phase or that has a liquid stationary phase coated on the walls of an open tube. The entire GC system, complex as it may be, supports this critical component and deficiencies in any part of the GC have an adverse effect on the efficiency of the column.

3.3.1 The GC column

The choice of column packing can be a daunting task and this will be discussed in some detail in Chapter 7. All early separations used some form of selective adsorption on natural materials including fire brick. The introduction of aluminium silicate in the form of Molecular sieves added a further physical separation mechanism. The bonding of layers of active stationary phase on inert supports saw the beginning of partition chromatography. The separation takes place due to selective solubility of the sample in the stationary phase. Solubility is well known to be temperature dependent, for example sugar does not dissolve well in cold tea. However, in gas chromatography we are looking at the solubility of a gas in a liquid and this is the opposite of what happens when dissolving solids in liquids in that the solubility of gases in a liquid improves by lowering the temperature of the liquid. Gas that was dissolved in a cold liquid will escape as the liquid heats up, as is evidenced by the bubbles rising from cold beer, champagne or carbonated soft drinks when they warm up after opening. In general, compounds tend to elute in order of their boiling points except when physical separation techniques are used.

3.3.2 The column oven

Apart from the required precision and stability of the oven temperature it must also be able to be programmed to heat up at a specific rate to follow a temperature program. The oven temperature was originally programmed purely in a linear manner using electro-mechanical devices but modern multi-ramp temperature programmers contain a dedicated microprocessor for this purpose. High precision temperature sensors in the oven feed the information to electronics that constantly control the temperature program. To ensure that there are no temperature gradients between

Fig. 3.12 A GC oven fitted with three packed and one capillary column used for gas analysis.

different parts of the oven a large fan is employed inside the oven to circulate the warm air. A photograph of a GC oven set up for gas analysis is shown in Fig. 3.12.

From the theory in Chapter 1 it is clear that there is an optimum flow, or rather linear velocity, for a column with specific dimensions and a chosen carrier gas. Having set the column flow to its optimum the only way to change the retention times of compounds in GC is by changing the column temperature. To elute all compounds in a reasonable time, the oven temperature is often set at a low temperature at the beginning of an analysis for good separation of the early eluting components and then increased during the run to a higher value to also elute the compounds that are more retained. Temperature increases of about 3–10 °C/min are typically used but some applications such as fast capillary GC require programming rates up to 100 °C/min. It is however important that the heating rate should not be too fast as that reduces the efficiency of the column. As a general rule for capillary columns, the rate should not exceed 1 °C per 10 m of column meaning that for a 30 m column the ideal heating rate would be around 3 °C/min. In practise much faster rates are often used as dictated by the elution behaviour of the compounds in the sample. This is a function of using a column that is too long for the required separation and users are reluctant to cut the column length. For samples containing both very volatile and less volatile compounds such as volatile organics in a gas mixture, it may be necessary to start the run at a low temperature for the gases and then ramp up to a much higher temperature to elute the organic compounds. Temperature programmes are usually set up on the data system and microprocessor controlled during the run.

At the end of the temperature programme, the cooling of the oven to the initial temperature determines the total cycle time. Some manufacturers offer a programmed cooling fan that speeds up the cooldown cycle. Most of the column materials used are poor thermal conductors and the oven needs to cool to the initial temperature and allow some time for the inside of the column to reach the same temperature as the oven before analysing the next sample. For this reason all manufacturers have an oven stabilisation time and it is recommended that the stabilisation time should not be shorter than 2 min which obviously extends the run time by this amount.

Because many of the detectors used are flow sensitive, such as the TCD, temperature programming is often limited in gas analysis. Also valves that have a low maximum temperature are often mounted in the column oven and multi-column systems may have columns with different maximum allowable operating temperatures. In practice it is desirable that gas separating systems are designed to operate isothermally. This helps in allowing analyses to be developed for process GC's which have limited programming capability and usually require fast cycle times. In isothermal operation the choice of the correct stationary phase is critical since the luxury of experimenting with different temperature programmes is not available. In multi-column systems the dimensions are selected such that all separations take place at the same isothermal temperature as far as possible. If this is not possible the different columns are installed in separate independently heated column ovens.

If temperature programming cannot be avoided in gas analysis it is recommended to minimise the range and speed of the program unless some compromise on stationary phases is purposely being made.

3.3.3 Sub-ambient operation

As discussed in Chapter 2, an option that is sometimes used in gas analysis is sub-ambient operation of the column oven. Many of the solid stationary phases used in gas analysis are capable of operation below −50 °C as opposed to liquid stationary phases that could have a minimum operating temperature of 40 °C although some allow operation down to −30 °C. Below the minimum temperature the liquid stationary phases solidify and are therefore not able to dissolve the sample components. Since gas analysis involves separation of very low boiling point compounds lowering the temperature increases retention and increases k′. In some applications a different stationary phase at sub-ambient temperatures can be used to separate compounds in a different elution order to allow much easier trace impurity determination by eluting this before the matrix peak. For example the elution order of hydrogen, argon, oxygen, and nitrogen on Molecular sieve 5A becomes hydrogen, nitrogen, oxygen and then argon on Hayesep Q. A good separation is only achieved at −70 °C but this would offer an alternative for trace determination of oxygen in pure argon.

All instruments use either a liquid nitrogen cryogenic valve or a liquid carbon dioxide cryogenic valve. They are not interchangeable since the liquid nitrogen system operates at approximately 200 kPa and −196 °C and the liquid carbon dioxide operates at full cylinder pressure and at −78 °C. Liquid nitrogen cools by being spread in the oven by the fan while liquid CO_2 cools by expansion through a restrictor utilising the Joule-Thompson effect.

Although sub ambient operation offers an extra dimension in gas separations, there are several reasons why this is not used in robust and routine methods. The main disadvantage of sub ambient oven operation is the fact that mechanical noise is added by the switching of cryogenic valves which can adversely affect the noise level of some detectors such as TCD’s. For routine operation, it requires automated filling of liquid nitrogen or automated switchover for liquid CO_2 cylinders. Depending on the temperature that is needed for analysis, one could easily have the situation where a liquid CO_2 cylinder must be replaced every few hours. Because liquid CO_2 cylinders require adequate cylinder pressure to deliver its contents as a liquid, one may end up with a laboratory full of cylinders that are still partially full. Liquid CO_2 also tends to contain particulate impurities that must be filtered before the cryogenic valve and can cause blockages.

Vortex coolers are also available for cooling the oven but will only cool about 10 °C below ambient temperature. These devices use vast quantities of compressed air to cool and are rather noisy.

In conclusion sub ambient oven control is ‘nice to have’ especially for non-routine applications, but makes the overall system somewhat less user-friendly. However, many systems do use sub-ambient cooling to speed up the oven cool down from say 50 °C to 30 °C after temperature programmed runs to increase sample throughput.

3.4 Detectors

The fourth module of our generic gas chromatograph contains one or more detector, which is usually situated in its own oven. If there are any connections between the column and the detector, the volume must be very small, there must be no unswept volume and must be maintained at a temperature above that of the maximum temperature the oven will reach during analysis to ensure that no sample condenses in the tubing or detector. This means that a separate conduit heater may be necessary. Any condensation of sample will seriously increase detector noise which will also reduce the detector response and adversely affect the accuracy of the results. The detector oven is usually controlled by its own temperature controller and is operated isothermally since most detectors show some sensitivity changes with temperature. The output from the detector is analogue and is then electronically converted to a digital signal that can be processed by the computer, which calculates the results, prints out a report or sends the results directly to another computer system.

Detectors can be either mass dependent or concentration dependent. Detectors are also either universal or selective in their response. A choice must be made from the wide range of detectors available depending on the concentrations and components to be measured. Each detector has unique operating parameters and performance characteristics. These are discussed in detail in Chapter 8.

3.5 The data system

The final unit of the separation system is probably external to the gas chromatograph but essential to operation of the whole analysis. Today most GC's use PC based chromatography data systems whose software has been designed specifically for calculating chromatographic data as well as chromatographic performance. Most data systems calculate noise as well as peak widths which allows the correct setting for peak threshold. It is good practice in complex separations to monitor performance such as resolution between two closely eluting peaks. Although chromatography is a separation technique, we claim to 'identify' components by their retention times relative to known standards. Most applications require quantitation and results are quoted in amounts or in concentrations. Several different quantitation techniques can be used depending on sample type. This is further discussed in Chapter 9 on data handling. Another important function of the data system is that it provides the numerical data required for method validation and uncertainty estimations.

Unattended operation may require the system to start the whole sampling process including selection of the standard or correct sample stream at some predetermined time, analyse the samples, calculate results, send these to a central database and communicate error conditions to the appropriate authority. In such a case it will probably be necessary to connect the PC on a network system. In general, it is always recommended to operate chromatography data systems from dedicated PC's that are not networked. Full administration rights are normally required to change methods while limited access can be given to users of the systems. A problem that is quite common in large organisations is that the IT department implement a new network strategy that requires renaming the PC. This may immobilise the chromatography data system if the computer name is stored in the system registry file.

3.6 The ideal GC for gas analysis

Probably the most frequently asked question is 'what GC should be bought for my application?' As far as the authors are concerned the market is so competitive that many instruments are comparable in performance. However, when considering their suitability for gas analysis there are significant differences. GC's used for gas analysis differ significantly from the ones used for liquid injection volatiles analysis. Apart from one or more gas sampling valves it is quite common to have valves for the selection and redirection of parts of the sample, either at the start or after passing through the first of two or more columns. More than one channel is often needed with a number of detectors being used in parallel or series, depending on what must be analysed. A photograph of the top of a GC set up for gas analysis with multiple valves and detectors is shown in Fig. 3.13.

It is also important to consider the availability of controlled electrical switches for valve actuation or other external switching. When these facilities are insufficient for a particular application these events can be supplemented by switching directly from the data system.

There are several reputable companies that will supply custom built GC's for specific applications such as refinery gas analysis and natural gas analysis. Provided these suppliers are given the correct information, suitable systems can be custom made for just about any application. The advantage

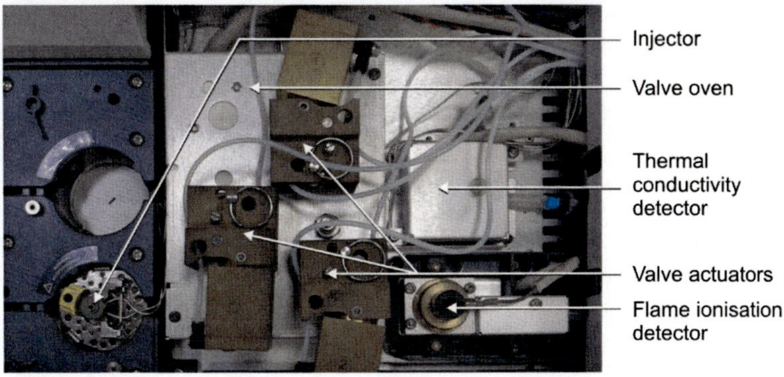

Fig. 3.13 Top of a GC set up for gas analysis showing the injector, valve actuators and detectors.

of the commercially customised systems is that the software is normally also customised to process the type of data generated by the instrument making it ideal for routine applications.

For a system used for method development in a R&D environment where sample compositions change frequently and where the analytical requirements vary, a generic gas analysis GC can be useful. However, such a system does not exist and its configuration will depend on what has to be analysed. The following chapters, especially those on detectors, multi-dimensional chromatography and special applications may offer more guidance and it will be useful to read through that before placing an order. Having said that, a system capable of simultaneously accommodating three injectors and three detectors will provide a much more flexible choice of configurations and is therefore preferred. Even if all the injectors are not used, the flow controllers that come with them can be useful for some multidimensional applications. The ability to accommodate capillary, micro-packed and packed columns will also ensure that the system is less limiting when it comes to the choice of columns. While valves are commonly used for stream switching, purchasing a system with a Dean's switch installed can be advantageous, especially if capillary columns are envisaged. Another useful feature available on most GC's is a separate valve oven and for some applications a separate GC column oven can be useful. In some specialised cases the GC hardware can be very simple although the application can be quite complex. For example we have built systems for trace analysis of gaseous impurities in electronic grade gases without using a GC. For this isothermal application an incubator was used to house the valves, columns and detectors while all flows were controlled by restrictors in the connecting tubing. Valve switching was controlled directly from the chromatography data system.

Many gas chromatographs are supplied with independently heated auxiliary zones which can be customized to heat valves, columns or post column reactors. These are particularly useful when the application requires trapping some components on a column at one temperature and then desorbing them at an elevated temperature. Some systems use external ovens for such applications but it is imperative that all temperatures and valve switching be done from a single system controller.

At this stage it is premature to discuss the benefits of fast, ultra-fast or Micro-GC systems for gas analysis and this will be introduced in Chapter 10.

References

[1] Valco Instruments Company Inc, Valco VICI Cheminert Catalog 65 INT, VICI AG International, Switzerland, 2012.

[2] A. Sensue, ChromaBLOGraphy: Which GC Injection Port Liner to Use for Gas Samples, October 9, 2014 [Online]. Available: Restek.com https://blog.restek.com/?p=13283.

[3] D. Rood, A Practical Guide to the Care, Maintenance, and Troubleshooting of Capillary Gas Chromatographic Systems, third revised ed., Wiley-VCH, New York, 1999.

CHAPTER 4

Standards, calibration and samples

Contents

Calibration standards can be a costly part of a gas analysis laboratory and may be very difficult to produce with sufficient accuracy. There are written ISO standards describing the preparation of gas standards and it is well worth getting these documents if gas standards are being prepared in-house. Especially useful are ISO 6141-6145 with the latter standard consisting of a series of documents; these standards are available from www.iso.org [1].

A practical guide to gas analysis by gas chromatography
ISBN 978-0-12-818888-0
https://doi.org/10.1016/B978-0-12-818888-0.00004-8

The book by Gary O. Nelson [2] is also an excellent source which is a lot more comprehensive than what we discuss here.

Standards in pressurised cylinders have a limited lifetime that depends on the components and their concentrations in the mixture. Good analytical practice dictates that multi-level calibration be done for quantitation. Unless an inlet system is used where multi-pressure injection or gas dilution is possible, several cylinders of varying concentrations will have to be purchased at least every six months to two years. An alternative way of producing standards is by permeation that works very well for low concentrations of one or two analytes but it is a little cumbersome if a nine component mixture is required. It also requires a relatively long time to reach equilibrium compared to cylinder standards. Standards can be produced by exponential dilution of a known quantity of pure gas injected into a vessel of known volume or by blending different gases using electronic mass flow controllers.

Applications that require low concentration standards of unstable or reactive compounds make the production of these standards at point-of-use imperative.

For standards that are manufactured at ambient pressure in a laboratory or production plant, any slight leakage will obviously lead to the ingress of air and the accuracy of oxygen, nitrogen, argon and possibly carbon dioxide can be badly compromised.

When blending gases it is very important to consider the compatibility of the gases used in preparing these blends. Some gases, such as acetylene, do not react at atmospheric pressure but may explode when pressurised. Any blends that contain a flammable gas as well as an oxidizing gas could be hazardous. While the dangers of mixing some gases such as oxygen and hydrogen are well known, gases such as carbon monoxide have enough oxidizing potential to form explosive mixtures with hydrogen. Reactions between hydrogen and halogen gases or gas mixtures containing compounds with unsaturated hydrocarbons will occur even at very low concentrations.

4.1 Preparation of static gas standards

There are many ways to prepare gas standards and some of them require specialised and expensive equipment that makes them impractical for in-house standard preparation. In this chapter we will however focus on methods that we used and that can be used in-house at a relatively moderate capital cost, but we will mention some others as well, if only to provide a few pointers on their use.

Since we are discussing the analysis of impurities in gases, the quality of gas used to blend standards and to dilute standards is of vital importance. In Chapter 2 we have already discussed how to purge regulators, how to plumb gas lines and how to install gas line filters. If impurities in the diluents gases are present this will show up as the calibration curve failing to have an intercept close to zero. When using nitrogen or helium as diluents the best results we achieved were using two getter systems in series to get the best quality gas and two getter systems in parallel with those to achieve excellent quality at high flows. Using such a system as shown in Fig. 4.1, allowed flows up to 2000 mL/min at good quality.

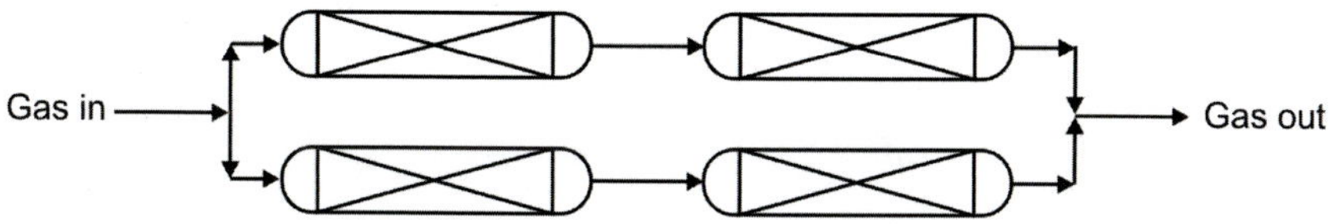

Fig. 4.1 Configuration of a four getter system.

The dilution system can reach the required concentration rapidly if small dilution ranges are used. On the other hand when traces of ambient gases are needed, the system may have to be flushed for many hours to achieve stable trace concentrations. In practice it is advisable to calibrate at the lowest concentration and increase the concentration stepwise so that stability is reached very quickly after changing dilution settings. The calibration curve in Fig. 4.2 was produced only after extensive flushing and

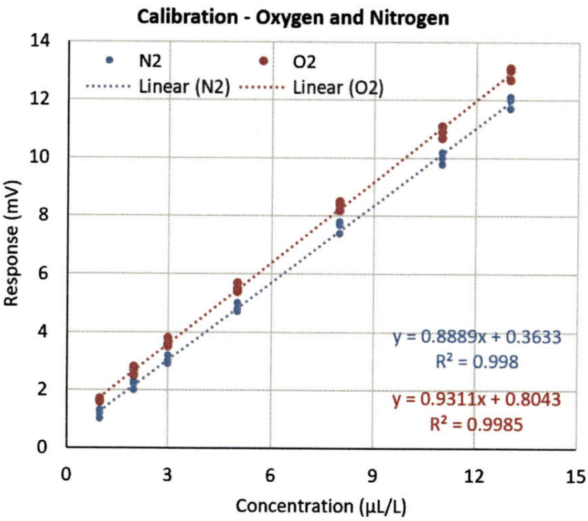

Fig. 4.2 Calibration curves obtained for the determination of trace impurities in an electronic grade gas.

both oxygen and nitrogen are linear and pass very close to the origin. Clearly this system is very leak-tight but when analysing trace amounts of oxygen and nitrogen in any other matrix it is always useful to keep an eye on the O_2:N_2 ratio; if approximately 1:4 it may indicate air ingress during sampling, storage, transport, sub-sampling or the GC pneumatics.

4.1.1 Gravimetric method

Gravimetrically prepared standards are prepared by gas suppliers and metrological institutes as they require very high precision large load balances that are expensive. In principle an evacuated cylinder is tared and the first component gas is added up to a certain mass. If more component gases are needed, they are added in turn, each time up to the required mass. When all the components have been added the cylinder is filled with the matrix gas up to a specified mass or sometimes a predetermined pressure. The exact concentration of each component can be calculated from the masses measured and is usually expressed in percentages or parts per million on a mole to mole basis. These cylinders can vary from small lecture bottles to full size cylinders and can be made of aluminum, steel or stainless steel. The internal surfaces of these cylinders should be lined with inert material or passivated using, for example, silation or fluorination. After filling, the contents are mixed usually by rolling the cylinder for a few hours using a special rig. Despite differences in gas density, the contents in the cylinder should remain mixed and will not separate, or stratify, within the period of validity of the standard.

Cylinder standards are arguably the most accurate with concentrations within 0.1−0.01% of the stated concentration and are certified, usually by analysis after preparation using standard reference materials. In this way traceability is provided and these standards are therefore preferred for critical analyses such as product certification and proving legal compliance. Generally higher concentration gravimetric standards have less uncertainty as the mass added to the cylinder can be more accurately determined. Some gas blends cannot be made as the contents may react with one another such as H_2S and CO and blends containing flammable gases above 50% of the lower explosive limit (LEL) should not be made. Despite the passivation of cylinders, many reactive gases will still react with the cylinder walls over time and standards containing these gases therefore have a very limited validity period. Examples include ammonia, the so-called SOx's and NOx's, the halogen gases F_2, Cl_2 and HCl. Some gases cannot be prepared at high concentrations or pressures as they can liquefy under pressure or when exposed to low temperatures. The suppliers of standards are usually well aware of all these potential problems and take care to avoid them.

Quite often the carefully prepared and expensive cylinder standards are not treated with sufficient care once they arrive at the point of use. We have already pointed out how a cylinder standard can be rendered useless, before it has been used, by failing to adequately purge the bull-nose or flexible hose that connects it to the regulator. Cylinder standards that are in use should never be exposed to extremes of temperature and therefore cannot be kept outside, unless you are situated in an area where there is no difference in temperature between summer and winter. In many places, safety standards require that cylinders may not be kept inside the building although there is no real danger if the cylinders are properly handled and secured. Two possible solutions to this apparent conflict between quality and safety would be to keep the cylinders in an air conditioned area outside but adjacent to the laboratory or to use explosive proof cabinets for the cylinders in the laboratory. When not in use, the cylinders, standards as well as other gases, should be kept out of direct sunlight in an area where the temperature will remain within 5–50 °C. These standards should never be used right up to the point where the cylinder is empty and also not after the expiry date. The suppliers will usually specify the validity both in terms of time as well as residual pressure.

4.1.2 Partial pressure method

Preparing standards using partial pressures requires a filling manifold that can be constructed using standard pipe fittings. The manifold would consist of several connections for component and matrix gases, one or more connections for the standard mixture, a connection to a good vacuum pump and connections to calibrated pressure gauges of various ranges. All these connections will have shut-off valves so that they can be opened and closed independently. To obtain better control a regulating valve can be installed between the section where the component and matrix gases are connected and the connections where the containers for the standard mixture will be connected. If such a system is constructed it is vital to ensure that the total system is vacuum tight using appropriate testing such as a helium mass spectrometric leak detector. A schematic of such a system is given in Fig. 4.3.

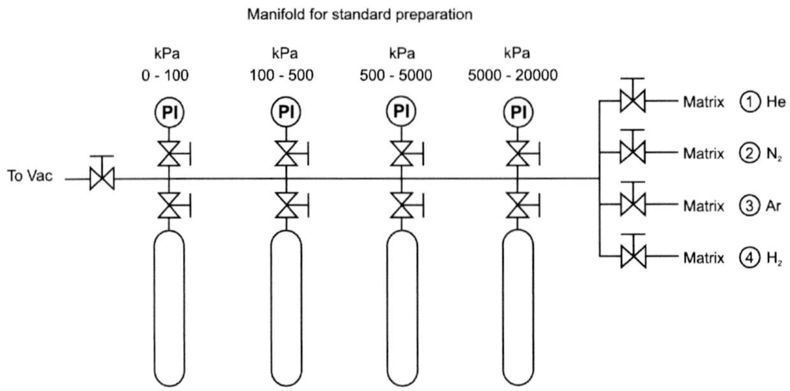

Fig. 4.3 Manifold for the preparation of standards using the partial pressure method.

To prepare a multi-component standard a sample container of suitable size and suitable pressure rating is connected to the manifold and evacuated. The component gases are connected to the manifold with all the valves to the manifold kept shut. When the sample container and manifold is evacuated, as seen on the vacuum gauge, the valve leading to the vacuum pump is closed. It is prudent at this stage to wait a few minutes and check that the vacuum does not deteriorate which would indicate a leaking connection. The valve to the vacuum gauge is closed and the valve to the first component gas is opened slowly to allow the component gas to fill the manifold and standard container up to the desired partial pressure as indicated by the pressure gauge, suitable for the pressure to be measured. All valves are closed and the manifold evacuated and then filled with the second component gas up to a pressure higher than the first component before the valve to the standard container is opened to prevent back-streaming of the first component. Again filling is done slowly to avoid temperature changes and to control the filling so that it can be stopped at the exact pressure required for the second component. The process is repeated for other component gases and finally the container is filled in a similar way with matrix gas up to the desired pressure.

The concentration of each component is calculated using Dalton's law which was discussed in Chapter 1. For those who forgot or skipped Chapter 1, the law simply says that the total pressure is equal to the sum of the partial pressures of all the component gases, including the matrix gas. Therefore:

$$P_{Total} = P_1 + P_2 + P_3 + \ldots P_{Matrix}$$

To get the concentrations of components 1, 2, 3, etc. in percent of the total the partial pressures are divided by the total pressure and multiplied by a hundred. Similarly concentrations in parts per million can be calculated by dividing the partial pressure by the total pressure and then multiplying by one million. The partial pressure of each component must be corrected for the purity of the component gas and the final pressure after adding the matrix gas should include the partial pressures of the impurities.

Standards prepared in this way are usually in-house standards and should have a relative accuracy of around 5%. Although commonly used for relatively high concentrations with a limited number of component gases it is possible to prepare blends in the low ppm range and containing many components by preparing a blend at higher concentration and then use the blend as a component gas in a subsequent blend to either obtain more dilution to lower concentrations or to add more components.

The accuracy of these blends can be considerably improved by including the compressibility of the different component gases. These values are available in many chemical engineering handbooks and web sites. It is usually wise to add the most compressible component first and also to keep in mind that every addition will cause an increase in temperature. This must be accounted for either by calculation or preferably by allowing the system temperature to return to the initial temperature after every addition, before noting the final pressure.

As with gravimetric standards there is a danger of condensation of condensable gases at high pressures and low temperatures. It is possible to include volatile liquid compounds into gas blends by evaporating them under vacuum before the other compound gases are added. These may be especially prone to condensation when the partial pressure exceeds the vapour pressure of that component at a specific temperature. It should also be kept in mind that when the pressure is released during use, the temperature of the gas may be reduced causing condensation of the condensable compounds in the lines leading to the GC.

4.1.3 Volumetric method

Arguably the easiest way to prepare a gaseous standard is by allowing a fixed amount of a pure gas, or volatile liquid, to expand into a fixed volume. Although various different sizes and types of containers can be used the most common are probably glass sample containers with valves at both ends and a short side tube fitted with a rubber septum in the middle. The major hazard associated with glass sample containers is that they may explode under overpressure or implode with the same result when evacuated, especially if the surface of the glass is scratched or chipped. Glass containers should therefore be handled with care to prevent scratches and chips and inspected before use. It is also prudent to use it while inside a secondary container or cover the surface with transparent plastic tape or plastic cling-wrap to contain pieces of glass in case of breakage. Even with these precautions, the valves at the ends of the glass bulbs will still fly off at speed should breakage occur under pressure.

Another disadvantage of glass sample bulbs is the difficulty of connecting them to 1⁄16″ tubing leading to the gas sampling valve without compromising the sample due to leaks and the volume of air contained in the connection. A useful way to counter this is to glue a 1⁄8″ to 1⁄16″ Swagelok connector into one end of the glass bulb with epoxy glue as shown in Fig. 4.4. The glass tube might have to be widened to accommodate this and the services of a glass blower could be required. The 1⁄16″ stainless steel tubing can then be connected without leaks and the connecting tubing can be evacuated prior to opening the valve on the glass bulb. If the sample size is sufficient, the connecting line can even be evacuated, filled with sample and evacuated again before the sample is injected. In this way contamination from ambient air is minimised.

Fig. 4.4 Photograph of glass sample bulb with a septum through which components can be added and equipped with a Swagelok fitting.

In principle the container is filled with the matrix gas to the desired pressure and the component gases are added through the rubber septum using a gas tight syringe. Even with a gas tight syringe equipped with a shut-off valve the needle will contain a certain amount of ambient air depending on the time taken to transfer the sample. Volatile liquids can be added using a microliter syringe and allowed to evaporate into the matrix gas. Some moderate heating may be required and an incubator is a handy addition to the ancillary equipment in a gas analysis laboratory. After addition the mixture should be allowed to mix thoroughly. There are a number of ingenious ways to accelerate mixing such as magnetic stirrer paddles and plungers inside the container but the easiest is probably to leave it for a few hours or overnight at a constant temperature and allow diffusion to take care of the mixing.

The accuracy of the gas blend is directly proportional to the accurate determination of the volume of the container used for preparing the standard. For the glass containers described above this is best done by weighing the container and then filling it with clean water and weighing again. The volume can then be accurate determined by the mass of water using the density of 1 g/cm^3. If necessary a correction can be made for the volume expansion of water at the temperature in the laboratory.

Equally important is the filling of the gas sample syringe used to introduce the component gases. Since it is impossible to fill a syringe in open air without introducing a considerable amount of ambient air as well, it is necessary to use a sampling device. A simple way to achieve this is to connect a piece of soft polymeric or rubber tubing with a flow restriction at the exit end to the cylinder containing the gas to be sampled. The gas from the cylinder is allowed to flow through the tube to flush out any air before the syringe needle is inserted through the tubing and the required amount of that gas is withdrawn without contamination with ambient air. Another way that works very well is to make a gas sampling divide consisting of a rigid tube that connects to the cylinder at one end and the laboratory ventilation system on the other end. In the middle of the tube a T-piece is installed with a GC septum instead of a ferrule below the nut of one of the legs of the T-piece as in Fig. 4.5. Gas can then be withdrawn from the tube with a syringe through the septum without contamination from ambient air but even more important, the sample gas cannot escape into the laboratory making it ideal for use with hazardous gases. There are also commercially available fittings for injecting or withdrawing samples from a gas stream also shown in Fig. 4.5.

The concentration of gas or vapour in the container after addition and mixing is calculated simply by dividing the volume of gas added by the volume of the container plus the volume of gas added. This ratio is then multiplied by 100 to give the percentage or multiplied by a million to give parts per million. When volumes of liquids are added the concentration is often expressed as mass per volume, using the density of the added liquid to

Fig. 4.5 A Swagelok T-piece with a septum fitted on the side for withdrawing of gas using a gas tight syringe (left) and a fitting for injecting or withdrawing gas from a sample stream (right).

calculate the mass corresponding to the volume added. In concentrations of less than 0.5–1% the volume added is often ignored as it is so small relative to the volume of the matrix.

4.2 Preparation of dynamic gas standards

Many of the problems associated with static standards can be overcome by using dynamically generated standards and it has therefore a much greater area of application. In essence it involves mixing two or more gas streams continuously while the flow of each stream is controlled accurately. This provides a continuous source of standard at the desired pressure when and where needed, not only as standards for gas analysis by GC, but also for instruments such as FTIR and direct reading gas sensors. It is also well suited for the generation of calibration standards in adsorption tubes used for thermal desorption GC and GCMS analysis.

Surface adsorption and even reaction with the tubing walls becomes less of a problem since equilibrium adsorption or passivation reactions are obtained reasonably quickly after allowing the gas blend to flow through the tubing. Having said that, some condensable gases, notably water, can take a long time to reach equilibrium especially when changing from a high to a low concentration.

One of the major concerns with dynamic standards is that in some systems it can lead to a considerable amount of wasted gas. The standard is generated continuously, but only used to fill the sample loop for a short time, before and after which the gas is wasted. Another concern is that the equipment can become quite costly especially if very low level standards must be generated. Given the difficulty in obtaining and manufacturing low ppb level certified static standards, the in-house generation of these standards becomes worth the cost of the initial capital outlay. Low concentration standards produced at the point of usage are by far the most reliable and reproducible.

4.2.1 Dynamic dilution method

Dynamic dilution refers to the continuous dilution of a standard by a diluent gas which, for gas chromatography, is preferably the same as the carrier gas. The standard can be a pure gas or a static gas blend from a cylinder or a permeation or diffusion source. Permeation and diffusion sources will be discussed in the next section.

When the dilution is done by blending the standard and diluent it is known as single stage dilution. This primary blend can however be used as the standard for a second stage dilution by adding another measured flow of diluent to get dual stage dilution. While theoretically possible, more stages can be added, however the combined uncertainty becomes unacceptably large considering that following three or more dilution steps the concentrations will become very low.

The concentration of any component can be calculated by dividing the flow of the standard by the sum of the standard and diluent flows. If the standard is already a mixture, then the concentration of each component is the concentration of the component multiplied by the standard flow divided by the sum of the standard and diluent flows. Once again multiplying by 100 will give the concentration in percent while multiplying with a million will give the concentration in ppm.

When a prepared cylinder standard is used with dynamic dilution to obtain different concentration levels for calibration, the standard should be specified to contain the highest concentration required. Of course the diluent need not be a pure gas and can also be a blend such as synthetic air or synthesis gas provided the quality requirement for diluent gas is met.

Provision must be made for proper mixing of the combined streams and the use of a mixing chamber is usually necessary. When we connected a cross piece in a dilution system, the measured values differed significantly from the calculated values as shown in Table 4.1. The reason for this is that the gases do not mix well but pass through the connection in a more or less

Table 4.1 Dilution of 28% methane in helium showing the expected and measured concentrations.

Dilution	1:1	1:10	1:100	1:500
% dilution	50%	9.09%	0.99%	0.19%
Calculated	14.0%	2.55%	0.28%	0.06%
Measured with mixing	13.94%	2.56%	0.29%	0.07%
Measured without mixing	18.13%	15.21%	11.24%	7.18%

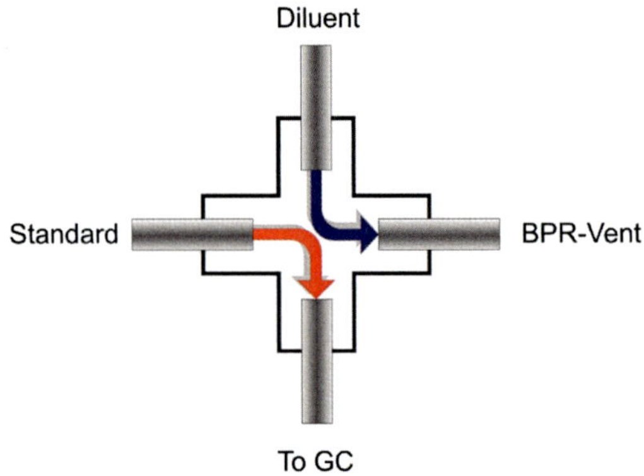

Fig. 4.6 The differences in calculated and measured results in dilution without a mixing chamber may relate to the laminar flow through the cross piece and therefore insufficient blending of the standard and diluent.

laminar flow as in Fig. 4.6. Adding a mixing volume solved the problem. Mixing volumes can be of many shapes and sizes but keeping it simple often works best. For example, when using 1⁄16″ tubing the mixing chamber could be a 20–30 cm piece of coiled 1⁄8″ tubing which we have found perfectly adequate for mixing most gas blends.

4.2.1.1 Mass flow controllers

Probably the best way to obtain the accurate flow rates needed for dynamic dilution is by using thermal mass flow controllers (MFCs) similar to those used for flow control in the GC as mentioned in Chapter 3. Digital mass flow controllers use a sensing tube and a flow restrictor to ensure laminar flow for accuracy and repeatability. They are relatively insensitive to temperature and pressure changes. Controllers can be programmed for various control functions but in chromatography we mostly need accurate flow control. All digital flow modules have a limited range and require regular calibration to ensure long term accuracy. Calibration is done over a range that uses 30 or more calibration points for each gas. When switching from one gas to another automated software calculations for change in compressibility, density etc. can be applied but for highest accuracy recalibration is preferred. For best accuracy it is advisable to use these controllers between 5% and 95% of their range as even a slight zero error could have a significant influence at a flow of 1 or 2%.

There are commercial systems available that will do dynamic dilution and some of them have electronic flow controls that allow a composition to be set on a data system and the mass flow controllers will be adjusted automatically to deliver the selected composition. Many of them also have the option to switch the injection valve to introduce the mixture into the GC and can be programmed to perform multi-point calibrations by linking to the GC data system. Another bonus is that most of these units also provide the uncertainty associated with the dilution. These units can be quite expensive but it is not overly difficult to buy mass flow controllers and do the plumbing in-house for a less expensive system. The downside is of course that the calculations will have to be done manually and that it is less easily automated. For routine applications the calculations needs to be done only once or alternatively a spreadsheet can be set up to perform the calculations.

The mass flow controllers must be calibrated for the gases used. Calibration is best performed by professionals but it is possible to do this in house as well provided the required instruments and skills are available. A less accurate, but useful way to approach the calibration of MFC's is to calibrate the diluent MFC for the GC carrier gas and the standard MFC for the matrix gas of the standard. Most manufacturers of standards prefer to use nitrogen as matrix gas but for GC gas analysis it might be better, albeit more expensive, to choose helium as the matrix. This will be suitable for most relatively low concentration calibrations and when high concentrations are needed conversion factors can be used to correct the MFC reading for gases that are different from the gas that the MFC was calibrated for. A table of such conversion factors for some common diluent gases is shown in Table 4.2. More detailed tables of conversion factors are available on the Internet, but there are small differences between the figures provided by different institutions [3,4].

Suppose the MFC was calibrated for nitrogen but the need is for using helium as diluent gas. If the MFC indicates a flow of 5 mL/min after

Table 4.2 Conversion factors relative to nitrogen for some common diluent gases.

Gas	Symbol	Density (g/L @ 0 °C, 101.325 kPa)	Heat capacity (cal/g K @ 20 °C, 101.325 kPa)	Correction factor (@ 20 °C, 101.325 kPa)
Nitrogen	N_2	1.250	0.249	1.00
Air	—	1.239	0.241	1.00
Helium	He	0.1785	1.24	1.41
Hydrogen	H_2	0.08991	3.44	1.01
Argon	Ar	1.784	0.125	1.40
Oxygen	O_2	1.429	0.222	0.98
Methane	CH_4	0.7175	0.568	0.76

connecting the helium to it, it will not be correct as it was not calibrated for helium. To correct the flow, the correction factor must be applied.

$$Flow_{He} = Flow_{N_2} \times \frac{CF_{He}}{CF_{N_2}} = 5\ \text{mL/min} \times \frac{1.41}{1.00} = 7.05\ \text{mL/min}$$

Therefore, the actual flow of helium is 7.05 mL/min and this value should be used for further calculation of the dilution. While this calculation can be made accurately, recalibration of the MFC using helium would be preferred.

With MFC's dilutions of 1:1000 or more can be made but it is prudent not to attempt large dilutions in a single step as the accuracy of MFC's at their lowest setting are usually not very good. For large dilutions of 1:10,000 or more it is therefore better to use a multiple stage dilution.

4.2.1.2 Fixed restrictors

The making of fixed restrictors was explained earlier and with them and a few valves a system can be constructed that allows dynamic dilution at much lower cost than MFC's, albeit with less accuracy and less flexibility. Narrow bore capillary tubes can also be used as flow restrictors with different lengths offering different flows. In Fig. 4.7 a schematic of a stream selector valve is shown with six different restrictors mounted on it. A photograph of this valve and the restrictors was shown in the previous chapter. The significant advantage of such a valve is that it can be switched using the vial number on the sequence table of most GC data systems allowing semi-automated multi-point calibration.

It is very important to keep the inlet pressure and the ambient temperature constant when using restrictor dilutors. A good quality regulator is

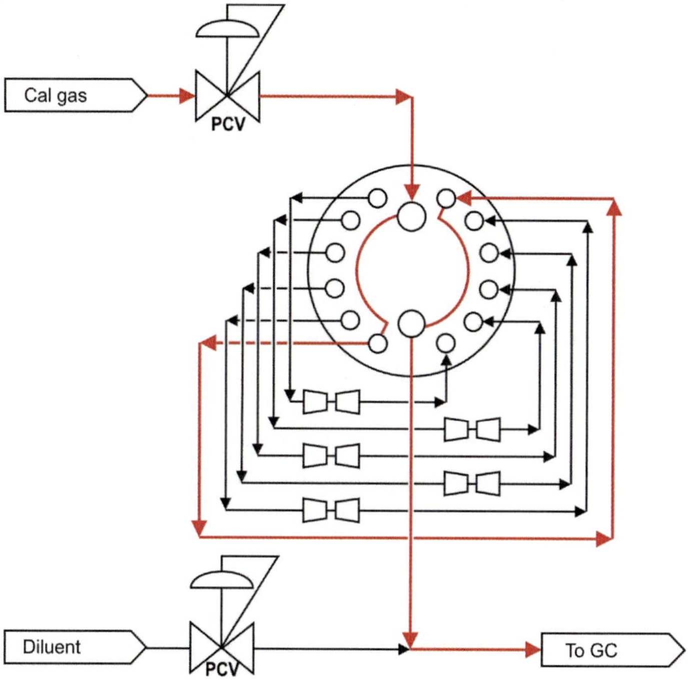

Fig. 4.7 Schematic of a restrictor based stream selector valve for delivering different flows of standard into a constant diluent flow.

therefore imperative and it is strongly recommended to mount the dilutor in a separate oven. Another consideration is the additional flow restriction resulting from the plumbing of the dilutor to the GC injection valve and it is therefore a good idea to have a flow meter at the injection valve outlet. A more detailed explanation of injection and stream selection valves is provided in Chapter 5 and the configuration of sample inlets and injectors are discussed in more detail in Chapter 6.

It is possible to calculate the flow through the different sized restrictors from the dimensions of the restrictors and ambient conditions of temperature and pressure as described in ISO standards 6145-5 and 6145-6 [1]. We found this laborious and not very accurate since the resistance of flow of the entire system contributes to the actual concentration delivered by the restrictor dilutor. A more practical way is to calibrate the GC using either different pressures or against a MFC based dilution system. The actual concentrations delivered by the different restrictors in the restrictor dilutor can then be determined from the calibration curve. Once the concentrations are established, the restrictor dilutor is 'characterised' and can be used for routine work. This dilutor system is small and is therefore easily portable.

For laboratories where many GC's need to be calibrated the cost of many MFC based dilution systems can become prohibitive which often leads to inadequate calibration of the GC's sometimes using only a single calibration point and zero to establish a calibration curve. The concerns relating to this type of calibration are discussed later in this chapter. With the restrictor dilutor a single MFC based system can be used to characterise the restrictor dilutors and from there on each system can be semi-automatically calibrated at multiple levels and at reasonable cost.

4.2.1.3 Bubble flow meters and rotameters

Similar to mass flow controllers, bubble flow meters or rotameters can also be used to measure the flows of two or more gas streams for blending to effect dynamic dilution. It is somewhat less accurate and conversion of the flow to standard reference conditions to compensate for the ambient pressure and temperature of the laboratory should be done [2].

For bubble flow meters this is done using the equation:

$$F_C = F_M\left(\frac{P}{101.3}\right) \times \left(\frac{298}{(T+273)}\right)$$

where F_C is the flow rate at reference conditions, F_M the measured flow rate, P the pressure in kPa and T the temperature in °C. The temperature and pressure are that of the bubble flow meter, usually the same as the ambient temperature. In Chapter 2 we have also said that the bubble is permeable and gases like hydrogen and helium will give lower readings for the same flow than more dense gases.

The height of the ball in a rotameter is a function of the gas density, viscosity and momentum. Since the ball tends to rise and fall even at fixed flow, it is difficult to get really accurate readings. For air the following equation is used to correct the measured flow to standard reference conditions [2]:

$$F_C = F_M\sqrt{\frac{P}{101.3} \times \frac{298}{(T+273)}} = 0.626 \times F_M\sqrt{\frac{P}{T}}$$

where F_C is the flow rate at reference conditions, F_M the indicated flow at the temperature T in °C and pressure P in kPa.

4.2.2 Permeation method

Earlier we mentioned that even metal tubing is permeable to gases and said that gas inside the tube will diffuse or permeate out of the tubing while ambient air will diffuse into the tubing. Polymeric tubing is much more permeable and this is the main reason why this should not be used when doing trace gas analyses. However, this property of polymeric tubing can be very useful for the preparation of low concentration gas standards. It turns out that the rate at which compounds permeates through the walls of a piece of tubing having a fixed length, diameter and wall thickness is constant provided the temperature and concentration gradient remains constant.

The simplest permeation device for gas standards consists of a long length of thin walled permeable tubing mounted in a purged container. The pure component gas is continuously flowing through the permeation tube and the holder is continuously flushed by the matrix gas. The concentration of the component in the matrix gas will depend on the length and nature of permeation tubing, the flows and the temperature which means that the permeation device must be calibrated against standards. This method is mostly used for non-condensable gases and is very difficult to characterise in-house without specialised equipment capable of micro gasometric or absolute pressure measurement methods. There are however excellent commercially available systems for this. Multi-component mixtures can be made by coupling several units in series with different gases flowing through each.

4.2.2.1 Making permeation tubes

Permeation sources work best for volatile liquids and gases that condense under their own vapour pressure such as CO_2, SO_2, H_2S, COS, butane, pentane and formaldehyde but can also be used for relatively less volatile compounds such as water, BTEX components, methanol, ethanol, acetic acid, acetaldehyde and compounds with similar boiling points. Care should be taken when halogenated solvents are used as they may cause swelling of some polymers and thereby change the permeation behaviour of the tubing. In general, permeation tubes will deliver the permeating compound at a constant rate as long as there are both liquid and vapour in the tube and the temperature and matrix gas flow remains constant.

Manufacturing permeation standards for volatile liquids and condensable gases is deceptively simple, need not be expensive and there are commercially available kits for manufacturing permeation sources. A piece of Teflon tubing is sealed at one end, the tube is filled approximately half to three-quarters full with the liquid and the open end of the tubing is sealed. For some compounds it might be necessary to cool the tube with liquid CO_2 or liquid nitrogen to keep the very volatile compounds in the liquid state while filling. This permeation source is placed in a container that allows gas to flow over it, such as the impingers used for environmental sampling. This is placed in a constant temperature oven or water bath and a constant, measured flow of matrix gas is allowed to flow through the container such that it flows over the permeation source. As the matrix gas passes over the permeation source the standard that permeated through the tubing wall is picked up into the matrix gas. The permeation source is taken out and weighed periodically and the concentration of the permeated compound in the matrix gas is calculated by the mass loss per time period. The flow of matrix gas is also known as a volume per time period and it is therefore relatively easy to calculate the mass loss per volume of matrix gas that passed over the tube in the same period, giving the concentration of the permeated compound in the matrix gas in mass per volume.

$$[Std_B] = \frac{R_A}{F_P} \text{ mass/volume}$$

where R_A is the permeation rate of compound A in mass per time, for example as µg/min or ng/h and F_P is the total volume flow of the permeation gas flowing through the permeation oven in volume per time, for example as mL/min or m^3/h. Note that the time for both parameters must be in the same units in order to cancel out and that the units for mass and volume are retained.

4.2.2.2 Types of permeation sources

The type of source described above is the simplest to make and characterise and is useful for moderately low concentrations. The length of the tubing can be varied to create sources that would give a fairly wide range of concentrations. To create tubes with a higher capacity the permeation tube can be attached to a stainless steel or glass reservoir filled with the permeating compound. These sources would obviously have a much longer lifetime than those consisting of a piece of tubing only. Reservoirs are also used to increase the lifetime of sources having very low permeation rates for which very short of pieces of tubing are used.

An extreme case of using a reservoir and a small surface area permeating membrane is the so-called permeation wafers. In these the reservoir consists of a piece of stainless steel tubing, sealed at one and containing a cap with a small piece of polymeric membrane at the other end. Given the small area through which permeation can take place, these devices can provide very low concentrations. They are also used for compounds that have high vapour pressures or low permeation rates. The downside is that they are very difficult to characterise in-house as using weight differences may require a few months of periodic weighing before the permeation rate is established and is therefore not really practical. It is much more practical to purchase certified permeation wafer standards that are commercially available. Examples of permeation devices are shown in Fig. 4.8.

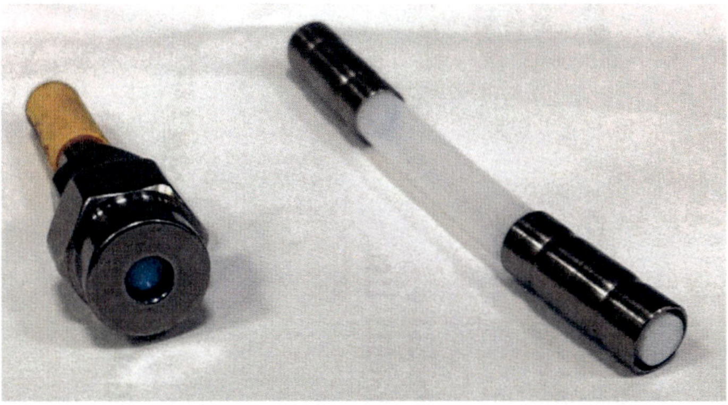

Fig. 4.8 Commercial permeation wafer (left) and an in-house made permeation tube (right).

Generally, permeation wafers will produce concentrations from 0.1 ng/min at 30 °C to about 10 ng/min at 80 °C while permeation tubes can produce concentrations from about 10 ng/min at 30 °C to 1000 ng/min at 80 °C. Of course this depends on the compound type and volatility, the tube materials and tube size. The permeation rate is normally not affected by the pressure of the matrix or span gas. Permeation rates are highly dependent on temperature and the permeation oven must have accurate and very stable temperature control. For concentrations higher than about 1 μg/min diffusion standards can be prepared in-house and are discussed next.

At the other end of the scale are permeation sources that will produce relatively large concentrations. In this case an inverse configuration is used.

A reservoir is used that has two sealable connections through which the permeation tubing can be fed. The permeation tubing is coiled inside the reservoir which is filled with the liquid and sealed. The matrix gas now passes through the permeation tube and picks up the standard compound that permeates into the tubing.

The tubing used is selected based on its inertness towards the compounds to be used in them. Many polymers have been used such as tetrafluoroethylene (Teflon), polyester (Mylar), polyamide (Nylon), polyethylene (Alathon) but fluorinated ethylene propylene (FEP Teflon) is the most common because of its superior inertness and durability. Diameters of between 1⁄16″ and 1⁄4″ are normally used and the length will depend on the concentration to be generated and can vary between 10 and 300 mm. The wall thickness of the tubing will clearly also determine the permeation rate.

4.2.2.3 Concentration calculation

The theory of gas permeation or diffusion through polymeric material is known and there are various sources in which this is explained in detail. Again we refer you to the book by Gary O. Nelson [2] and the ISO standards [1]. However such a discussion falls outside the scope of this book. For the practicing gas analyst the two important calculations are to know the concentration delivered by the permeation device and to get an idea of its expected lifetime.

The concentration of the permeated compound can be varied by changing the matrix or scavenger gas flow. However, at low concentrations and with 'sticky' compounds like water, the time required to reach equilibrium after changing the flow can be very long. It is therefore more practical to use dynamic dilution of the permeating gas to generate standards of different concentrations. The flow of permeating gas stays constant and the permeation rate is readily determined as described below.

The gas standard B, containing the permeating compound A, can be diluted by adding a diluent gas (first diluent) using a mass flow controller calibrated for the diluent gas.

The concentration of A in the resultant gas mixture C, is given by:

$$Std_C = \frac{R_A}{F_P} \times \frac{F_P}{(F_P + F_{D_1})} = \frac{R_A}{(F_P + F_{D_1})} \text{ mass/volume}$$

where F_{D1} is the flow of added diluent with R_A and F_P the same as before.

The gas standard C can be further diluted (secondary dilution) by blending all or part of it with a second diluent gas using appropriately calibrated mass flow controllers.

The concentration of A in the resultant gas mixture D, is given by:

$$Std_D = \frac{R_A}{(F_P + F_{D_1})} \times \frac{F_C}{(F_C + F_{D_2})} = \frac{R_A \times F_C}{(F_P + F_{D_1})(F_C + F_{D_2})} \text{ mass/volume}$$

With F_C the part of the flow of gas mixture C and F_{D2} the second diluent flow.

If the permeation gas is measured as a mass flow the mole fraction of the resultant gas mixture can be calculated using the molecular mass of the permeating compound and assuming the permeation gas is ideal:

$$Std_B = \frac{\frac{R_A}{M_A}}{\left(\frac{F_P}{M_P} \times 22.4\right)} = \frac{R_A \times M_P}{M_A \times F_P \times 22.4} \text{ mol/mol}$$

where M_A is the molecular weight of the permeating compound and M_P the molecular weight of the permeation gas. The prefix for the mass (rate) and volume must remain the same, for example, if the rate is in ng/min and the volume flow in mL/min then the mole fraction is given in nmol/mmol.

The permeation rate is given for commercial permeation sources and empirically determined for homemade devices. Since the permeation rate is strongly dependent on the temperature these rates are always determined and given at a specific temperature. Generally the permeation rate will change by nearly 10% for every 1 °C change in temperature. Therefore, if the permeation source must be used at a different temperature to that at which it was calibrated the permeation rate at the new temperature can be calculated using the formula:

$$\log Q_1 = \log Q_2 - 2950\left(\frac{1}{T_1} - \frac{1}{T_2}\right)$$

where Q_1 is the new permeation rate, Q_2 the calibrated permeation rate, T_2 the temperature at which the device was calibrated and T_1 the new temperature at which the permeation source will be used with the temperatures in Kelvin.

For example, a permeation source was calibrated at 25 °C (298 K) to give a permeation rate of 200 ng/min but a higher concentration is required, so the tube oven temperature is increase to 40 °C (313 K). The permeation rate at this temperature is therefore [5]:

$$\log Q_1 = \log 200 - 2950\left(\frac{1}{313} - \frac{1}{298}\right) = 2.7754$$

$$\therefore Q_1 = 596 \text{ ng/min}$$

Given the temperature dependence of permeation sources it is very important to control the temperature of the permeation oven very accurately. If the oven is controlled within ±0.1 °C, an accuracy of about 1% will be obtained. It is prudent, especially for critical applications to confirm the calculated permeation rate if the permeation source is used at a temperature that differs substantially from the temperature at which the source was characterised [6].

Most GC ovens are capable of isothermal temperature stability within this range. When replacing a GC with a newer model it is not a bad idea to keep the old GC and use it as a permeation oven. The source is then placed inside a glass impinger such as those used for sampling gases by bubbling it through liquids and the entire assembly placed in the oven. The inlet and outlet lines are easily fed through the holes where the injector or detector used to be. The GC oven temperature is stable enough for many non-demanding applications, however a commercial permeation oven with better temperature stability will be required for critical applications.

4.2.2.4 Some guidelines on the use and care of permeation sources

Permeation sources can be quite expensive and it is therefore important to care for them. It is sometimes recommended to keep the sources in a refrigerator when not in use to prolong their lifetime, while others believe that this cooling and reheating causes a sort of thermal hysteresis that changes the permeation rate. We found that keeping the permeation source in a secondary container, such as a piece of stainless steel tubing slightly longer and wider than the permeation source, with Swagelok end-caps, and avoiding temperature extremes, works fairly well. For permeation sources containing compounds that are reactive the secondary container can be flushed with an inert gas before closing.

Several permeation sources for different compounds can be placed in the same permeation oven as long as there is adequate flow of matrix or span gas through the oven. If the flow is interrupted for some time, cross contamination of the sources could occur. When using different sources in this way it is very important to ensure that the vapours of the different compounds are compatible.

The relative simplicity of permeation sources offers many advantages that do not exist with other types of gaseous calibrants. It is ideal for the calibration of field instruments such as air pollution analysers and its small size and weight makes it even more attractive for field work. For low levels of about 200 ppb to 200 ppm its accuracy is arguably the best and it is therefore suitable for calibration of GC's with very sensitive detectors such as ECD's, PFPD's and PDHID's. Another major advantage is that it is suitable for the preparation of standards of some corrosive, flammable and other hazardous compounds. On the downside not all compounds permeate fast enough to make it useful and by its nature it is only suitable for fairly volatile compounds that do not attack or react with the tube materials.

4.2.3 Diffusion standards

When a volatile liquid evaporates it diffuses into the gas above it and if this gas is constantly removed, a gas stream containing the vapour of the volatile liquid can be obtained. If the flow rate of the gas, the temperature, the concentration gradient and the geometry of the system remains constant the diffusion rate will be constant and this property can be used to create standards. Diffusion standards will generally have a higher concentration than permeation standards, typically from about 5 μg/min at 30 °C to above 1000 μg/min at 80 °C depending on the properties of the compound and the configuration of the system.

The flow of the matrix gas is probably best controlled using thermal mass flow controllers although fixed restrictors work very well provided the pressure is kept constant. To obtain different concentrations the flow of gas can be varied, as with permeation devices but it can take a considerable time for a new equilibrium to be established after each change in flow. A quicker way is to add a secondary diluent gas after the diffusion or permeation device. The primary flow is then kept constant and diluted to different ratios using different secondary diluent flows. This method of dynamic dilution was already described in some detail earlier.

4.2.3.1 Making diffusion standards

To make a diffusion device is probably the easiest way to make in-house standards as practically any small narrow necked container with a fixed length through which diffusion can take place can be used. For example a 5 mL volumetric flask without a stopper will do and if the diffusion rate is too fast the neck can be narrowed by inserting a stopper with a small hole drilled through it. Vials with screw caps and Teflon lined stoppers such as those used with headspace analysers works well as a 40–100 mm length of wide bore 1⁄16″ stainless steel tubing can be pushed through the septum to make the diffusion source. The replacement needle of a syringe also makes a reliable and repeatable diffusion device. There are many different diffusion source designs given in the literature and some experimentation might be necessary to determine which works best for the required standards. Once gain the book by Gary O. Nelson [2] and the ISO standards [1] provide valuable information on the preparation and use of many different types of diffusion standards. To the best of our knowledge diffusion sources are not commercially available.

The diffusion source can be placed in a glass impinger which is placed inside an oven or water bath or several sources with different compounds can be placed in a small desiccator modified such that the incoming gas goes in at the bottom and flows upwards past the diffusion source to the outlet at the top. Diffusion sources can also be connected directly to the source gas line using T-pieces although care should be taken that the gas flow is not fast enough to create the Venturi effect and thereby withdraw the diffusing compounds faster than the diffusion rate. When a number of diffusion sources must be used together to create a standard containing more than one compound all the compounds should be kept in separate diffusion sources that are separately connected to the matrix gas. Mixing two components in the same diffusion source does not work as the more volatile compound will evaporate faster causing the relative amounts of the compounds to change continuously and therefore making it impossible to determine the concentration of the standard. When using multiple diffusion standards for creating a multicomponent standard it is also advisable to keep the diffusion sources separated when not in use. Leaving several sources in the same impinger without a flow of matrix gas may lead to cross contamination of the diffusion sources.

As with the permeation rate the diffusion rate is strongly temperature dependent and it is very important to keep the diffusion sources at a constant temperature and also to ensure that the matrix gas is at the same temperature when it gets to the diffusion source. One way to do this is to make a coil of a meter or more of tubing in the matrix gas tubing after the flow controller and to keep the coil in the same oven or water bath as the diffusion source.

4.2.3.2 Calibration and concentration calculation

As with permeation sources the concentration delivered by a diffusion source can be calculated using first principles such as the diffusion coefficient and the molecular weight combined with terms for the minimum diluent gas flow rate and time taken for obtaining steady state equilibrium. These calculations seldom provide an accurate concentration and it is much easier and more accurate to calibrate the diffusion source gravimetrically.

For gravimetric calibration the diffusion source is weighed and placed inside the impinger in the oven or water bath with a constant measured flow of gas flowing through the impinger. After an initial equilibration time of 6–24 h, the diffusion source is taken out, weighed and returned to the impinger. This is repeated every 24–48 h, or longer, depending on the diffusion rate until a constant mass loss over time is obtained. Using the average mass loss over time and given that the flow rate in volume per time is known and therefore also the total volume that passed over the diffusion source in the same time as the mass loss occurred, the concentration of the diffusing compound in the outlet stream can be calculated.

$$Q_1 = \frac{M}{T} \times \frac{T}{V}$$

where Q_1 is the concentration of the diffusing compound, M the mass loss in a specified time T and V the volume that passed over the source in the same time, T.

For example, assume that a mass loss of 10 mg was observed over a 24 h period with a flow of 100 mL/min flowing over the diffusion source. Converting everything to the same units gives a mass loss of 10 mg or 10,000 μg in 1440 min. At a flow of 100 mL/min the total volume of gas that passed over the diffusion source in the same 24 h period will be $1440 \times 100 = 144{,}000$ mL which gives a concentration of 10,000 μg/144,000 mL or 0.0694 μg/mL.

Unlike permeation sources there is no universal direct relationship between diffusion rate and temperature as the rate depends on so many varying factors. Probably the best way to determine the rates at different temperatures is to construct a graph of diffusion rate versus temperature for each compound and device combination. The diffusion rate at each temperature must be determined for this purpose and although this can be done gravimetrically, it can be extremely time consuming. A more efficient way is to calibrate a GC or FTIR and to use the calibrated instrument to determine the diffusion rates at different temperatures. At first glance this may look like a catch 22 situation where you need the calibration standard to calibrate the instrument with which you need to determine the concentration of the standard. However, it is not impossible to do. First the concentration delivered by the diffusion source is determined gravimetrically at its maximum operating temperature. The source is then used at that temperature and combined with secondary dilution to construct a calibration curve for the instrument. Once calibrated, the instrument can then be used to determine the concentrations delivered by the diffusion source at lower temperatures allowing the construction of the diffusion rate versus temperature graph.

With the diffusion rate known in mL/min at the desired temperature the concentration delivered by the diffusion standard can be calculated by:

$$C = \frac{F_{diffusion}}{F_{diluent}}$$

where C is the concentration, $F_{diffusion}$ the diffusion flow and $F_{diluent}$ the diluent flow rate with all flow rates in mL/min. To get the concentration in ppm (v/v) the above is simply multiplied by one million.

4.2.4 Exponential dilution standards

For exponential dilution a vessel of known volume is filled with a liquid or gas. The vessel should have an inlet and an outlet and some means of mixing the contents, for example using a mechanical stirring paddle. A diluent is added at a known rate, allowed to mix with the contents of the vessel and the mixture is then allowed to leave the vessel at the same rate as the incoming diluent. The mixture in the vessel will become more diluted with time at an exponential rate [7].

If the volume of the vessel is V (mL) and the starting concentration of the fluid or gas in the vessel is C_0 (g/mL) and the flow rate of the diluent is F (mL/s), then the concentration C at different times, t (s) is given by the fundamental equation of exponential dilution:

$$C = C_0 \times e^{\left(-\frac{Ft}{V}\right)}$$

It is therefore theoretically possible to calculate the concentration at any time (t) and therefore a whole range of standards of different concentrations can be generated for calibration purposes. By selecting a slow enough flow rate of the diluent, the times at which the concentrations needed for calibration are generated can be matched with the run time of the instrument.

This is however, quite tricky and given that the relationship between concentration and flow is exponential the concentration depends significantly on the accuracy of the timing. Another problem with exponential dilution is that it assumes that mixing is complete and through the years many designs for exponential dilution vessels have been proposed. Some of these used mixing plates, baffles or double paddles turning in opposite directions and were reasonably successful. However, when the effluent of these vessels is continuously monitored there always seems to be some deviation from the theoretical concentrations. The concentrations are sometimes lower than the calculated concentrations at the beginning of the dilution and higher than the theoretical concentration towards the end. As with diffusion standards, it is probably best to determine the concentrations experimentally rather that to rely on the theoretical calculations.

Exponential dilution was quite popular some time ago and is still being used in specific applications. It is useful if the amount of standard is limited or cumbersome to prepare, for example as in the preparation of VOC standards in air. It has, however, to a large extent been replaced by more robust methods such as dynamic dilution using mass flow controllers.

4.3 Calibration

There are probably few areas of analytical chemistry in which more invalid assumptions are made than in calibration. It has been said that never before has it been possible and with so little effort to produce bad results. Very often the assumptions associated with calibration are to blame for these bad results. All too frequently standards are assumed to be valid past their expiry date, calibrations are assumed to be valid without confirmation and proper calibration is neglected quoting time and cost as reasons. Strangely enough, while there is often not enough time to do it properly first time around, there seems to be always enough time to do it over. The question is therefore, what constitutes a proper calibration and the answer is quite obvious; it has to be fit for purpose. Not all analyses will require the same kind of calibration; for some determinations minimal calibration can be enough while others, such as those required for product certification, clinical and legal applications might require extensive calibration using traceable standards and an unbroken audit trail. Calibration should therefore be the first step in method validation to prove that the calibration is fit for purpose.

One of the reasons why calibration in gas analysis is often neglected may be the fact that it is a bit more difficult and a lot more time consuming to prepare standards in-house compared to standards prepared in liquid matrices. The latter can also be injected using an autosampler and without operator intervention while the same is not possible in gas analysis except with an automated dilution system linked to the GC. Another reason that is sometimes presented is that the error in gas sampling is much greater than the analytical error and that careful calibration is therefore not all that necessary which is, of course, pure nonsense.

4.3.1 Single point calibration

The easiest calibration is to analyse a standard mixture of known concentrations of the compounds to be determined. The peak areas of the individual compounds divided by their concentrations provide the responses of the detector towards the compounds in the standard. This is useful for the monitoring of processes where the concentrations of the compounds do not change much over time. It has a very limited validity and is actually only really valid if the concentrations of the compounds in the standard are the same or very close to the concentrations in the sample. A common mistake is to use a single calibration point to construct a calibration curve by including zero as a calibration point. This is simply not valid and the measurement error would increase greatly the further away the result is from the single calibration point. In fact, a single calibration point in GC is not really a calibration at all and cannot be used to determine a range of concentrations with any degree of accuracy as can be seen in Fig. 4.9.

4.3.2 Multi-point calibration

There is no doubt that multi-point calibration over the range to be analysed is, and always will be, the method of choice for determining analytes with any degree of accuracy. Some of the ways in which this can be achieved are listed below but there are many more given in the literature. With a bit of creativity the calibration method can be customised to suit the available facilities as well as the analyses for which it is needed.

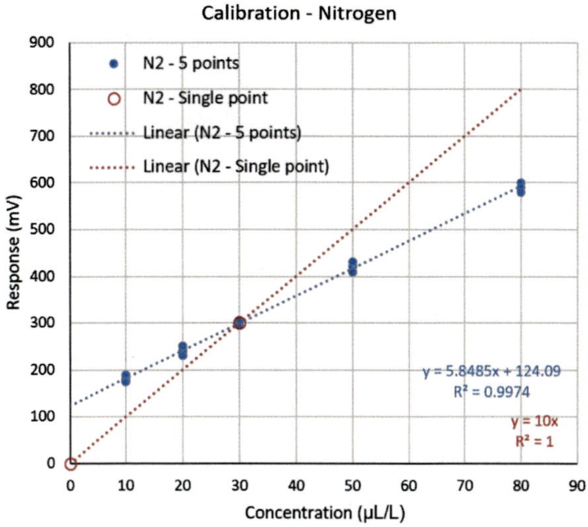

Fig. 4.9 Calibration curve done by calibrating at 5 concentrations compared with a single point forced through zero showing that the latter is only valid for measurements very close to the calibration point.

4.3.2.1 Calibration using dynamic dilution

Arguably the best way to get accurate calibration curves is to use dynamic dilution to prepare standards at different concentration from a cylinder containing a gas blend or from a permeation or diffusion source. When purchasing a standard gas blend for this purpose it is best to purchase a standard containing the maximum concentrations of the analytes to be used for the calibration. Using the standard undiluted then gives the highest calibration point. The other calibration points are generated by stepwise dilution of the standard. Dynamic dilution using mass flow controllers is probably the most flexible and gives the possibility to generate many different concentration levels. However, for routine applications a dilution system made with fixed restrictors can be a lower cost alternative without affecting the accuracy significantly and can be automated as described earlier. Most commercial dilution systems based on mass flow controllers can be automated and would obviously be the method of choice for the routine gas analysis laboratory.

4.3.2.2 Calibration using differential pressures

Probably the cheapest and simplest way to obtain points of different amounts for calibration is to use different pressures of the standard in the sample loop. To do this, a pressure sensor and needle valve is installed in the vent line after the sample loop. Using the needle valve the pressure in the sample loop can be adjusted to different values. Given that the volume of the sample loop stays the same, injecting different pressures is the same as injecting different amounts. For example, a standard containing 5% of a compound is injected at 100 kPa to give the lowest calibration point. Next pressures of 150, 200, 300 and 400 kPa are injected that will correspond to calibration points at 7.5%, 10%, 15% and 20% respectively. The sample is then injected at 100 kPa and the concentration can be determined from the calibration curve. By connecting a vacuum pump to the vent line, the range below 5% can be done at sub-ambient pressures.

This method of calibration results in calibration curves that may deviate slightly from a perfectly straight line but is usually sufficiently linear for most applications, especially given the fact that the range over which one can calibrate using differential pressures is fairly limited. Another potential problem develops when injecting pressures that are higher than the carrier gas head pressure. Upon injection the sample expands into the carrier gas line causing broad, flat topped peaks. A sample inlet that will allow injection of different pressures is explained in more detail in Chapter 6 on sampling.

4.3.2.3 Calibration using static dilution

While performing static dilution is relatively easy, the difficulty comes in introducing these standards into the sample loop as the standard is usually at atmospheric pressure. The simplest way is to use a gas tight syringe, withdraw the standard through the septum in a glass sample container and then to inject the sample through the injector septum. The main problem with gas tight syringes is the difficulty in injecting the same amount and the repeatability is usually substantially worse than using sample loops and injection valves. When gas tight syringes get older, they are prone to leak and contamination can cause significant carry-over. Even when the syringes are fitted with shut-off valves, there will always be a certain amount of air in the needle as the sample is transferred. Using a larger gas tight syringe to overfill the sample loop usually gives much improved repeatability. It is also possible to set up a configuration in which the sample loop and line connecting it to the sample container is evacuated before opening the sample container. The vacuum is then closed off, the sample valve opened and the sample drawn into the sample loop. The other way to get the standard or sample from the glass sample container into the sample loop is to connect the one end of the container to a water source. By holding the container upright and allowing the water to fill it from the bottom, the gas is displaced at the top connection which is connected to the sample loop that is filled with the sample gas. Obviously the sample should not contain compounds that are highly soluble in water and using cold water can lead to more gas dissolving in it as gases love cold liquids. Using warm water will add a significant amount of moisture to the gas and may reduce the efficiency of some columns such as the Molecular sieves. This technique will not easily give satisfactory results for very low level concentrations of oxygen or nitrogen and the ultimate disaster would be to inject a sample loop full of water into the GC system.

4.3.3 A good calibration

To get a good calibration it is best to start with the lowest level first and then to work up to the highest level. The integration parameters must be optimised so that baseline allocations are identical for all levels of concentration. This will be discussed in detail in Chapter 9. It is always rather difficult to decide how many calibration levels would be enough and also how many replicates should be done at each level. A good starting point during method development is to have at least seven levels for each two orders of magnitude, with five replicates at each level, for example levels could be at 1, 5, 10, 25, 50, 75 and 100. The five replicates allows for removing outliers at each level using statistical outlier testing and still remain with a reasonable number of replicates. Following systematic method validation the number of levels and replicates can usually be reduced substantially for routine methods. It is also prudent to have more calibration points at the lower concentrations where the uncertainty is usually the highest. It should also be borne in mind that the high concentration values will determine the slope of the calibration line to a larger extent than the lower concentrations. It is useful to enlarge the lower end of the calibration curve to see if it still passes through all the points. In some cases weighted calibration curves can be useful to get better results at low concentrations. If the calibration range is large and spans several orders of magnitude, consideration should be given to use one calibration line for the low concentration range and another for the higher concentration range. It is also prudent to add more calibration points at, and either side of, critical concentrations such as the certification and specification limits. Although there is some debate about this the authors believe that calibration curves should never be forced through zero. The single most compelling argument against the inclusion of zero is that it is not a measurement that was made. Results should only be reported for values that are between the lowest and highest calibration points. Beyond that constitutes extrapolation of the calibration curve which is at best guesswork since there is no proof that the line remains straight beyond the calibration points. If concentrations lower than the lowest calibrated point must be measured, then more calibration points should be added to encompass the values to be determined. Similarly, more points should be added to the high end of the calibration if values higher than the calibrated maximum must be determined.

Once a calibration curve has been constructed it is customary to view the calibration as good if the regression coefficient, r^2, is close to 1. In fact the more nines there are the better it is believed to be. This coefficient indicates how well the data conforms to a straight line but does not show that the correct values can be read from it, or calculated using the regression equation, over the entire range. There are other statistical measures that can be used in addition to r^2 to evaluate the quality of the calibration line and it is well described in method validation literature. The easiest is to reprocess the data files used for calibration as if they were samples and compare the results obtained with the calibration line equation with the known concentration values. If the differences are acceptable then the calibration line is probably good enough but if not, then the calibration is clearly not fit for use.

Lastly it is not wise to assume that calibrations, no matter how carefully done, would remain valid forever. A calibration check should therefore be done at regular intervals. The calibration check sample should have a concentration matching that of typical analyses that are done which would usually also be somewhere in the middle of the calibration range. If widely varying concentration samples are analysed it may be prudent to have more than one calibration check samples for say the low and high ranges. For compliance testing, the calibration check would usually be equal to the critical value, for example if a product must be certified as containing less than 2 ppm of an impurity, then a calibration check sample containing 2 ppm of the impurity should be used. For routine applications the calibration check results should be plotted on a control chart on which decision limits can be plotted at a multiple of the standard deviation (usually 2sd or 3sd) of repeated quality checks. If the QC sample falls within these limits the calibration is still valid and when outside the limits, the instrument should be recalibrated. By doing this, calibration will be done only when necessary saving time as well as standards.

There is a belief that the more frequently you calibrate, the better the results would be, leading some laboratories to calibrate daily. There is, however, a danger in doing that as there will then be no indication of drift, whether caused by the instrument or changes in the calibration standard. Most GC instruments are remarkably stable and provided a suitable QC is done, it can remain stable for several weeks, depending of course on the number and type of samples analysed. A stable laboratory environment is valuable in assisting in the stability of the instrument calibration.

References

[1] ISO, Standards Catalogue, ISO/TC 158 – Analysis of Gases, International Standards Organisation [Online]. Available: http://www.iso.org/iso/iso_catalogue/catalogue_tc/catalogue_tc_browse.htm?commid=53314.

[2] G. Nelson, Gas Mixtures Preparation and Control, Lewis Publishers, CRC Press LLC, Boca Raton, London, NY, Washington, 1992.

[3] MKS Instruments, Gas Correction Factors for Thermal-based Mass Flow Controllers, MKS Instruments [Online]. Available: https://www.mksinst.com/docs/ur/MFCGasCorrection.aspx.

[4] Brooks Instrument, Gas Correction Factors, Brooks Instrument [Online]. Available: https://www.brooksinstrument.com/en/resources/sizing-tools.

[5] B. Boyle, Owlstone Whitepaper: Generating Calibration Gas Standards with OVG-4 and Permeation Tubes, Owlstone Nanotech Inc.

[6] J. Susaya, K.-H. Kim, J. Cho, D. Parker, The controlling effect of temperature in the application of permeation tube devices in standard gas generation, Journal of Chromatography A 1225 (2012) 8–16.

[7] J. Ritter, N. Adams, Exponential dilution as a calibration technique, Analytical Chemistry 48 (3) (1976) 612–619.

CHAPTER 5
Valves

Contents

The main difference between a gas analysis GC and a GC used for liquid samples is in the extensive use of valves for many varied purposes such as selecting streams, introducing samples, switching between different columns or injectors and more. Valves are used for sample injection because

A practical guide to gas analysis by gas chromatography
ISBN 978-0-12-818888-0
https://doi.org/10.1016/B978-0-12-818888-0.00005-X

they are the most precise method of introducing gas samples. They can also be used to change flows or flow directions which allows different columns to be used in a single analytical system. Therefore, before discussing the topics in which valves are used we felt it prudent to dedicate this chapter to the different types of valves and how to use, control and maintain them. For some column switching applications a Dean's switch can be used instead of a valve and the functioning of the Dean's switch is included in this chapter. Not only are sampling valves available but any configuration from 3 to 14 ports is available for a large variety of sampling and multidimensional column switching applications. Detail on the application of the different types of valves for different types of sampling and analyses will be covered in the following chapters while only sample injection will be covered in this chapter.

5.1 Two position slider valves

Early gas sampling valves were typically slider valves using elastomer O-rings sealing on to a metal slider. Although there were several limitations with respect to lubrication and sample compatibility this was seen as a major advance in reliable gas injection.

Slider, or sliding plate valves are still popular in many production environments and are often used in dedicated on-line analysers. Their popularity in this environment stems from the simpler actuation mechanism that requires linear rather than rotary motion which means it can be switched directly using a solenoid whose shaft is linked to it. It can also be switched manually or using an air actuator.

They typically have stainless steel bodies and polymeric, usually Teflon, sliders. Both the body and the slider can be machined to precise internal volumes allowing the production of valves with internal sample loops similar to the liquid injection valves as well as valves with external sample loops.

Slider valves are two-position valves and their operation is very similar to that of the rotary valves. Comparing Fig. 5.1 with Fig. 5.6 clearly shows the similarity in operation between the slider and rotary valves. In the load or stand-by position carrier gas flows through the channel connecting port 1 and port 4, while the sample flows in at port 6, out at port 5, through the sample loop, then entering the valve again at port 2 and out at port 3 that is connected to vent. In the inject or actuated position, sample flows through the channel connecting port 6 and port 3 to vent, while the carrier gas flows

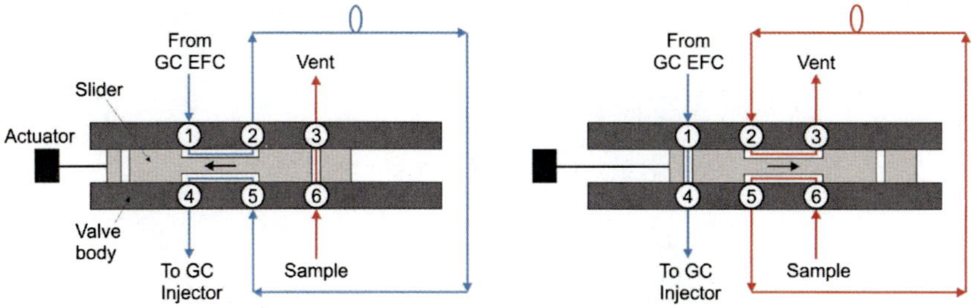

Fig. 5.1 Schematic diagram of a six port slider valve configured for sample injection and showing the connection between the ports in the load (right) and inject (left) positions.

in at port 1, out at port 2, through the sample loop, then back into the valve at port 5 and out to the GC column at port 4.

The ten port version of the slider valve can be used for the same applications as the ten port rotary valve. Since the slider valves also rely on the friction between the polymeric slider and metallic housing to seal, they require the same care in handling as described for rotary valves. Any scratches on the inside of the housing or slider will severely compromise its ability to seal between the ports and may render the valve useless. As with rotary valves, slider valves should also be tested for leaks in both standby and actuated positions.

5.2 Two position rotary valves

There are a number of valves available on today's market but our experience is predominantly in the application of two position rotary valves, an example of which is shown in Fig. 5.2.

Valves typically have two functional parts, the valve body or stator in which ports are situated to allow the connection of lines and a rotor that allows connecting ports adjacent to one another as shown in Fig. 5.3. There

Fig. 5.2 A picture of a six port valve with a pneumatic actuator [1]. ©VICI AG International. *(Reproduced with permission, courtesy of VICI AG International.)*

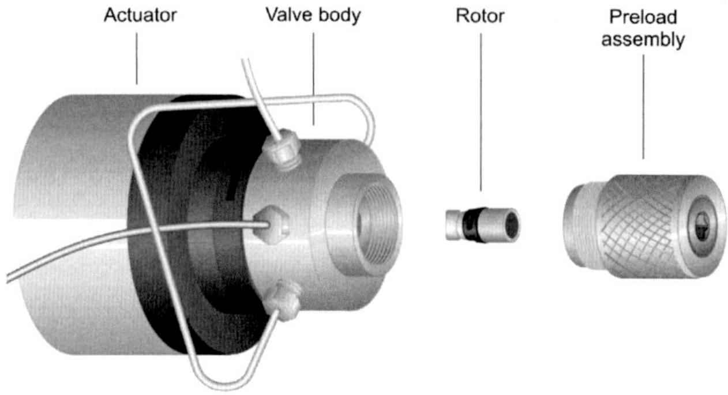

Fig. 5.3 An expanded view of a six-port, two position valve showing the connections and the rotor that is used to connect different pairs of ports [1]. ©VICI AG International. *(Reproduced with permission, courtesy of VICI AG International.)*

is also a retaining nut to keep it together and standoffs, mounting rings, mounting brackets and actuators which will be discussed in more detail later.

As the name indicates, a two position valve has two positions, normally designated as 'load' and 'inject' in reference to using the valve to inject samples. Sample valves are typically installed such that the counter-clockwise position is used as the 'Load' position which implies that clockwise rotation to the 'Inject' position introduces the sample.

When these valves are used in other applications it is easier to refer to the two positions as 'clockwise' and 'counter clockwise'. To understand how it works, refer to Fig. 5.4 showing diagrams of a four and a six port valve in both positions.

For the four port valve, ports 1 and 2 as well as ports 3 and 4 are connected when the valve is in the counter clockwise position. This means that gas that enters at positions 1 and 3 will exit at ports 2 and 4 respectively as indicated by the solid lines in the figure. When the valve is turned to the

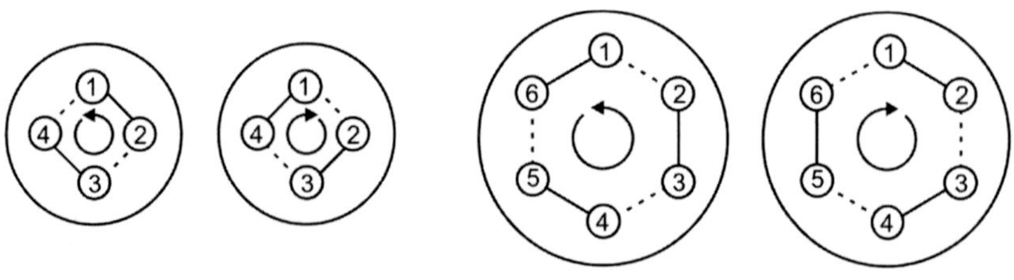

Fig. 5.4 Diagrams of a four and a six port valve in the counter clockwise (off or load) and clockwise (on or inject) positions.

clockwise position port 1 will be connected to port 4 and port 2 will be connected to port 3 as indicated by the dotted lines in the figure. In the same way the six port valve will allow the connection of three incoming gas lines and three outgoing gas lines with the possibility of directing gas coming in at port 1 to either port 2 (clockwise position) or port 6 in the anticlockwise position. Similarly gas entering at port 3 can be channelled either to port 2 or port 4 while gas entering at port 5 can be directed to port 4 or port 6, depending on the direction of rotation of the valve switch.

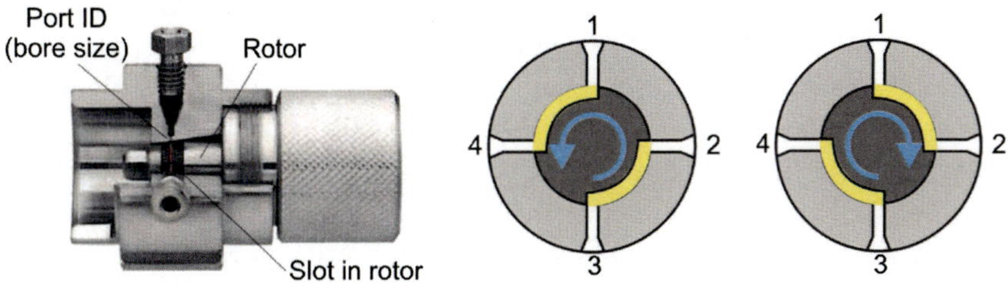

Fig. 5.5 A valve cut open to show the connecting port and the groove in the rotor on the left [1] and a schematic of a two position four port valve showing how the groove in the rotor connects adjacent ports in the clockwise and counter clockwise positions. ©VICI AG International. *(Reproduced with permission, courtesy of VICI AG International.)*

Inside the valve the switching between the ports is done by the rotor with small grooves in it. The length of each groove is exactly the same as the distance between two adjacent ports on the valve as illustrated in Fig. 5.5.

When the sample is completely in the gas phase at room temperature and at ambient pressure, an external sample loop of known volume is normally used for sample introduction. In Fig. 5.6 the sample loop is connected between ports 1 and 4, the line from the sample container is connected to port 6 and port 5 is connected to a suitable vent line in order to get rid of any

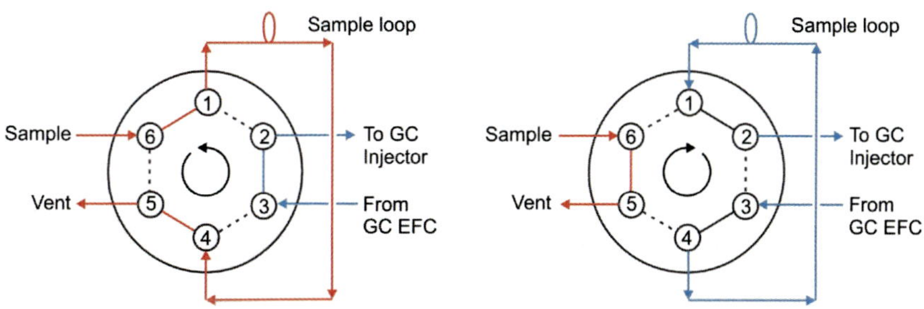

Fig. 5.6 A six port gas sampling valve plumbed for injection of a gas sample.

excess sample. The carrier gas line from the GC that normally leads to the injector is now interrupted and connected to port 3 and port 2 is connected to the GC injector so that the valve is in series with the flow controller and injector. In the 'load' or counter clockwise position the sample flows in at port 6 and exits at port 1 where it flows through the sample loop to port 4 where it enters the valve and exits at port 5 to the vent. In this position the carrier gas flows from the GC, in at port 3 and out at port 2 to the GC injector. When the valve is switched clockwise to the 'inject' position the sample flow enters at port 6 and exits at port 5 to the vent. The carrier gas entering at port 3 now exits at port 4 and pushes the sample out of the sample loop and into the valve at port 1 from where it exits at port 2 and continues on to the GC injector or column. For two position valves, the degree of angular rotation is given by 360 degrees divided by the number of ports so that a 4 port valve rotates 90 degrees while a 10 port rotates 36 degrees.

The size of any analyte peak is directly proportional to the loop volume and the loop pressure at a constant temperature. This will be discussed in more detail in the next chapter. When a 10-port valve is used for sample introduction, some rather special applications are possible and these will also be discussed in the next chapter.

When the sample contains components that are not all gas at room temperature and pressure a representative sample is best obtained by injecting it at elevated pressure to ensure that a liquefied sample is injected. For this type of application a liquid sampling valve is used as shown in Fig. 5.7. Although these valves have only four ports it may be drawn as a six port

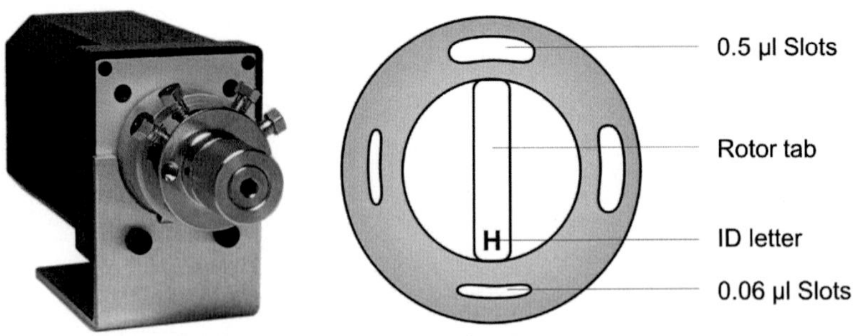

Fig. 5.7 Photograph of a liquid injection valve (LSV) showing the four connecting ports (left) and a schematic of a LSV rotor showing the two different sized engravings for injecting different sample sizes depending on the way in which the valve is assembled (right) [1]. ©VICI AG International. *(Reproduced with permission, courtesy of VICI AG International.)*

valve because the sample loop is an internal engraving in the rotor of the valve. Two different volumes can be engraved in the same rotor so that the way the valve is assembled determines the volume injected. Note that the volume of liquid injected is very much less than the gas volume injected from an external sample loop. However, a liquid sample valve rotates 90 degrees, the same as a 4 port gas valve.

5.3 Valve handling and maintenance

When a valve is delivered there is a protective tape wrapped around the valve body to cover the valve ports. This should not be removed until you are ready to connect the tubing to the valve. It is also important to cover the ports when valves are removed and stored for any length of time. Open ports invite the ingress of particulate material that will cause the valve body (stator) to be scored. Clearly the old adage that prevention is better than cure holds good for valves as well. The most common cause of valve failure is scoring of the valve body from particulate material. All tubing used must be chemically cleaned and all cuts must be burr-free and mechanically cleaned. For this reason we prefer to use pre-cut cleaned and polished tubing when building a system. Also it is imperative that any particulate material be removed from the sample by installing in-line filters, as shown in Fig. 5.8, in the sample line. These filters can have fixed or removable frits of typically 2 μm but filters as small as 0.5 μm are also available. The small price difference between fixed and removable filters hardly justifies the hassle of replacing the frits.

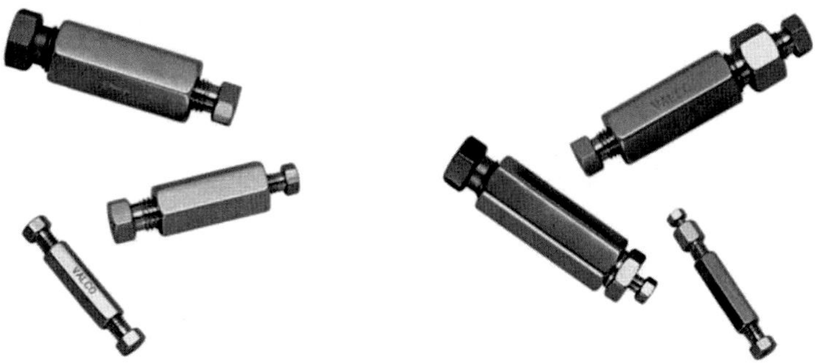

Fig. 5.8 Photograph of Valco in-line filters with fixed (left) and removable frits (right) [1]. ©VICI AG International. *(Reproduced with permission, courtesy of VICI AG International.)*

5.4 Valve actuation

5.4.1 Manual actuation

This is achieved using a handle to turn the valve manually from one position to the other and is adequate provided the valve switch is not time dependent, for example, switching from vent to vacuum prior to sample injection. Simplicity and cost are the main advantages. An option is available to allow the switching of the valve to give a contact closure signal to start the data system or other device. Apart from the obvious fact that this cannot be automated, the disadvantage of manual actuation is that there may be some slight off-axis twisting of the rotor. An optional position feedback module is available (Fig. 5.9).

Fig. 5.9 Photograph of a manually actuated valve.

5.4.2 Air actuation

Air actuation is the preferred way of automation in any environment where a spark could cause an explosion. Valves with air actuators are supplied with stand-offs of various lengths typically from 50 to 200 mm (2–8 inches) long; an example is shown in Fig. 5.10. The length must be chosen to ensure that the air actuator is not heated by heat conductance from the oven in which the valve is mounted. If space permits, rather err by having a longer stand-off than one which is too short. The maximum operating temperature of a standard air actuator is 50 °C and a high temperature air actuator is 100 °C. Air actuators can be installed in any orientation which can be advantageous and allows one to mount the actuator in the position where it is least affected by the elevated temperatures of the oven, injector and detector. While the actuator can be placed in any orientation, it should be mounted so that any potential leakage of liquid from the valve or fittings would flow away from the actuator.

Fig. 5.10 A photograph of an air actuator connected to the standoff with a mounting ring. The second clamping ring is for mounting the whole assembly. The valve body was removed [1]. ©VICI AG International. *(Reproduced with permission, courtesy of VICI AG International.)*

Air actuators must be matched to the degree of rotation of the valve on which they are mounted. A 60 degrees actuator cannot operate a 4-port valve as the 4-port valve requires 90 degrees of actuation. Conversely, while a 90 degrees actuator is theoretically capable of rotating any other valve, damage to the mechanical mechanism will result when the stop pin hits the end-stop in the valve stator. Never attempt to actuate a valve with an air actuator that is not the correct one for that valve. As stated earlier, angular rotation of a valve is obtained by dividing 360 degrees by the number of ports, for example a 6-port valve rotates 60 degrees, an 8-port valve rotates 45 degrees and a 10-port valve 36 degrees. Air actuators are marked A90 or A60 or A45 or A36 depending on their angular rotation.

The recommended way to operate these 2 position valves is to apply air only when switching the valve to alleviate potential problems associated with continuous air pressure. This is normally achieved with a pair of 3-way solenoid valves which are activated for approximately 2 s. The pulsed operation simulates switching by hand but provides full automation capability. Switching is usually controlled by timed events from the microprocessor-based instruments, data systems, or valve programmers. Obviously two separate events per valve must be available. A 4-way solenoid uses only one external event, but there is pressure on the actuator at all times. A digital valve interface is available which sends a 2 s pulse of air to the actuator after which the pressure is automatically released. It also has a start output signal which can be used to start the data system. The digital valve interface also allows timed event signals in the method to be translated into air pulses for switching valves during a run in multidimensional applications.

It is also recommended that bottled air or nitrogen be used for actuation but when plant air is used, an oil trap and a dehumidifier may be necessary. The optimum actuator air pressure depends upon the type of valve in use. Beyond that, variances between like valves and actuators mean that some turn easier than others. Although the actuators can withstand pressures in excess of 500 kPa, the optimum pressure for any particular valve-actuator combination is that pressure which yields a reasonable switching time (nominally 0.5 s). Typically one would start at a pressure of 300 kPa and observe the valve rotation. If the valve turns slowly increase this pressure slightly, but do not exceed 400 kPa as leakage across the solenoid valve may occur. It is probably better practice to run at a pressure that is a little above the optimum pressure rather than being a little below optimum.

When high switching speed is required for partial loop injections or micro-bore chromatography, an accessory is available to reduce the switching time from approximately 500 ms to <10 ms using helium. This also involves the conversion of logic level signals or contact closures into pneumatic pulses for the operation of the actuator.

5.4.3 Electric actuation

Two position standard electric actuators may be operated manually by a toggle switch or automatically by any data system with momentary contact closures or 5 V DC negative true logic outputs. Since the actuator is designed for room temperature use, valves which will be mounted in ovens require a standoff assembly so that the actuator is located out of the heated zone. As with air actuators, the actuator's rotation must be properly matched to the valve. For example an 8 port valve requires an E45 electric actuator. Electric actuators have largely been replaced by micro-electric actuators which are better equipped for automation and are also capable of being set for any degree of rotation required.

5.4.4 Micro-electric actuation

A micro-electric actuator is self-adjusting from 30 degrees to 90 degrees so that no valve alignment is necessary. When an actuator is installed on a valve by tightening the clamp screw of the stepper motor assembly, the actuator senses the positive stops within the valve and memorises them within the first four cycles which are relatively slow. The manual override is a very convenient way to switch the actuator these four times. When the correct positioning has been detected, there is an audible difference in switching time and the correct switching is more rapid. If there is no change after four to six cycles, check that the actuator is tightly mounted on the valve and repeat this procedure. To reset the stored positional information, disconnect the cable to the actuator and cycle the controller four times again. This clears any stored information in the controller. After reconnecting the actuator cable, the actuator position can again be sensed and memorised.

Micro-electric actuators also require two contact closures per valve. However, when there are insufficient events, a single pole double throw relay can be used to switch the actuator to both positions with a single event. The actuator is supplied with a 24 V DC power supply. If a number of micro-electric actuators are used in one system, it may save considerable space by using a power supply capable of supplying sufficient current for all the actuators. Each micro-electric actuator requires a current of 2 A.

The latest in this range are the universal and modular universal actuators that can operate virtually any Valco valve, whether stream selectors or two position valves.

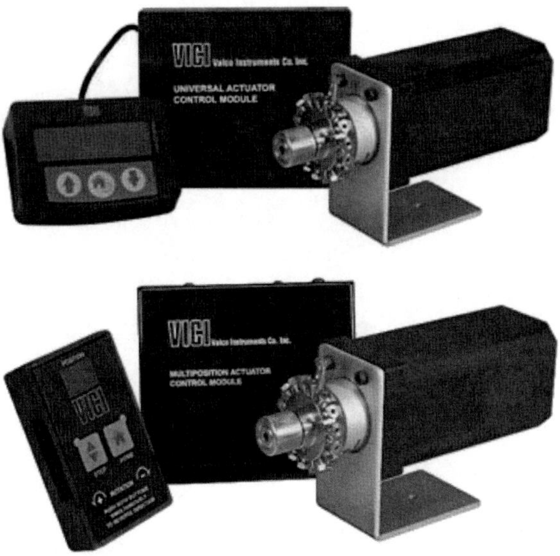

Fig. 5.11 A micro-electric actuator mounted on a valve (bottom) with its controller and manual control switches and a modular universal actuator with its controls (top) [1]. ©VICI AG International. *(Reproduced with permission, courtesy of VICI AG International.)*

Although the actuator can be placed in any orientation, it should be mounted so that any potential leakage of liquid from the valve or fittings would flow away from the actuator. The control module should similarly be installed in a position where there is no chance of liquid leakage falling onto the controller. To allow for different configurations there are different lengths of standoffs and also a right angle drive available. Microelectric actuators must be at room temperature. Images of the micro-electric and modular universal actuators are shown in Fig. 5.11.

5.5 Identifying Valco™ valves

Valves can be actuated manually, by air actuators, by electric actuators or by micro-electric actuators as discussed. The first letter used in the part number defines this. An 'A' means air actuated an 'E' means electrically actuated. Micro-electric actuators use an additional letter after the E to specify the actuator speed or torque. Manually activated valves have no code.

The stand-off length must be chosen and this length in inches is the first number of the part number.

The next letter defines the size of the fittings where a prefix of N is used for 1⁄32″, C is used for 1⁄16″ and if there is no prefix it indicates 1⁄8″ fittings. The next number specifies the number of ports, for example 4, 6, 8 or 10.

As a general rule the valve ports should be matched to the chromatographic system so that restrictions or void volumes can be avoided. For packed column systems 0.75 mm port sizes are typically used while 0.4 mm and 0.25 mm ports are used with capillary and micro-packed columns. The letters used are W for 0.25 mm ports on valves with 1⁄32″ fittings and 0.40 mm ports on valves with 1⁄16″ fittings, UW for 0.75 mm ports on valves with 1⁄16″ and 1⁄8″ fittings and MW for 4.0 mm ports valves with 1⁄4″ fittings.

Valves have different temperature ratings based on the polymer material of the rotor; 'P' valves are used from ambient to 175 °C; 'E' valves from ambient to 220 °C and 'T' valves from 150 °C to 330 °C. Liquid sampling valves have a maximum of 75 °C. The valve code is therefore made up of the number of ports, the diameter and rotor temperature rating; an example is shown in Fig. 5.12. The serial number is stamped on the stator. The pressure rating of these valves is not part of the identification number and must be obtained from the literature for each valve.

The valve body is made from Stainless Steel (Nitronic 60) but, if corrosive gases are present, a special body made from Hastelloy C is

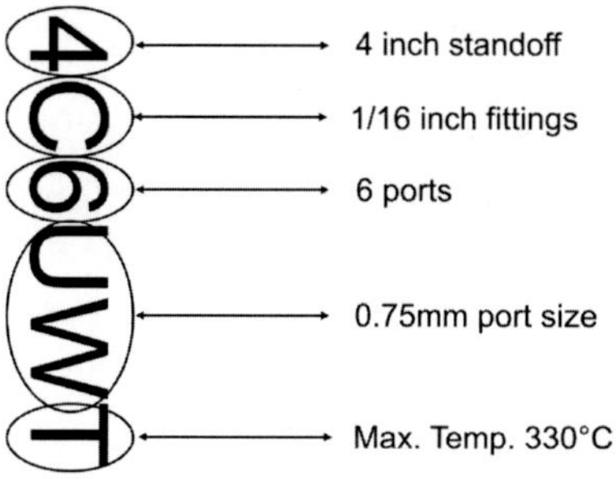

Fig. 5.12 A typical valve code with an explanation of the symbols used.

typically used. Other materials including Monel, Nickel, Titanium and Zirconium are also available for difficult samples. This will be stamped on the stator and will cause the part number to change.

For example, a 4C6UWT valve is manually actuated, has a 4″ standoff with 1⁄16″ fittings, 6 ports with 0.75 mm port diameter, has a temperature minimum of 150 °C and maximum of 330 °C and a maximum pressure rating of 2 MPa. Since this body will be Nitronic 60 the stator will be stamped N6. If this was fitted with an air actuator the catalogue number would be A4C6UWT.

If one looks at an assembled valve it is possible to gauge the port size from the outside diameter of the valve body as W type valves will have a diameter of 32 mm and UW valves a diameter of 38 mm. When looking at any two position valve from the handle side, port number 1 is always diametrically opposite the cut-out for the rotor pin. In the counter clockwise position ports 1 and 2 are connected on 4 and 8 port valves while port 1 and the highest number port are connected on 6 and 10-port valves. Stated in another way, if the number of connections divided by 2 is even, then ports 1 and 2 are connected in the counter clockwise position and if odd, port 1 is connected to the highest port number of the valve. This is clearly illustrated in Fig. 5.4 where a four port valve (number of ports divided by 2 is even) and a six port valve (number of ports divided by 2 is odd) have port 1 connected to ports 2 and 6 respectively. The same applies to valves with more ports.

5.6 Troubleshooting valve problems

The first priority is to determine whether there is a leak at a connection or whether the valve has an internal leak. Finding and fixing leaks in the external connections is done similar to what was described earlier for tubing connections in Chapter 2. Although it is recommended that one should never loosen or tighten any connection while the system is pressurised, this is not always practical. If a small leak is found at a nut on the valve, the nut should be slightly loosened then quickly retightened. This allows the ferrule to reposition itself more easily. Obviously the use of toxic or hazardous substances can make venting of the system mandatory before such work may commence. It is useful to fill the lines to be tested with helium and then locate the leaks with a helium leak detector. Soap solutions should never be used for leak detection on valves. Fittings using one-piece ferrules do not require the use of thread lubricant and the dangers of the lubricant contaminating the analytical system far outweigh any potential benefit. It is important to ensure that the system is not being operated at higher pressure or temperature than the components can handle.

5.6.1 Internal leaks

The symptoms of an internal leak are very difficult to define. If the leak is on a sampling valve, changes in retention time will probably occur. Internal leaks on valves used for column switching are impossible to characterise as different configurations will give totally divergent information. Therefore make sure that all other possible sources of error are eliminated before deciding that there is an internal valve leak. If possible the valve should be replaced to confirm internal leakage but if no spare is available the valve must be disassembled. Although it is possible to disassemble the valve in situ, generally it is more practical to remove the valve from the system. Sample loops can be left on the valve body. Since small particulate material can damage the valve, it is imperative that this work be done in a dust-free environment and the open ports covered with adhesive tape. Loosen the large knurled nut, called the preload assembly, at the rear of the valve. Using a permanent marker, make a mark on the body and rotor to ensure correct alignment upon reassembly. The rotor can be extracted with a magnet, but if this is not available tap the body lightly so that the rotor will drop onto a tissue paper. If the rotor is really tight it may be necessary to push it out from the actuator side. Using a magnifier, visually inspect the interior of the valve body. The inner surface should appear highly polished. Look for scratches or ridges between the ports. Also visually check the rotor to ensure that the engravings are clean and the correct length. If there are any scratches inside the body, the valve needs to be returned to the factory for regrinding and polishing. It will be fitted with an appropriately over-sized rotor.

If the body looks polished and without scratches, the rotor is field replaceable. Firstly clean the valve body with a suitable solvent on a cotton swab. The solvent of choice is isopropanol provided this is compatible with the analytical system. After swabbing the interior, blow dry with a clean dry gas to remove any lint that may be left by the swab.

If the old rotor appears to be in good condition and the grooves clearly visible, it too should be cleaned before reassembly. Grasp the rotor with a pair of long nosed pliers on the actuator end and immerse it in the solvent. Using a clean tissue, gently wipe the rotor polymer dry. Do not touch the rotor or inside of the body with bare fingers. Make sure that the rotor is fitted such that its tab lies within the cut-out, as it will be aligned later. Assemble the valve and tighten the round knurled nut. This nut does not require force to tighten, finger tight will usually be sufficient.

The valve can now be returned to service and reconnected and tested. The chance of successfully refurbishing a valve in a system is excellent for a four port valve but less predictable as the number of ports increase. High temperature valves require conditioning after the rotor is replaced. The valve needs to be plumbed with a flow of oxygen-free carrier gas through all the ports and rapidly heated to 325 °C. At this temperature the valve is cycled at least 10 times and then cooled to allow the plumbing to be returned to normal.

5.6.2 Actuator problems

Mechanical misalignment between the valve and actuator is a common problem and usually results in restricted or blocked flow in one or both positions. This will normally be noticed by changes in the retention times of some or all component peaks. Restricted flow in a sampling valve will result in increased retention for all components whereas restricted flow on a switching valve between columns would only affect those components eluting through this valve. Increases in column head pressure are often an indication that some restriction is taking place when using constant flow or constant velocity modes.

Another symptom of misalignment can be the appearance of multiple peaks or doublets. The sweeping motion of the rotor appears as a multiple injection in certain valve configurations. However this symptom is difficult to isolate from multiple injection effects caused by internal leakage between ports.

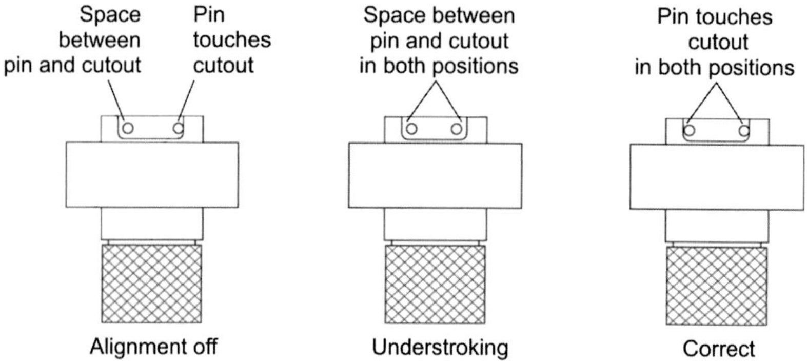

Fig. 5.13 Schematic showing the pin positions when the rotor is aligned and misaligned [2]. ©VICI AG International. *(Reproduced with permission, courtesy of VICI AG International.)*

Confirmation of misalignment must be done visually by observing the position of the pin in both positions as shown in Fig. 5.13. This is relatively easy with standard valves but those with a purged housing require the housing to be dismantled.

For air and electric actuators mechanical realignment is required. A valve with air or electric actuator is shipped from the factory as a unit which is accurately aligned and ready to use. However, installation of the valve usually involves loosening the screw in the clamp ring to fit the valve through a bracket or an oven wall. Before loosening the clamp ring, use a soft pencil or felt tipped pen to make a mark on the standoff in line with the slot in the clamp ring. Reassembling the actuator on the valve with these lined up should closely approximate factory alignment. The alignment is not changed as long as the screw in the clamping ring is not loosened. However, any time the clamp ring screw is loosened, the alignment must be checked. The clamp ring uses either a 9⁄64″ or a 7⁄64″ Allen key and the use of metric Allen keys will cause damage to these heads. A set of Imperial hexagonal keys is essential.

One method that has given good results is as follows. When connecting the actuator to the valve ensure that the actuator is switched to the counter-clockwise position and the valve is switched to its counter-clockwise position. Now hold the square on the shaft of the air actuator with a 5⁄16″ spanner and gently push it in the counter-clockwise position while tightening the clamp screw.

If the valve and actuator are not properly aligned, the rotor pin will strike one side of the valve body cut-out with considerable force, but will not go all the way to the other side. If the rotor does not reach the end of the cut-out in both positions, there is a different problem, such as an actuator which doesn't throw far enough or a rotor with engravings deformed from overheating.

All actuators create some mechanical vibration so that it is quite common for a clamp ring screw to loosen, allowing the valve to move out of alignment. Sometimes it will be possible to see the valve moving relative to the actuator when this problem occurs. After some time the thread of the clamp ring will become dry, especially at elevated temperatures, and a little lubricant on the thread will allow further tightening. If this fails to give a tight connection, the clamp ring and screw may need to be replaced.

5.6.2.1 Air actuators

Air actuators can fail to rotate the specified amount due to internal leakage as the O-rings deteriorate. To check for this, remove one of the air inlets to the actuator and apply pressure to the other inlet. If air escapes from the non-pressurized inlet, the O-rings need to be replaced. If no leak is detected, reverse the air fittings to repeat the test with pressure from the opposite direction. O-rings are field replaceable, but make sure that the replacement O-rings have the same temperature rating as those being replaced. Should this not restore the total swing of the actuator factory repair will be required.

One source of trouble with air actuators is inadequate pressure or flow of the actuation gas. Most valves will switch at 300—400 kPa, but some high temperature or high pressure valves may require more pressure. If there is adequate pressure at the source and the actuator still does not cycle the valve completely, check the solenoid air valves for leakage or failure. Also check that the tubing is not being clamped or restricted anywhere which will reduce the flow of air, thereby causing slow or insufficient switching. The solenoid valves used to switch the actuators can also be a source of leaking air which can often be heard as a hissing sound. Depending on the manufacturer, these have a rating of about 400 kPa and excess pressure may cause damage to the solenoid valve.

5.6.2.2 Electric actuators

Apart from mechanical alignment electric actuators have no user serviceable parts. To realign a valve with a micro-electric actuator requires reprogramming the controller as described in Section 5.4.4.

Fig. 5.14 Photograph of a diaphragm valve [1]. ©VICI AG International. *(Reproduced with permission, courtesy of VICI AG International.)*

5.7 Diaphragm valves

These valves, as in Fig. 5.14, have a much longer life expectancy than the two position valves discussed above. Actuation is always by a built-in air actuator. Diaphragm valves are only available with 1⁄32″ or 1⁄16″ fittings and the standard bore is 0.4 mm although 0.75 and 0.25 mm are optional. Only 4 port internal sample valves or 6 port external sample and switching valves or 10 port multifunction valves are available.

When no air is applied to the actuator, the valve will be in the 'standby' or 'off' position and the springs force the upper and lower pistons together. The plungers on the lower piston force the diaphragm against the cap, which creates a seal between ports 1 and 2, ports 3 and 4, and ports 5 and 6, while flow is permitted from port 2 to port 3, from port 4 to port 5, and from port 6 to port 1. When the gas pressure is applied to the actuator the upper and lower pistons are pushed apart. Plungers on the lower piston retract from the cap, permitting flow between ports 1 and 2, ports 3 and 4, and ports 5 and 6. At the same time, the upper piston is pushed up, forcing these plungers against the cap and creating a seal between ports 2 and 3, ports 4 and 5, and ports 6 and 1. This is illustrated in Fig. 5.15. Since the 'off' position is the default position to which the valve will return in a power outage, a well-designed system will have the valve in the 'off' position most of the time.

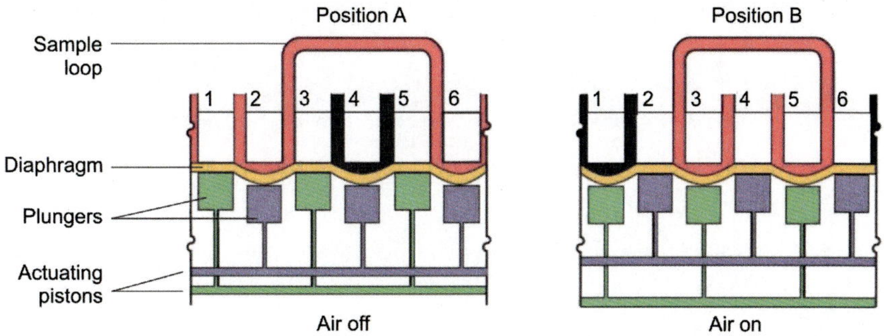

Fig. 5.15 Schematic of a diaphragm valve showing the 'on' and 'off' positions [1]. ©VICI AG International. *(Reproduced with permission, courtesy of VICI AG International.)*

When used as a sampling valve, the sample loop is fitted at ports 3 and 6 on a 6 port valve, or ports 3 and 10 on a 10 port valve. Loops are available in volumes as small as 2 μL.

Unlike the two position valves previously discussed, the actuator only holds the valve in the 'on' position while there is a supply of actuator gas. A 3-way solenoid valve is used to switch the actuator air or pure gas. The 6 port and 4 port valves require 350 kPa for actuation while the 10 port valve needs 400 kPa.

The valve mounting orientation does not affect performance so that these valves can be installed either vertically or horizontally. The internal volume sample valves have a maximum temperature rating of 50 °C while the external sampling and switching valves are limited to a maximum of 200 °C. These valves can handle pressures up to 2 MPa.

Although it is possible to purchase the individual plungers, diaphragms and springs, it is recommended that diaphragm valves are refurbished by the factory when necessary.

5.8 Purge housings

Many trace gas analysis systems are mounted in a cabinet that is flushed with a suitable gas to minimise any ingress of air from the environment. This uses a considerable amount of purge gas and if helium were used it would be very costly. However, each valve that is critical to the application can be fitted with two purge chambers connected in series and supplied with a low flow of inert gas which shields both sides of the valve from atmosphere. Another use is to ensure personnel safety by isolating the valve against leaks into the atmosphere when pyrophoric, toxic, or carcinogenic materials are

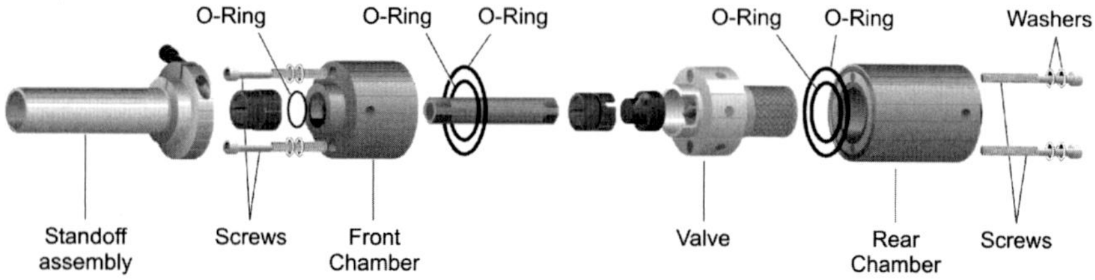

Fig. 5.16 Exploded view showing how the purge housings shield the valve from ambient air [1]. ©VICI AG International. *(Reproduced with permission, courtesy of VICI AG International.)*

present in the sample stream. An exploded view of a valve fitted with purge housings is shown in Fig. 5.16.

Valves with purge housings are strongly recommended for all trace analyses. It is still prudent to mount the valves in a gas tight enclosure, especially if potentially hazardous gases are being analysed. If several valves are used in the application, the purge housing can be connected in series and the outlet of the purge housings can end in the cabinet so that it will, in time, help to purge the cabinet. External purge housings reduce the maximum operating temperature of any valve to 175 °C. Another method that is also available is internal purge which has a space-saving advantage but is only available on valves with 1/8″ and 1/16″ fittings. This is not an option on smaller bore valves.

5.9 Multi-position valves

The rotors of two-position valves typically have the grooves in the rotor in the same plane and have only two switching positions. However, it is possible to have valve cores with multiple planes of ports and this allows a variety of multi-position configurations which may be used for stream selection, column selection, or trapping. Furthermore these valves can be turned through a full circle and therefore have many possible positions with many possible applications. As with two position valves, a wide range of port sizes, actuators, stand-offs and materials is available. Multi-position valves are available for up to 16 streams, requiring 34 ports, and are typically used for selecting different sample streams onto one analytical system.

The temperature rating of these valves is somewhat lower than for two position valves. Multi-position valves having 1/8″ and 1/16″ fittings are rated at 200 °C maximum temperature and a maximum pressure of 1.2 MPa while valves with $^1/_4''$ fittings are rated to 75 °C. Higher temperature valves or with other body materials can be manufactured to specification. High pressure versions can handle pressures up to 34 MPa but their temperature is then limited to 75 °C.

As discussed with two position valves the major cause of valve failure is scoring of the stator by particulates. Each and every stream must be equipped with particulate filters prior to the stream selector valve. It is still recommended to use a further filter between the stream selector and sampling valve.

5.9.1 SD or dead ended valves

SD valves, Fig. 5.17, select one stream to an outlet port while all other streams are dead-ended. The selected stream is then directed to the analytical system. When a new stream is selected it is necessary to purge that stream so that a representative sample can be analysed. These are available for 4 to 16 streams. This valve can also be used to direct one stream to a number of outlets such as collection of discrete fractions from a single stream.

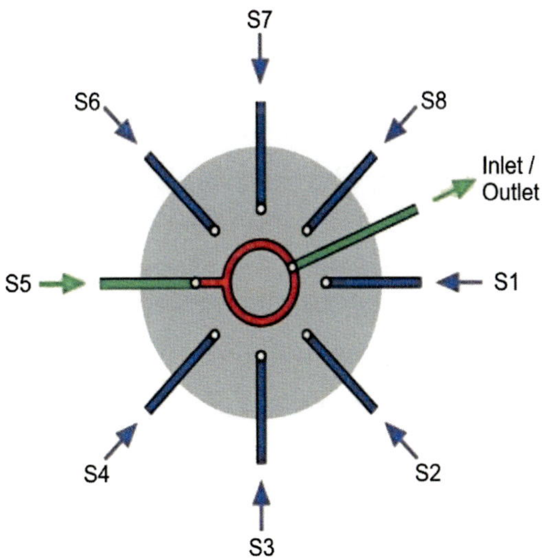

Fig. 5.17 SD valve with several incoming streams that are all dead ended except the selected stream [1]. ©VICI AG International. *(Reproduced with permission, courtesy of VICI AG International.)*

5.9.2 SC or common outlet valves

The SC valves, Fig. 5.18, are similar to SDs, except that all non-selected streams flow to a common (waste) outlet. This configuration has the

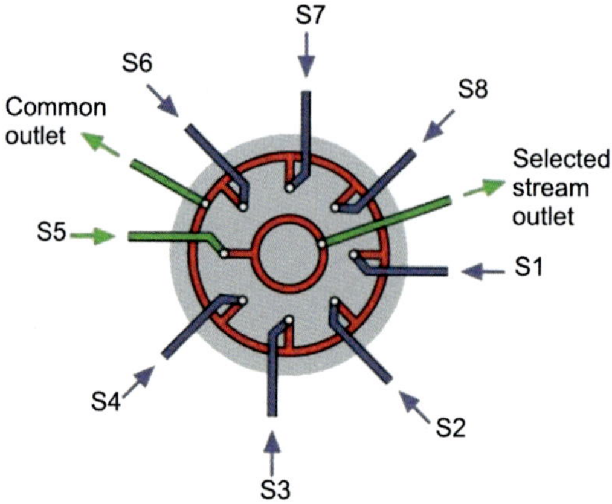

Fig. 5.18 SC valve with one stream selected while all non-selected streams are combined to a common outlet [1]. ©VICI AG International. *(Reproduced with permission, courtesy of VICI AG International.)*

advantage of having a continuous flushing of each stream but it does consume a considerable amount of sample. Care must be taken to avoid any hazard that could occur when the various streams combine in the waste line.

5.9.3 SF or flow-through valves

SF valves, Fig. 5.19, select one stream similar to SDs and SCs, sending it to the common outlet. However, non-selected streams flow through individual

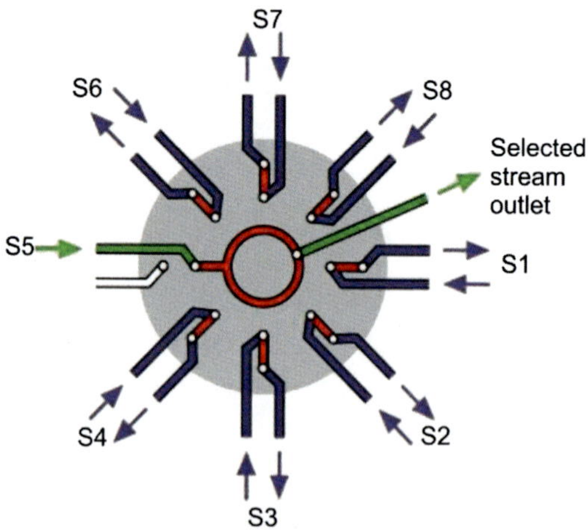

Fig. 5.19 SF valve with one stream selected and all non-selected streams returning the excess sample to its own destination [1]. ©VICI AG International. *(Reproduced with permission, courtesy of VICI AG International.)*

outlets back to each individual source instead of a common waste outlet. In this way there is no waste of sample while a representative sample is available at all times for analysis.

5.9.4 ST or trapping valves

When reaction changes happen faster than the analytical cycle time it is possible to store samples for later analysis using ST (trapping) valves, Fig. 5.20. They can also be used for multi-column operations. For example, a rapid reaction requires sampling every 2 min while the analysis time is 10 min. The valve is then used to store several samples which will then be analysed once all the samples have been trapped. This system must be very leak-tight and inert to ensure that the samples do not change composition before analysis.

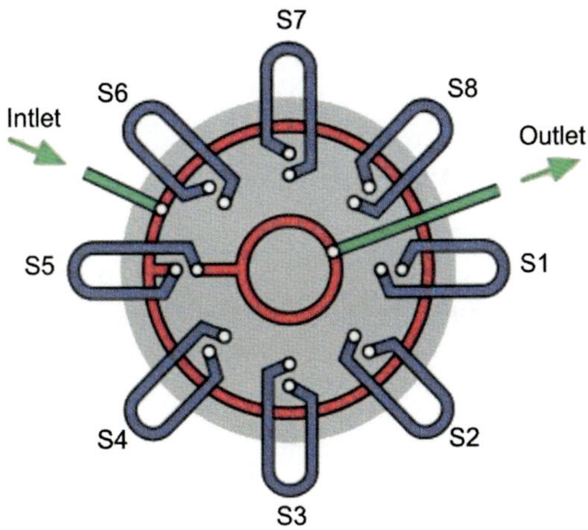

Fig. 5.20 ST valve in which samples can be trapped and stored in loops for later use [1]. ©VICI AG International. *(Reproduced with permission, courtesy of VICI AG International.)*

5.9.5 STF or trapping, flow-through valves

For maximum flexibility the capabilities of the SF and ST are combined in the STF valves, Fig. 5.21, where the non-selected streams can be returned to their individual sources while trapping takes place on the selected stream.

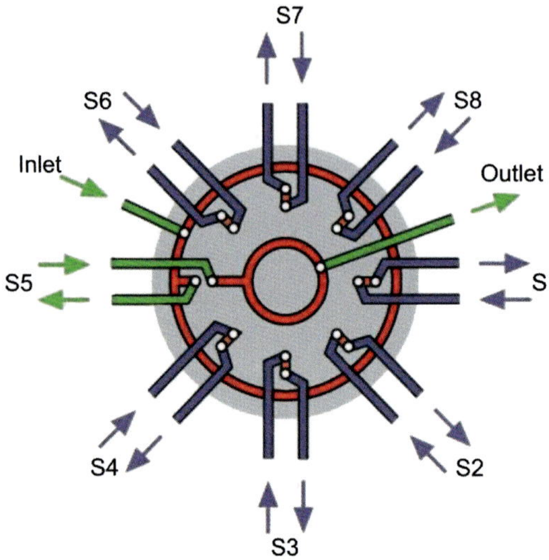

Fig. 5.21 STF valve that traps the selected stream while non-selected streams are returned to individual destinations [1]. ©VICI AG International. *(Reproduced with permission, courtesy of VICI AG International.)*

5.10 Multi-position valve actuators

Actuation of multi-position valves can be manual, air, electric and micro-electric. Clearly the more streams, the closer the ports and therefore the more critical the alignment accuracy of the valve and actuator becomes. When using 6 streams or more, it is highly recommended to use micro-electric actuation.

Multi-position valves move in incremental steps through continuous revolutions, unlike the back and forth switching of two position valves that has been discussed. Manual, air and electrically activated actuators allow these valves to move in one direction only while micro-electric actuation allows the direction of rotation to be changed. Micro-electric actuators are also self-aligning.

5.10.1 Manual actuators

Manual actuation is used when automated stream selection is not required or desirable. The manual actuator consists of a handle using a ratchet and pawl mechanism that rotates the valve the required amount in only one direction. An optional position feedback module is available for manual and air actuated valves.

5.10.2 Air actuators

Air actuation must be used in areas where there is any danger of explosion due to electrical sparking. All the requirements for air actuation have already been discussed under the section on valve actuation of 2 position valves.

Mechanical alignment is different in that the actuator and valve have a mechanical keying system. After resetting this alignment it is advisable to connect a test gas to the valve and step the actuator through all positions and confirm that all ports are sequentially selected. Clearly the alignment becomes more critical as the number of streams increase i.e. the port spacing becomes closer.

5.10.3 Electric actuators

Electric actuators have also been described for two position operation. For multi-position actuation they have a manual override to step the valve through its positions. There is however no feedback of current position unless an optional accessory has been installed. For advanced automated operation the micro-electric actuator offers a lot of additional advantages.

5.10.4 Micro-electric actuators

When a high degree of automation as well as reliability is required, the actuation of choice is the micro-electric actuator. One actuator can be used on any multi-position valve as the number of streams can be programmed. The first requirement is to set the number of streams in the memory of the control module. Simultaneously depress and hold the STEP and HOME buttons for about 8 s. The position display lights will begin to flash, as will the current setting for the total number of positions. The STEP button will increase the number of streams while the HOME button will decrease them. The new number of streams value is saved to memory by pressing and holding both buttons simultaneously until the LED's stop flashing.

Before changing the switching direction, check which LED is illuminated then depress and hold the STEP and HOME switches simultaneously for a second or two. This change of direction will be confirmed by observing that the other direction LED is now illuminated.

These actuators can be stepped from a simple TTL contact closure but cannot be controlled directly. They can be driven by a serial port directly from a PC using VICI's VCOM software. However, most chromatography software packages allow the control of the actuator either as a valve number or as a vial number. The actual protocol used is often Binary Coded Decimal (BCD). Macros can be written to start the sequence at predetermined times without operator intervention.

The time it takes to step a valve from one position to another depends upon the actuator model, the number of positions for which the actuator is set, and the total amount of rotation involved.

When the application requires more streams than can be handled by one multiposition valve, several can be coupled to increase the number of streams that can be selected. For example, nineteen streams can be handled by two 10 position valves that are connected such that streams 1–9 are on positions 1–9 of the 'master' valve and position 10 is used to select streams 10–19 using the second ('slave') valve. Control is done using a standard 'Daisy chained' serial communication cable to a serial (COM) port on the PC.

5.11 Mounting

As with the two-position valve actuators, the actuators for multi-position valves can be placed in any orientation. It should be mounted so that any potential leakage of liquid from the valve or fittings would flow away from the actuator. Control modules should similarly be mounted safely. Multi-position valves are usually not mounted on the instrument, but in a separate valve enclosure to allow for all the connections and ancillary equipment such as in-line filters, pressure and flow controllers that may be needed. Depending on the sampling requirements such an enclosure could be heated separately from the injection valve oven.

5.12 Utilising timed events

Depending on the choice of actuator and implementation one or two events will be required to switch a valve as described in the sections above. When using a valve only for sampling, we recommend that the valve stays in the inject position for the duration of the run since the loop is kept clean with the flow of carrier gas and is switched to the 'load' position only at the end of the run. On an external sample loop valve there is a constant flow of sample in the inject position so that a fresh sample is immediately available when the valve is switched back to the load position. An internal volume sample valve will require flushing before the next injection.

On some applications it is important not to inject the sample immediately that the GC run begins. For example, to allow the sample inlet pressure to drop to atmospheric pressure before injection, it may improve precision by opening a vent valve at time 0.01 min to equilibrate the pressure and then injecting at 0.3 min. This would effectively give the sample pressure nearly 18 s to equilibrate to ambient pressure.

The setting of switching times of two position valves for multidimensional GC is covered in Chapter 10 on Multidimensional Chromatography. The use of signal switching and integration events is covered in Chapter 9 on Data Handling.

Air actuated and electrically actuated stream selectors advance one step per event in the same direction. Because of the mechanical ratchet, the event must be deactivated after each activation to reset the ratchet for the next movement. For example, if the application requires that the stream selector be advanced every 10 min the appropriate event would be switched on at 10.00 min and switched off at 10.2 min.

Systems that use BCD (binary coded decimal) control of multi-position stream selector valves allow the system to identify each stream as a 'vial' which would be equivalent to having a liquid auto sampler. At the start of each run the sequence file will determine the correct vial number and set the stream number to the value. The micro-electric actuator rotates the stream selector the shortest route to the required position and does not have the limitation of only rotating in one direction.

Many data systems operate their external events on TTL (transistor-transistor) logic rather than a contact closure. In such cases it is necessary to build a small circuit with a transistor and relay. Alternatively a small TTL relay circuit board can be purchased.

5.13 Troubleshooting stream selector valves

As with two position valves, the most common reason for failure is scratching of the valve body due to particulate material entering the valve. All streams must have in-line filters installed as was discussed under sample valves. We also recommend a further in-line filter between the stream selector and sampling valves.

Mechanical misalignment accounts for most of the problems associated with stream selection. If there is a serious misalignment there will be no flow through any of the ports and this will be noticed by all streams having poor results. However, slight mechanical misalignment may only affect certain of the streams. As has already been discussed the greater the number of ports, the closer the ports and therefore the more critical the alignment. Flows can be checked by stepping the actuator through its cycles and ensuring that there are flows for all connected streams. For flow checking it is convenient if at all possible to connect the gas flow to the common outlet of the stream selector and check flows on each inlet port as the valve is stepped through its range of positions. This can also help to backflush any dust particles to waste. Alignment procedures for the different actuators have been discussed above.

5.14 Dean's switches

The so-called Dean's switch is not a valve but offers an alternative way of switching flows in a capillary column system. It is named after David R. Deans who pioneered this technology for use in multidimensional capillary chromatography in the 1960's. It will be discussed in more detail in Chapter 10 on multidimensional gas analysis but the principle will be introduced here as its application resembles that of valves.

Fundamentally the Dean's or fluidic switching is based on the principle that flow in an enclosed channel will be in the direction of least resistance which translates into the least restriction or lowest pressure. When a set of three T-connectors are plumbed in series as in Fig. 5.22, the direction of the incoming flow through the central tee can be changed by increasing the pressure on either side of it. Note the characteristic colour of the passivated T-pieces.

Increased pressure on the left side of the central tee compared to the pressure on the right side of the central tee will effectively prevent any flow from right to left and direct the flow from the central tee to the right. Increased pressure on the right side tee compared to the left side, will

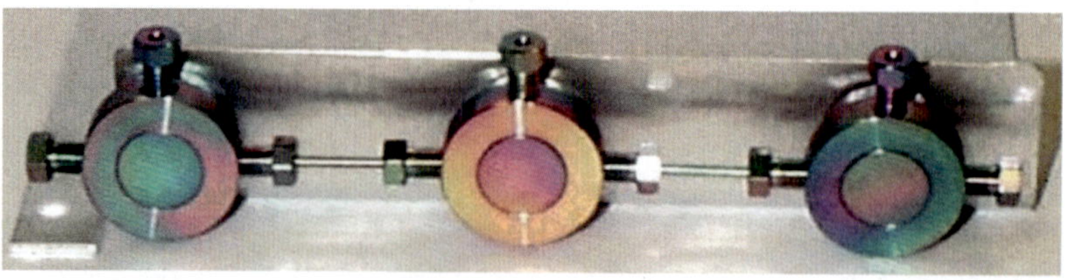

Fig. 5.22 Three interconnected T-connections used as a Dean's or fluidic switch [3]. ©Agilent Technologies, Inc. 2003, 2006. *(Reproduced with permission, courtesy of Agilent Technologies, Inc.)*

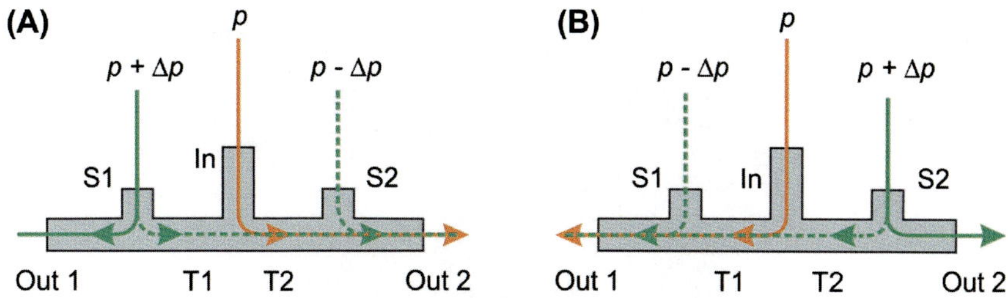

Fig. 5.23 Schematic diagram of a Dean's switch.

reverse the situation and direct the central flow towards the left. This is illustrated in Fig. 5.23. This type of application is made possible by the developments in electronic and mechanical systems that make up modern electronic pressure control (EPC) systems [4].

In practice a small amount of the gas that is used for the switching will mix with the sample flow coming in from the central tee but with careful selection of pressures the dilution of the sample can be kept to a minimum. The constant flow through all three connections prevents back streaming of sample and also constantly purges the internal channels.

Applications using the rotary valves described above are usually somewhat more robust and easier to maintain than the Dean's systems and also do not add any additional gas that may dilute the sample stream. On the other hand, the rotary valve actuators have a slower switching speed; the valve rotors have temperature limits and mechanical valves wear out after some time [4]. The Dean's switches do not suffer from these disadvantages but are less flexible than rotary valves in that some multidimensional applications such as column sequence reversal cannot be done with Dean's switches. On the other hand, Dean's switches may be more susceptible to blockages than rotary valves. Stream selection using fluidic switches is theoretically possible but would be much more difficult and cumbersome to implement than using a stream selector valve.

While the functioning of the Dean's switch was explained with respect to three interconnected T-pieces, commercially available units are integrated with very precise internal channels having very small volumes and highly inert surfaces. The characteristic discolouration of the metal in the example in Fig. 5.24 is due to the treatment to make the metal highly inert. The example in Fig. 5.25 clearly shows the configuration inside the unit connecting the two columns, detector and auxiliary carrier gas connections to the two switches.

Fig. 5.24 Example of a microfluidic Dean's switch from Agilent [5]. ©Agilent Technologies, Inc. 2013. *(Reproduced with permission, courtesy of Agilent Technologies, Inc.)*

Fig. 5.25 The microfluidic Dean's switch from Trajan Scientific and Medical. ©Trajan Scientific Australia Pty Ltd 2018. *(Reproduced with permission, courtesy of Trajan Scientific and Medical.)*

There are many configurations available and the exact configuration will be dictated by the specific application which can be simple backflushing of unwanted components or even sophisticated comprehensive GC × GC systems. Retrofitting Dean's switches to an existing GC usually requires additional EPC modules and other bits of hardware and it is best to consult the instrument supplier to establish the exact requirements. Software for calculating the different flows and pressures for the columns that are used is usually available from the supplier of the specific instrument, the Dean's switch and the ancillary modules. There is also a handy calculator using Microsoft Excel that was developed by Dr Peter Boeker at the University of Bonn [6].

References

[1] Valco Instruments Company Inc, Valco VICI Cheminert Catalog 65 INT, VICI AG International, Switzerland, 2012.
[2] VICI AG International, Technical Note 101. Valve Alignment Troubleshooting, Valco Instruments Co. Inc. and VICI AG, Switzerland, TN-101 2/10.
[3] Agilent Technologies, Inc, Agilent G2855A Deans Switching System. Installation and Operation, vols. G2855-90100, January 2003, p. 61.
[4] J. Hinshaw, Valves for gas chromatography Part III: fluidic switching applications, LCGC North America 29 (11) (November 1, 2011) 988–994.
[5] Agilent Technologies, Inc, Capillary Flow Technology: Deans Switch. Increase the Resolving Power of Your GC, vols. 5989-9384EN, June 10, 2013.
[6] P. Boeker, Research - AltraSens. Deans Switch Calculator V 1.0, 2014 [Online]. Available: http://www.altrasens.de/research.html.

CHAPTER 6

Sampling and sample introduction

Contents

Sampling is more often than not the weakest link in the analytical process. Contrary to the laboratory where good laboratory practices are in force, the sampling is quite often less controlled. Apart from environmental factors and accessibility of sampling points which can make sampling very

A practical guide to gas analysis by gas chromatography
ISBN 978-0-12-818888-0
https://doi.org/10.1016/B978-0-12-818888-0.00006-1

complicated, one of the biggest contributing factors to the integrity of the sample is the person who does the sampling. While the laboratory utilises trained scientists and technicians, the sampling is often shifted onto the shoulders of non-scientists with little or no analytical background. One just has to look at the EPA methods for sampling of air and stack emissions to realise that sampling is a science of its own and requires specially trained scientists and technicians to ensure that samples are representative and that the integrity of the sample is preserved from the time it is taken to the time it gets analysed.

Samples submitted to the laboratory should be large enough so that repeat measurements can be made but should also be small enough to avoid disposal problems. Arguably the next weakest link is getting a representative sub-sample from the sample container submitted to the laboratory into the instrument without contamination or losses. Furthermore, this sub-sample should be large enough so that the instrument can detect all the components but also be small enough to prevent overloading of the instrument. Given that the approximate concentrations in the sample container submitted for analysis are often not known, this can be rather difficult and some detective work may be required. It is useful to know where and when the sample was taken, which analytes must be determined and up to what level, why the analysis is necessary and what components, other than the ones that must be analysed, are or could be in the sample.

Given the above, it is therefore not surprising that sampling and sample introduction usually contributes most to the combined uncertainty of analytical methods. In this chapter it is not our intention to describe all the possible ways of sampling and how it should be done. To do so would fill a number of volumes, so we rather shared some of our experiences on sampling with the understanding that it is neither comprehensive nor conclusive. Application specific sampling methods are described in detail in various documents published by the API, ASTM, ISO, USEPA and others, depending on the field in which the sampling and analyses are done. In many cases the sampling procedures are described in regulatory documents such as the environmental regulations that are in force in most countries.

6.1 Sampling

Broadly speaking, sampling can take one of three forms that we will call, for the lack of more descriptive terms, spot, continuous and integrated samples. Spot samples are taken at a single point by filling a sample container from the source to be sampled. This could be ambient air in a room, a process pipe, a smokestack, a process vent, a storage tank or any similar source. The sample is taken into a suitable container and transported to the laboratory for analysis and is therefore termed off-line analysis as there is a considerable time lapse between taking the sample and analysing it. This type of sample is representative only of that sample point and only at the time when the sample was taken and is often called a 'grab' or 'spot' sample.

The opposite of spot sampling would be continuous sampling where the analyser is situated in the stream to be monitored and takes measurements continuously. This is also known as in-line sampling and the only time difference between the sampling and reporting of the result is the time it takes the instrument to process and relay the information. This is often called real-time reporting of results. One of the main problems with this type of analyser is that the results depend on the accuracy and continuous performance of the analyser since there is often no way of readily checking the results. However, advances in technology are making in-line sensors more specific, sensitive and affordable and it is therefore becoming more and more popular and these have, in recent years, replaced many complex instruments in many production facilities. Since the gas analysis GC, no matter how fast it can be made, needs time to separate components it is not capable of real-time analyses and is therefore not used for in-line analysis. Some GC detectors can, however, be used for continuous monitoring such as an FID to monitor the total hydrocarbon content of a stream or the use of a TCD in hand-held leak detectors. In these applications no separation is involved.

Somewhere in between spot and continuous sampling, gas analysis GC's are most often used at-line. In at-line sampling a shunt is installed into a process line to be monitored and a small part of the gas is continuously flowing through the shunt to a vent, flare or, for high value products, an area of lower pressure in the plant. The GC sampling valve is installed in the shunt and samples analysed at regular intervals with the result that there is always a time lag between taking the sample and reporting the results. This is however much shorter than spot sampling and off-line analysis, but still only representative of the stream at the time the sample was taken, provided the flow velocity through the shunt closely matches that of the process stream being monitored. Although not as immediate as an on-line analyser, at-line analyses are usually more extensive and since the reliability and accuracy can be checked at regular intervals, they often provide more reliable results.

Integrated sampling requires more extensive sampling equipment as sampling takes place over extended time periods. For gases, this usually involves a system that will take spot samples at regular intervals into the same sample vessel over a time period. Another method is to fit an evacuated sample vessel with a fixed restrictor that will allow a very small flow into the vessel. Once opened, the vacuum in the sample vessel will draw the sample in over a long time period as the rate of flow into the vessel is determined by the restrictor. Commercial restrictors for sampling of air into canisters are available for this.

A popular and very practical way of obtaining average samples is to pass a known volume of gas through a tube containing a suitable adsorbent. This is not really suitable for trapping gases, but rather for the trapping and concentration of volatile components in the gas, as is needed in the detection of hazardous atmospheric pollutants in ambient air for example. In the same category, although also not really useable for permanent gases, are the diffusive samplers, where a container containing a suitable adsorbent is exposed to the air to be sampled for several hours or days during which the substances in the air will diffuse into, and be adsorbed onto the sorbent. The sorbent can then be taken to the laboratory where it can be chemically or thermally desorbed into an analytical instrument. We have not included detailed sampling of volatile organic compounds onto thermal desorption tubes in the following discussion as it falls outside our definition of gas analysis.

6.2 Sampling vessels

One of the early ways that low pressure gas samples were taken and transported was in the rubber bladder of soccer or rugby balls. After the sample was taken the neck of the bladder was folded closed and clamped with anything from a Mohr's clamp to a Bulldog clip. It was soon discovered that gases such as hydrogen disappeared as it diffused through the side wall. Also reactive gases like sulphur gases disappeared due to reaction with the inner surface. These shortcomings lead to the development of non-permeable sampling bags.

When selecting a sampling vessel it is of the utmost importance to consider the sampling system design and intended application from the sample source up to the point of analysis to ensure safe, trouble-free sampling and preservation of the sample integrity. Things to consider include the function, material compatibility, adequate pressure and temperature ratings, proper installation, operation, maintenance, schedules and always the safety aspects.

Irrespective of who does the sampling or where it is done, good analytical practice dictates that the sample containers must be checked from time to time for cleanliness and sample carryover. This could take the form of a container being flushed with an inert gas such as nitrogen and analysed for the impurities in the purge gas.

6.2.1 Sampling bags

Gas sampling bags, Fig. 6.1, are used for spot sampling of permanent gases and are also effective in sampling ppm levels of volatile organic compounds in air. Compared to other sample containers they are relatively cost efficient and are used in many standardised methods such as EPA method 0040 and NIOSH 6603. The most common bags are made of Tedlar™ which is a polyvinyl fluoride polymer and these bags are used for most general purpose samples. Other polymers are also used, such as Restek's ALTEF™ bags, but not all are suitable for permanent gases and may be incompatible with certain chemicals such as ketones and sulphur gases. For more demanding samples, multilayer bags can be used, especially for permanent gases and low molecular weight compounds such as methane, carbon monoxide and hydrogen sulphide.

Since these bags are made of polymers they are not completely gas tight and gases will permeate through the bag walls. The multilayer foil bags are made of up to five different layers of different polymers and aluminium foil to reduce the permeation effect. Even so, sampling bags are not suitable for

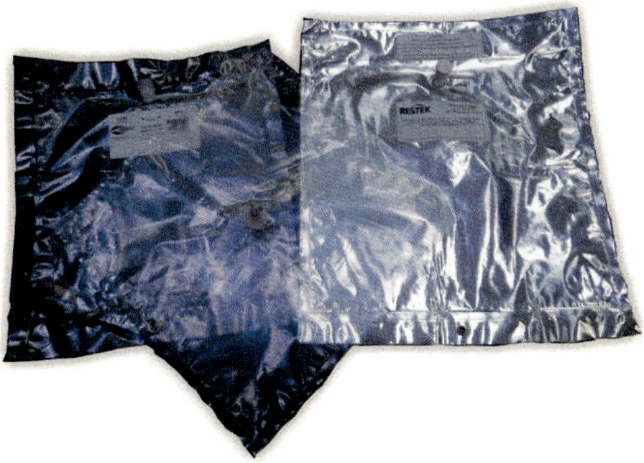

Fig. 6.1 Photograph of Tedlar™ bags on the left and multilayer foil bag on the right.

the very light gases such as hydrogen and helium unless analysed within a short time after sampling. Gas samples will only retain their integrity in the bag for a relatively short time due to the permeation of not only the gases inside the bag to the outside, but also of the ambient air into the bag. Sample bags are used extensively today for gas sampling, but it should not be used for critical or trace analysis or for samples containing hydrogen or helium. The bags are available in sizes from 0.5 L to 25 L and are intended for taking a single sample. They should not be re-used as it is virtually impossible to clean them and this may make them less cost efficient than reusable sample containers. Some companies supply sampling bags that are bigger than 25 L, such as the 100 L bags but they are rather cumbersome to use, in fact, all the bags bigger than 5 L are somewhat impractical for most applications and other types of sample vessels should be considered for large volume samples.

Sampling bags are equipped with a valve, Fig. 6.2, through which filling and subsampling can be achieved. Most bag valves incorporate a septum through which samples can be withdrawn using a gas tight syringe. One of the frustrations associated with sampling bags are the variations in the valves that are used on bags from different manufacturers. There seems to be no standard for this and it is often difficult to obtain a leak tight seal between the outlet of the valve on the bag and the inlet of the GC. Any leak at this point will contaminate the sample with ambient air and compromise the integrity of the sample. We would therefore not recommend sampling bags for any sample where low concentrations of air components must be quantified. There are bags available with Swagelok connections which overcome this to a large extent but of course not everyone has the same preferences.

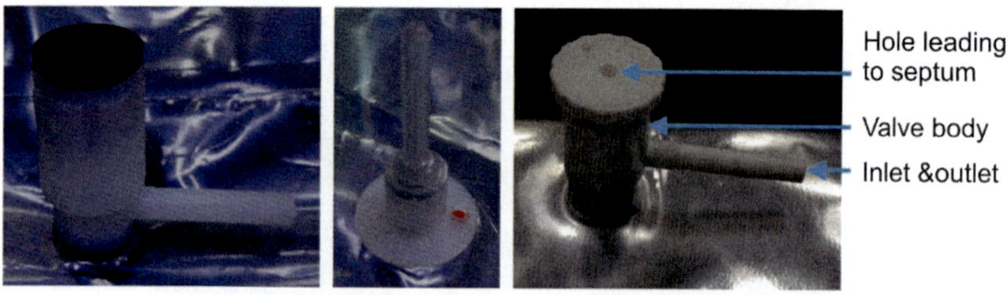

Fig. 6.2 Some typical sampling bag valves.

When storing unused sample bags it is prudent to keep them in a clean environment in a sealed secondary containment bag to prevent contamination due to adsorption. Before use, fill the bag with clean nitrogen to ensure that it has no holes in it and that the valve seals properly. When emptying the bag it also rinses out any volatiles that may be in the bag. Of course, if the sample to be taken contains nitrogen and it must be determined, another pure gas should be used for rinsing the bag.

Before sampling, check the chemical compatibility of the sample to be taken with the bag and valve material as well as the tubing that will be used to connect the bag to the source. Also ensure that the latter is clean. If the source is above ambient pressure the bag can be filled using the source pressure, but the bags should not be filled more than 80% of its capacity and not be filled too rapidly in order to retain control of the filling process. Samples at or below ambient pressure will require a sampling pump for filling. These can be hand operated or electrical but should be suitable for the purpose and not contaminate the gas being sampled. Another way to fill the bag with gas at ambient pressure is to place the bag in a so-called vacuum box sampler. This is a sealed container in which the bag is placed with the valve outside the container. The bag valve is opened and the container evacuated which causes the bag to inflate as it draws in the sample from the outside.

After sampling the bags should be protected from direct sunlight and stored above 0 °C to prevent condensation. The bags are fragile so when necessary to transport them, use a rigid container and protect against sharp objects that may puncture them. Sample bags should normally be analysed within 48 h of sampling although this time might be less for some applications.

6.2.2 Glass bulbs

Glass has long been the scientist's material of choice for storing chemicals and for doing chemical reactions because of its superior inertness. This is still true to some extent although glass can be very reactive towards certain chemical substances and the liners and glass wool used in GC injectors are therefore deactivated, usually by silation. For gas sampling it works very well provided chemicals that will react with glass, such as gas streams containing HF are avoided. For most of the more benign chemicals the reactivity of glass usually only becomes a problem when used for low concentrations, especially sulphur gases and gases that will adsorb onto the glass. Glass surfaces are highly reactive towards sulphur gases unless passivated in some way. It is possible to take a sample of a sulphur gas in a glass container and not detect anything by GC although it is clearly present at the source due to the distinctive smell. Taking many successive samples from the same source can eventually result in the glass becoming sufficiently deactivated and allow the detection of sulphur. The problem is how to know when the glass is sufficiently passivated to detect amounts that are representative of the source. This surface adsorption problem can be lessened significantly by silating the glass vessels to increase the inertness of the inner surfaces.

The major practical problem with glass is that it is very fragile and it does not take much effort to accidentally snap off a valve and lose the contents. Although glass can handle both pressure and vacuum depending on the wall thickness of the bulb, small, almost invisible scratches on its surface can cause it to explode when pressurised or implode when evacuated. In both cases the result is the same and is associated with considerable risk of injury. That may be reason why glass gas sampling bulbs are known colloquially as glass bombs — the description is not far off. All commercially available glass bulbs are oven annealed to remove stress and will resist damage during transportation and on-site use, but scratches will appear after using them for some time. For peace of mind it is a good idea to wrap the vessel with a few layers of transparent plastic or adhesive tape to prevent bits of glass from flying all over when an accident does occur. Safer still is to use foam lined metal boxes with lids that can be latched and keep the glass containers in them, especially when being evacuated or pressurised.

Glass sampling bulbs such as shown in Fig. 6.3, are designed for the collection and short term storage of samples such as environmental samples and are better than sampling bags as far as the sample integrity upon storage is concerned. They may contain valves at either end that allows connection

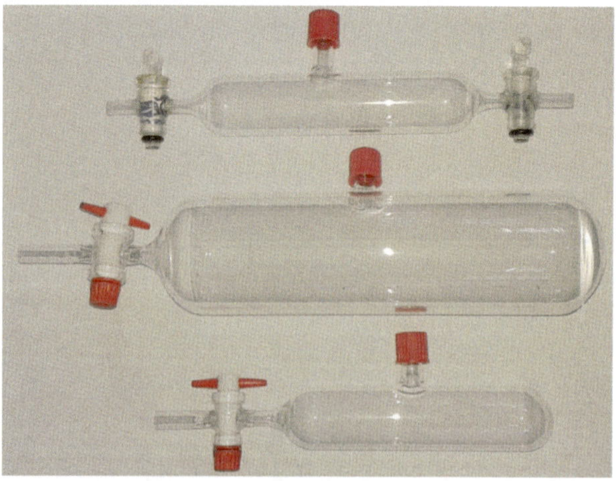

Fig. 6.3 Glass sampling bulbs.

to the sample point. The sample can then either be drawn into the glass bulb that has previously been evacuated or the sample can be allowed to flow through the bulb with both valves open to ensure the air inside is replaced with the sample and then closed. Always ensure that the valves are a good fit and it is good practise to keep the valves with the sampling bulbs it came with when purchased during cleaning and storage. Although glass stopcocks are also available the Teflon valves are usually much more user friendly as they do not get stuck so easily and requires no sealing grease. Most of these sampling bulbs contain a side opening fitted with a Teflon lined silicone rubber septum through which aliquots of the gas can be withdrawn with a gas tight syringe or through which standards can be added to the contents of the glass bulb. Since the glass bulbs are used for many samples there is always a possibility of contamination from previous samples or cleaning solvents that either remain on the glass surface or that are adsorbed onto the Teflon stopcock or side port septum. Purging with an inert gas and analysing the blank is good practice, especially for critical and trace level analyses.

6.2.3 Gas tight syringes

Gas tight syringes are very popular due to their ease of use but are also arguably the worst of the sample containers discussed here. Their popularity is probably due to the fact that they are very quick and easy to use, re-usable, easily transported, require no specialised sampling equipment and are available in many different sizes. This seems to make them ideal for spot sampling. Syringes are not without problems of their own as they are prone to developing leaks and very susceptible to contamination, especially by compounds that will adsorb onto, or react with, the syringe walls or the sealing polymer.

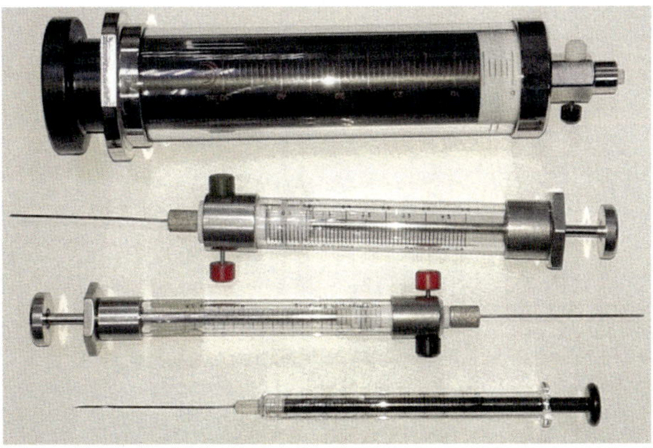

Fig. 6.4 Gas tight syringes of different sizes.

Most gas tight syringes are made of borosilicate glass and use a stainless steel plunger with a precision machined Teflon plunger tip; examples are shown in Fig. 6.4. These syringes are available in the microliter range with volumes between 10 μL and 500 μL and in the millilitre range between 1 mL and 100 mL. Very large volume syringes as shown in Fig. 6.5, with a capacity of 50 mL up to 2 L are usually made of clear acrylic and are suitable for sampling of air and benign gases that do not contain components that will react with the syringe material. The larger syringes are usually used for sample collection and the smaller ones for sample introduction or standard addition. Some large volume syringes also have a septum port through which sub-samples can be withdrawn with a smaller syringe or standards can be added to create mixtures.

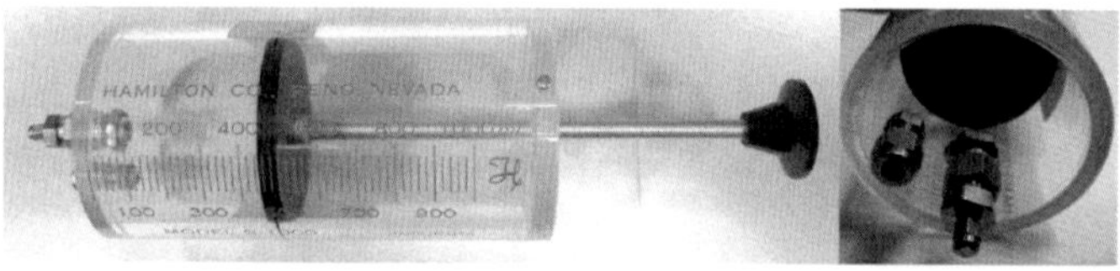

Fig. 6.5 Large volume acrylic sampling syringe (left) with a septum port for adding standards next to the centre sampling connection (right).

Some syringes require wetting of the plunger to get a proper seal whereas liquid should be avoided with some others. For most gas tight syringes replaceable needles, barrels and plungers are available. Syringes are usually fitted with needles which can be fixed or removable with the most common probably the syringes with a luer fitting to allow the attachment of luer lock needles. Gas tight syringes are normally fitted with a syringe valve, Fig. 6.6, with which the contents can be sealed in. If not closed the integrity of the syringe contents will be rapidly compromised by diffusion with ambient air.

Fig. 6.6 Syringe valve.

When using gas tight syringes it is important to remember that it is intended for low pressure samples and can handle pressures up to 300 kPa although there are specialised syringes that can handle higher pressures. When the syringe needle is inserted into a pressurised area such as a sample container or the GC injector the pressure can cause the plunger to fly out and become a mini-missile. Before use always ensure that the plunger moves easily, check that the needle is not blocked, that the valve seals properly and that there is no condensation or other contamination in the syringe. Gas tight syringes do not normally need lubricant for the plunger. Do not force the plunger if it seems stuck as it can damage the plunger seal, bend the plunger stem or even crack the syringe barrel. If the plunger does not move smoothly, it can be cleaned with an appropriate solvent taking care not to contaminate it with your fingers and to avoid scratches to the plunger tip. Ensure that no residues of cleaning solutions remain after cleaning the syringe. Avoid bending the needle by not using force and rather replace bent needles than trying to straighten them.

The accuracy of syringes may be as good as one per cent of the measured volume. As with all volumetric glassware, volumetric syringes should be used at room temperature as their reproducibility specifications are usually determined at 25 °C. For improved precision in dispensing volumes with a syringe, as in standard addition, there are mechanical and electrical devices (syringe pumps) that can be used. When using the syringe for grab sampling only, this is less critical, but it is important to thoroughly flush the syringe with the sample by filling and emptying it several times with the sample. Samples in gas tight syringes are typically stable for a day depending on the gases sampled. For light gases such as hydrogen the samples should be preferably be analysed within an hour or two while for air samples containing VOC's the analysis can be postponed for a day or more. While syringes can be used to crudely dilute a sample, typically using air as diluents, this is by no means quantitative and should only be used for qualitative analyses. It is possible to inject a sample aliquot directly from the sampling syringe but the accuracy will be poor. Much better results will be obtained if the syringe is used to flush and fill a sample loop on a gas injection valve, allow the pressure to equalise and then switch the valve to inject the sample.

Fig. 6.7 Canisters for air sampling typically used for USEPA TO methods. Deactivated Siltek® treated canisters on the left and electro-polished and passivated canisters on the right [1]. ©Restek Corporation. *(Reproduced with permission, courtesy of Restek Corporation.)*

6.2.4 Canisters

The so-called TO-canisters as shown in Fig. 6.7, derive their name from the USEPA TO methods for air sampling. Although primarily used for sampling of air for VOC analysis, they offer an excellent means for sampling and transporting permanent gas samples. These stainless steel canisters have electropolished and passivated inner surfaces, offer very good inertness and can be pressurised up to 275 kPa. There is also a special range of canisters that have silated inner surfaces which offers superior inertness for active compounds such as sulphur-containing and polar compounds. They are fitted with a two or three-port diaphragm valve with metal to metal seats and the valves have a temperature limit of 250 °C. The valve is situated inside a bracket that protects it from damage during use and transport and a pressure gauge can be fitted to the valve to check the pressure in the canister. Canisters are available with 1 L, 3 L, 6 L and 15 L volumes. Canisters are usually filled with nitrogen prior to delivery to clients. When receiving a new canister there should be a release of nitrogen when it is opened for the first time; if not, then the canister is leaking and should be returned to the supplier. It is also prudent to check the cleanliness of new canisters using methods TO-12, TO14A or TO-15 if possible.

Canisters are evacuated before use and are filled by drawing the sample in, either as a spot (grab) sample or as an integrated sample. For the spot samples the canister is simply opened where the sample needs to be taken. For sampling at specific points, a length of tubing can be connected to the

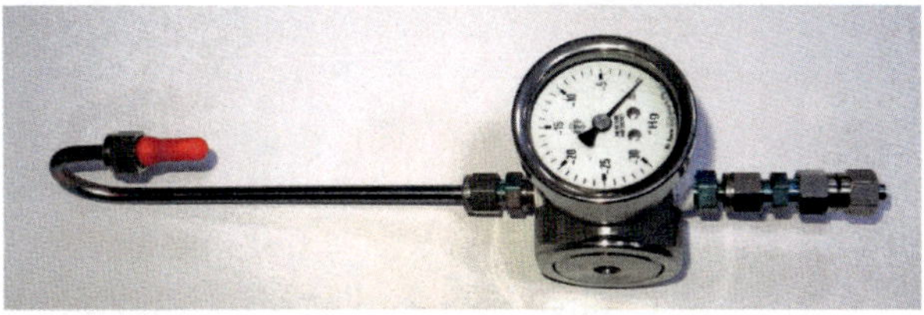

Fig. 6.8 Critical orifice sampler with vacuum gauge.

canister and the tip of the tube held at, or connected to, the point where the sample must be taken. This also allows the canister to be filled over several minutes and the sample may therefore be more representative. For integrated samples there are two accessories that can allow sampling to be integrated over a pre-selected time of between 4 min and 24 h depending on the size of the canister and the sampling controller used. Using a mechanical mass flow controller the flow into an evacuated canister can be controlled to achieve the desired sampling interval. This device has the advantage that it opens up as the pressure differential between the inside of the canister and the sample point decreases and therefore ensures a constant flow into the canister over the entire sampling time. Another way which is more suited to field sampling is to use a critical orifice device which is a restrictor such as a length of capillary tubing or a commercially available restrictor, in the sampling line allowing only a small flow through; an example is shown in Fig. 6.8. This is a passive device and does not compensate for the changing pressure of the canister and the sampling therefore favours the beginning of the sampling period with less sample being taken as time progresses.

Since the amount of sample collected in canisters relies on the difference between the initial vacuum and the sample point pressure it is important to verify the initial vacuum as well as the ambient pressure at the sampling point. For example, when sampling ambient air, a 6 L canister may contain 6 L of air at sea level but will only contain about 5 L at 1500 m above sea level, about 4 L at 3000 m and less than 1 L and the top of Mount Everest (8848 m above sea level).

Sampling using canisters is almost intuitive and it is relatively easy to get good repeatability provided common sense prevails. Canisters must be clean, the valve must be functioning properly and the pressure gauge must be correct. When sampling, use mechanical line filters to prevent particulates from entering the canister, do not over-tighten the valve as it can be damaged and cap the valve inlet when not in use. Before reusing canisters ensure that they are cleaned properly. Cleaning normally requires filling the canister with humidified nitrogen to just above atmospheric pressure and the heating it at 80 °C (or up to 250 °C if the gauge is removed) for an hour and then evacuate for an hour. These steps may have to be repeated several times if the canister is badly contaminated. Most canister suppliers offer a reconditioning service to restore used canisters to a working condition.

6.2.5 High pressure sample cylinders

The strength of steel makes it ideal for the manufacturing of pressure vessels; it is used in almost every industry for storage and reactors and of course for supplying pressurised gas in cylinders. For sampling, the one problem was always that steel surfaces are active and will adsorb some compounds, especially moisture and sulphur containing compounds. Considerable research into passivation has led to the introduction of metal containers that are highly inert and do not have the fragility of glass and are able to handle high pressures. High pressure sample containers are available in various shapes and sizes ranging from 10 mL to 3.8 L capable of withstanding pressures up to 34 MPa. Two examples are shown in Fig. 6.9. Sample cylinders are manufactured from different stainless steel alloys or aluminium for various applications and it is therefore important to match the cylinder specifications with the chemical and physical properties of the sample to be taken in them.

These sample cylinders are manufactured from seamless tubing and have a consistent wall thickness. The internal surfaces are smooth and the ends have thick walls to resist flaring when connecting valves and fittings to it. There are single and double ended containers available and they are usually supplied with a pressure test certificate. It is however, important for the end user to have the containers recertified from time to time to ensure the containers remain safe for use. Pressure vessels are susceptible to metal fatigue and should never be used unless it has been certified as safe.

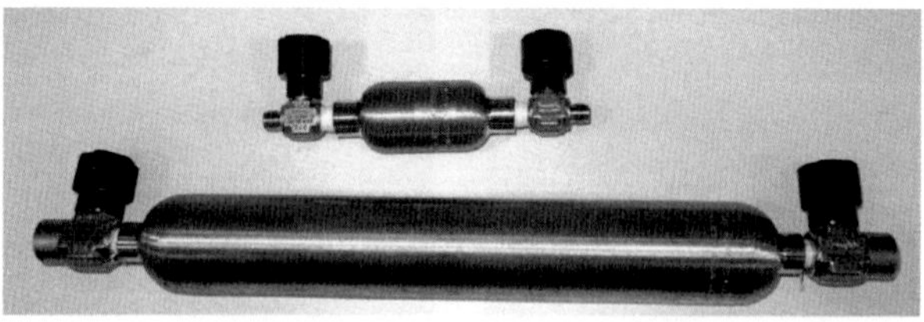

Fig. 6.9 Two stainless steel high pressure sample cylinders of different capacities. There are many more with capacities up to as large as 3.8 L.

In addition, keep the safety considerations that were discussed in the Chapter 2 in mind as it applies equally to sample cylinders as to high pressure gas cylinders in which gases are supplied to the laboratory.

The sample cylinders are typically fitted with valves capable of withstanding the pressure in the cylinder and fitted with a rupture disk of suitable rating. The rupture disk will burst when excessive pressure is applied and will allow venting of the cylinder contents to the atmosphere. The rupture disks can be replaced but the sample will be lost, so it is best to ensure that the maximum certified pressure of the cylinder, valve and rupture disk are not exceeded. Also keep in mind that increased temperature during storage or transportation can cause an increase in pressure which could lead to bursting of the rupture disk. Using high pressure cylinders without overpressure protection is irresponsible as a sample cylinder will become a deadly missile when a valve fails at high pressure.

The cylinders can be electropolished on the inside for increased inertness and, if required, the inside can be coated in Teflon. For special applications such as the sampling of sulphur containing compounds, silated cylinders are available similar to the silated canisters; see Fig. 6.10. There are also various accessories available such as end caps, plugs and carrying handles. Outage tubes can be used to provide a vapour space of a desired volume inside the cylinders when sampling liquefied gases and special valves are available for extreme pressures and severe service applications.

Fig. 6.10 Silated cylinder showing the characteristic discolouration as a result of the passivation.

The single ended sample cylinders are used similar to canisters in that they are evacuated and can then be filled by opening the valve at the desired location for spot sampling. If required a sample pump can be used to compress the sample in the cylinder to the desired pressure. For sampling of product streams in production or processing plants the double ended cylinders are preferred as they will allow the cylinder to be purged with several volumes of the sample gas before the valves are closed and the sample contained at the same pressure as the line being sampled. It is usually necessary to perform this type of sampling using a sample shunt or similar purpose designed sampling point as it may be necessary to reduce the pressure or temperature of the stream to be sampled prior to filling the cylinder.

When a high pressure sample arrives in the laboratory there is generally no way of knowing what pressure is in the cylinder because adding a pressure gauge adds an unswept volume. For high pressure samples this is less serious compared to canisters as the amount of gas in the gauge and its connections is small relative to the amount of gas in the cylinder. It is therefore prudent to insist that each sample is submitted with sampling information which should at the very least include the pressure in the cylinder. If the pressure is not known a pressure gauge and another valve can be connected to the cylinder valve. The second valve is opened and the connection evacuated. The second valve is then closed and the cylinder valve opened to get a pressure reading of the cylinder contents.

When connecting a regulator to a sample cylinder it is very important to use a high integrity regulator and to take special care when connecting it to the cylinder to ensure that the air in the connections does not contaminate the cylinder contents. We found it works well to evacuate the regulator and all connections before opening the sample cylinder briefly to fill the connections, evacuate again and only then open the sample cylinder fully. Depending on the analytical requirements and the internal volume of the regulator and connections this may have to be repeated a few times to ensure all air is purged from the regulator and connecting tubing. This is essentially the same procedure that should be used for the connection of high purity gases as described in Chapter 2.

From an analytical viewpoint, high pressure samples are seldom needed. Modern GC instruments are very sensitive and depending on the application it is possible to perform several replicate analyses on as little as 10 mL sample at just above atmospheric pressure with a properly designed, minimum volume sample inlet.

A summary of the different sampling vessels is given in Table 6.1.

Table 6.1 A comparison of some popular sampling vessels.

	Sample bag	**Glass bulb**	**Syringe**	**Canister**	**Cylinder**
Sample type	Spot	Spot or Integrated	Spot	Spot or Integrated	Spot or integrated
Technique	Active (needs Pressure)	Active or passive	Active (negative pressure)	Passive – no pump	Active or passive

Table 6.1 A comparison of some popular sampling vessels.—cont'd

	Sample bag	Glass bulb	Syringe	Canister	Cylinder
Sensitivity (max)	ppm	ppm	ppm	ppb	ppb
Analytes	Perm gas, VOC, Not H_2 & He	Perm gas, VOC	Perm gas, VOC	Perm gas, VOC	Perm gas, Liquefied gas
Pressure	Ambient	Sub-ambient to 200 kPa	Ambient	Sub-ambient to 275 kPa	Sub-ambient to 34 MPa
Durability	Single use	Reusable	Multiple use	Reusable	Reusable
Stability	48 h	2–5 days	1–2 h	30 days	30+ days depending on contents
Volume	0.5–25 L (100 L)	100 mL–5 L	1 L	0.4–6 L	10 mL–3.8 L
Sampling time	Minutes	Minutes	Minutes	Minutes–day	Minutes–days
Cost	Low	Low	Low	Moderate	High

6.2.6 Other gas sample containers

Apart from the containers described above there are many other ways of taking and keeping gas samples. The simplest is to take a length of stainless steel tubing and add valves on either end as shown in Fig. 6.11. The internal volume will depend on the diameter and length of tubing and the sample tubes can therefore be tailor made for a specific application. For ease of handling such a tube can be bent in a U-shape and the valves mounted in a block to make the assembly more rigid. These tubes can then be used for passive or active sampling and depending on the tubing material and valves used can be designed for high pressure sampling if needed. Cleaning can be done by heating and evacuation or if severely contaminated it can be disassembled and cleaned with chemical cleaning agents. If required, the tubes can be passivated with silation or fluorination agents or coated on the inside with a protective coating suitable for the type of samples to be taken in them.

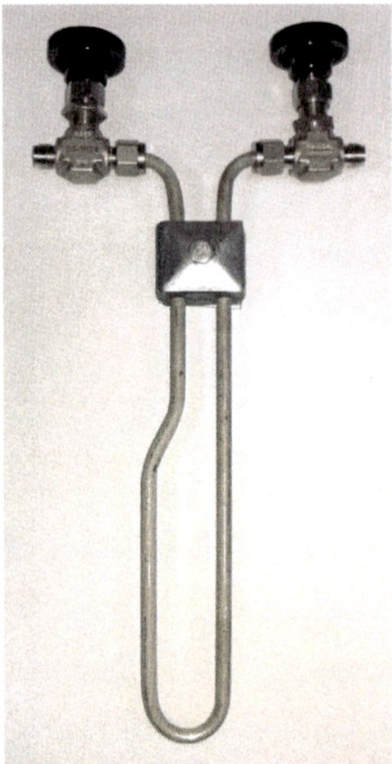

Fig. 6.11 Sampling tube bent in a U-shape with valves and mounted in a block to make it more robust.

Another useful sampling container is the humble aluminium aerosol can. Empty, evacuated cans can be obtained from the manufacturers and are very easy to use for spot sampling. Aerosol cans are lightweight, easy to use and disposable. The can is simply held at the sample point and the valve depressed to draw the sample into the can. In the laboratory a $^1/_8''$ Swagelok fitting with a Teflon ferrule can be used to connect the can to the sample inlet. The valve is depressed to open the can and the sample is drawn into a previously evacuated sample inlet. Obviously this works well only for less demanding applications as it is particularly prone to contamination by ambient air during sampling and analysis and the inner surfaces of the cans are not passivated. With a bit of ingenuity a sampling system using aerosol cans can be constructed that will provide reasonable results.

6.3 Sample introduction

Assuming a gas sample was taken with sufficient care to ensure that it is fully representative of the gas at the sample point and that its integrity remained inviolate throughout storage and transport, the next obstacle is to get a representative aliquot of that sample into the GC for analysis without contamination or unreasonable losses.

6.3.1 Static versus dynamic sampling

There are fundamentally two sub-sampling techniques relating to the way in which the sample is introduced to the valve or syringe with which the injection is done. In static sampling the sample is drawn into an evacuated sample volume such as a sample loop on a valve or a syringe with the plunger withdrawn to a certain point. The sample is then isolated from the source and injected into the carrier gas stream and carried on to the column. In dynamic sampling the sample flows continuously through the sample loop and is injected after the sample loop on the valve has been purged with the sample. It is not possible to do dynamic sampling with syringes. Both methods have their inherent advantages and disadvantages.

Dynamic sampling is the method of choice if sufficient sample at a high enough pressure is available. For trace gas analysis, the best results will be obtained with dynamic sampling as the surface adsorption effects in transfer lines are in equilibrium and therefore the sample composition remains constant. For at-line applications dynamic sampling ensures that the analysis results reflect the sample composition at the time of sampling as closely as possible. Static sampling is arguably easier to achieve and can be done using gas tight syringes. Static sampling is useful when the amount of sample is limited or if the sample is at ambient or sub-ambient pressure. Static injection using valves is done by evacuating the sample loop and drawing the sample into it or if sufficient pressure is available in the sample container, the sample can be allowed to purge the sample inlet.

6.3.2 Syringe injection

In many gas analyses, samples are still injected through the GC septum using a gas tight syringe. This has serious shortcomings such as sample carry-over and poor reproducibility. However, it can be good enough in some cases, for example if hydrocarbon gases are analysed by GC-FID and the results are normalised.

Repeatable manual syringe injections are difficult enough when doing liquid sample injections and repeatability is much more difficult to achieve when doing gas sample injections. Where the liquid injections have the advantage of using auto samplers to get more repeatability there are no auto samplers for gas samples, apart from those used for headspace and similar types of analyses. The first problem originates in the ability of the analyst to accurately draw up a constant volume into the syringe, in other words withdrawing the plunger to exactly the same position every time. During this process it is also fairly easy to draw in some ambient air with the sample. During injection the sample should be delivered consistently injection after injection. This means that the speed at which the plunger is depressed should be consistent as the gas can be compressed in the syringe barrel and therefore affect the amount of sample that is delivered over the injection period. If a split injection is done, the rate of sample delivery will affect the amount that is split and therefore the amount that ends up in the column. Injecting large gas volumes too quickly can create overpressure in the injector and cause some of the sample to be pushed into the carrier gas line and lead to excessive tailing of the peak.

There are many advantages in using syringes though, mainly related to their relatively low cost and ease of use. When sufficient care is taken in the filling of the syringe and the injection technique, results can be obtained that may be good enough for some purposes. In qualitative analyses and establishing the retention times of various component gases during method development, the syringe injection is very useful, especially since two or more pure gases can be mixed by drawing them into the syringe one after another. However, this will never be quantitative and syringes should not be used for analyses where the quantitative values are critical or in trace gas analysis. It is always prudent to validate the method used and determine whether it is fit for purpose. There are always many ways that seems to work, but not all work well enough to satisfy the accuracy requirements of the method.

Gas tight syringes are available with a choice of needle points and the correct style must be chosen for the required application [2]. For the injection of gas samples, the thinnest needle will ensure the longest septum lifetime. However, very thin needles will bend easily and limits the speed of sample delivery. The length of the needle should also be compatible with the type of injector used on the GC. A side port needle of 23 gauge and 50 mm length is typically used for gas analysis.

6.3.3 Valve injection

Using two position valves for introducing gas samples into the GC offers the most accurate injections and therefore the most repeatable results. Injection valves are usually installed in the carrier gas line in series with the GC injector, if fitted. If both gas and liquid injection valves are used in the same application, the liquid sample valve is installed between the gas injection valve and the GC injector. Since these valves can be readily automated they are also used in a large variety of on-line and at-line applications.

6.3.3.1 The sample loop

When injecting gas phase samples with a gas sampling valve an external sample loop of known volume is used. Clearly the size of any analyte peak

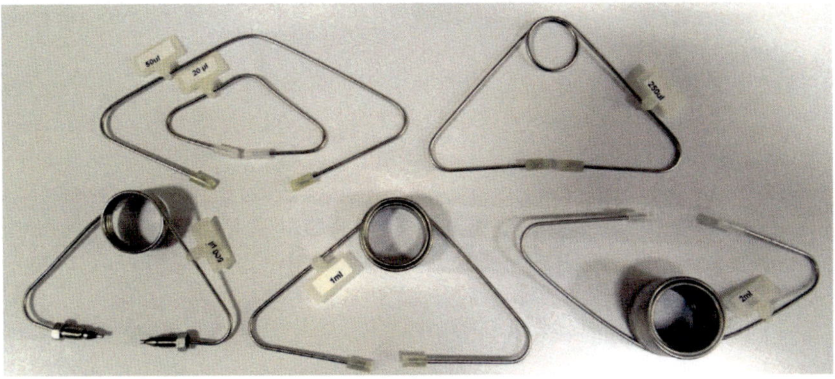

Fig. 6.12 A collection of sample loops with different volumes.

is directly proportional to the loop volume and the loop pressure at a constant temperature. Where possible valves are mounted in an area of controlled, and preferably fixed, temperature.

The volume of the sample loop chosen will depend on the analysis that is to be done. Loops are available from 5 to 5000 μL using tubing of various diameters; a selection of sample loops is shown in Fig. 6.12. For general purpose operation a loop of 250 μL is typical. For use with capillary columns a split injector is normally used to reduce the sample volume transferred to the column. High concentrations can be injected with a small loop such as 10 μL. Trace analyses require large samples and typically a 2000 μL loop is used. Loops larger than this have 1⁄16″ legs on 1⁄8″ tubing which leads to broader rather than higher peaks.

Since the actual loop volume is rarely used in gas analysis, sample loops can be made in a few minutes from pre-cut cleaned tubing. For example, a 50 cm length of 1⁄16″ tubing with and internal diameter of 0.75 mm has a volume of 228 μL and this loop would be usable in many applications.

For improved precision the sample can be vented to atmospheric pressure for a fixed time before injection. The best inlet system would be equipped with a digital pressure transducer and needle valve that can be used to control sample loop pressure. The least flow disturbance will be achieved when the sample pressure is the same as the column head pressure.

6.3.3.2 The sample inlet

Depending on how the sample is delivered for analysis some inlet system will be required. In all cases there must be some form of injection valve whether it is situated inside the GC or not. Based on the amount of sample available a decision must be made of whether static or dynamic injection will be best for the specific application.

For the introduction of a sample gas into the carrier gas stream, the general purpose sample inlet should satisfy some minimum requirements. It should allow for precise static sampling of small gas volumes as well as precise and repeatable dynamic sampling of samples at, above and below ambient pressure. The sample inlet should allow selection of either calibration gas or sample gas using valves to minimise the need for connecting and disconnecting the calibration gas line. The loop pressure must be known accurately and some form of flow indicator is required. An example of such an inlet is shown in Fig. 6.13.

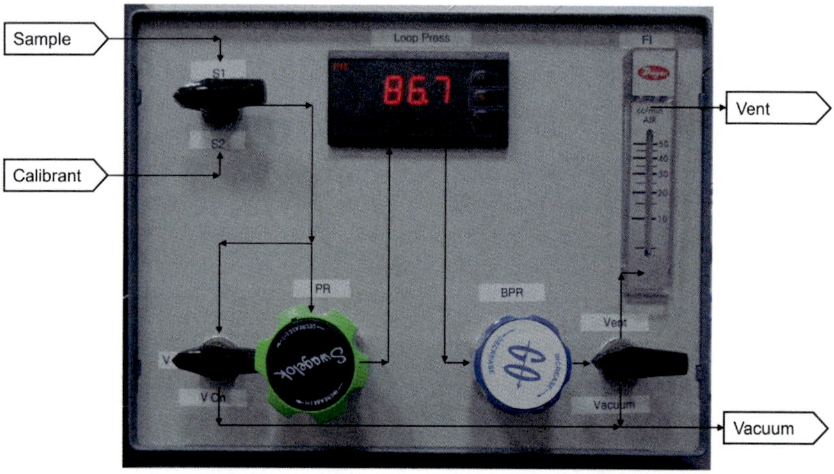

Fig. 6.13 A photograph of a general purpose gas sample inlet with the flow direction indicated. The sample loop pressure is regulated between a pressure regulator and a back-pressure regulator.

Firstly the potential for contamination from leaks before the sample loop should be minimised by using as few connections as possible. To accurately determine the sample pressure in the loop, a precise, calibrated and traceable pressure indicator, must be mounted after the sample loop and as close to it as possible. The precise and fine adjustment of the sample pressure can be done by installing a needle valve or back pressure regulator after the pressure indicator. The pressure indicator, needle valves or regulators must be downstream of the sample loop to prevent contamination by back diffusion. To allow the analysis of small gas samples and minimise waste, the internal volume of the inlet should be minimised by using $^1/_{16}''$ tubing and the plumbing should be free of dead-legs and unswept volumes.

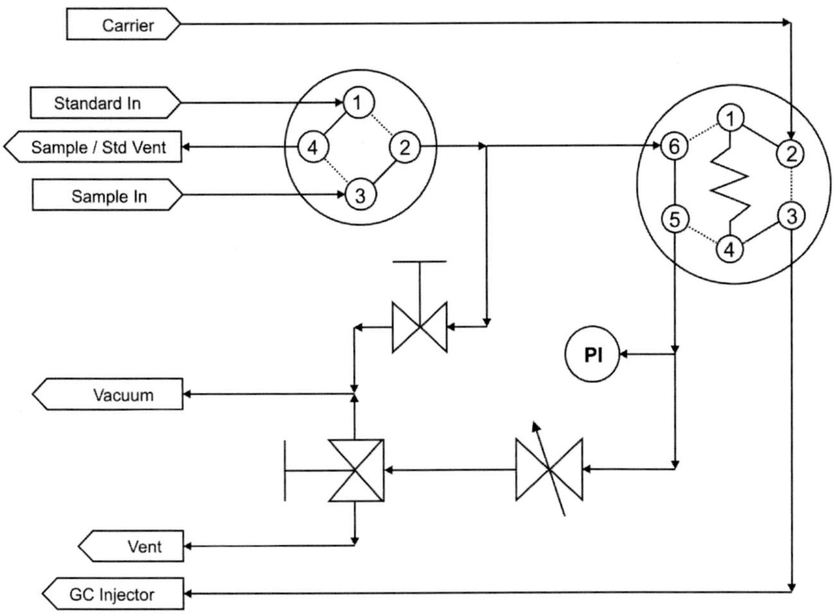

Fig. 6.14 A diagram of a general purpose sample inlet.

For trace analysis, the integrity of the inlet system should be ensured by using multi-port valves with purged housings, purging the entire sampling inlet with an inert gas and using gold plated ferrules for all connections.

A plumbing diagram for a general purpose gas inlet that satisfies the above requirements is shown in Fig. 6.14. The four port valve is used to select between the sample and standard. When used with permeation standards or a dynamic dilution system the standard stream should flow continuously and this is achieved by using the four port valve. Similarly if used for dynamic sampling of the sample line, continuous flow is also required and that is also possible with this configuration. If continuous flow of the standard and sample is not required then a three way valve can be used for selecting between standard and sample. The selected flow is then channelled through the sample loop mounted on a valve. After the valve, a pressure transducer is used to monitor the pressure in the loop and a needle valve, regulating valve or back pressure regulator is used to adjust the pressure in the loop to the desired pressure. Next a three port valve is used to select the outlet as either vacuum or vent. An additional open-close valve in the vacuum line is used to quickly evacuate the system as well as the connecting tubing to the standard or sample source when needed.

When used in conjunction with the dynamic dilution system a very broad calibration range can be covered from a single standard of relatively high concentration. Incidentally, higher concentration gravimetric standards are usually preferable since they have less uncertainty as the mass added to the cylinder can be more accurately determined. With this inlet, static and dynamic calibration and sampling can be done at, above and below ambient pressure.

In many sample inlets, the sample loop is vented to atmospheric pressure for a fixed time after loading the loop but before injection to ensure the sample loop pressure remains the same for repeated analyses. This time is typically a few seconds to ensure that diffusion of ambient air into the sample loop does not occur. The advantage of the inlet system in Fig. 6.14 is that the analysis pressure can be chosen and it allows sample injection at any pressure between a few kPa and several hundred kPa. Being able to load different pressures in the sample loop allows multi-point calibration from one standard mixture analogous to injecting different volumes.

Samples are often received at atmospheric pressure such as integrated samples taken in canisters and samples in glass bulbs. Having a system with which the loop can be evacuated and the pressure regulated will allow the sample to be drawn into the loop. The pressure is noted and the results can be corrected to the pressure used when the system was calibrated once the analysis is done.

It is also useful to be able to adjust the pressure in the sample loop to a value that is close to the carrier gas pressure as the least flow disturbance will be achieved when the sample pressure is the same as the column head pressure. If the pressure in the loop is much higher than the carrier gas pressure, the sample will expand into the carrier gas line when the valve is switched. This will result in a much broadened sample application, similar to using a very large sample loop and therefore the peaks will become broad and flat-topped but not higher as illustrated in Fig. 6.15.

Loading a very high pressure sample in the sample loop and then reducing the loop pressure, for example by connecting to vacuum, may show some discrimination against the early eluting compounds can occur as can be seen in Fig. 6.16. It is therefore preferable to inject samples at pressures as close as possible to the carrier gas pressure.

The pressure can be measured by digital pressure transducers, Fig. 6.17, and such transducers are also used extensively in sample and sample preparation lines. Typically a piezo-resistive element is used to accurately measure absolute pressure. Each transducer covers a certain range but is not

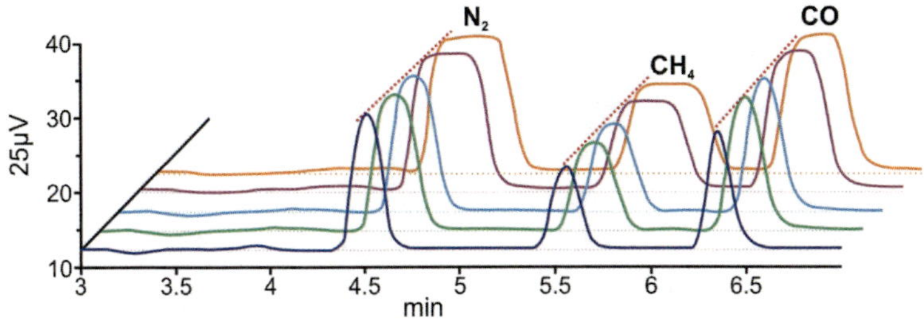

Fig. 6.15 Chromatograms with the sample loop loaded at the carrier gas pressure (bottom) and then from bottom to top at 50, 100, 150 and 200 kPa above the carrier gas pressure.

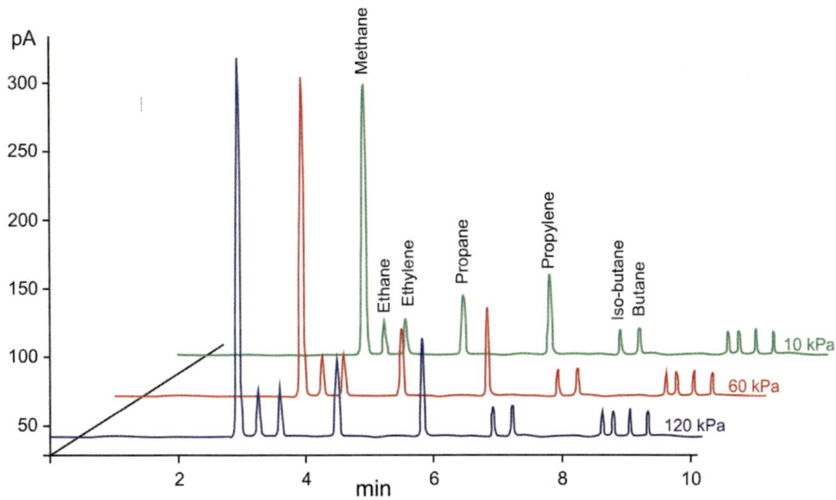

Fig. 6.16 Discrimination against early eluting compounds due to a sample loaded at high pressure and reducing the loop pressure to below the carrier gas pressure prior to injection. The chromatograms were scaled to give the same peak size for propylene. The methane and early eluting peaks reduce in size relative to propylene while later eluting peaks remain constant.

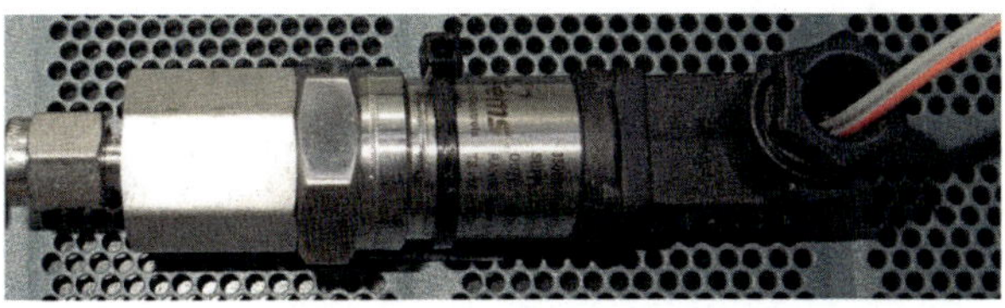

Fig. 6.17 Picture of a digital pressure head.

specific to the composition of the gas measured. When using digital pressure indicators they may be calibrated for absolute or gauge pressure and care must be taken to ensure the correct usage of pressure units. Since analogue pressure gauges nearly always showed zero at ambient pressure, we became accustomed to units such as psig (pounds per square inch gauge) but it is always much better to stay with absolute pressures and SI units and avoid any confusion and misunderstanding. Similar to the analogue pressure gauges described in Chapter 2 these also form an unswept volume that requires careful flushing between samples. This is also the reason that the pressure measurement is done just downstream of the sample loop.

6.3.3.3 At-line sample inlets

For at-line sampling, the gas inlet should not sacrifice any of the functionality of the off-line inlet described above. In addition it must be possible to automate the sampling and injection as well as selection between multiple sample streams without compromising the integrity of the sampling. Sampling and injection should require minimum attention from the analyst and unattended operation must be possible, including periodic automated calibration checks. To ensure control of the sample pressure, regardless of which line is selected, we found that the best repeatability is obtained when the sample loop is situated between a pressure regular and a back pressure regulator. We also found that evacuating pressure- and back-pressure regulators from both sides and drawing samples into the manifold using sequential closing of vacuum points works well to draw the sample into the inlet manifold quickly. Using stream selector valves as discussed in Chapter 5 is better as it requires fewer connections and uses thinner transfer lines resulting in less of the sample stream being wasted. When designing a stream selector, remember to include a connection for the calibration gas. A flow indicator in the vent gas line is a good way to check that there is continuous flow through the system and the sample loop. Examples of plumbing diagrams using on-off and stream selector valves for at-line sampling are given in Fig. 6.18.

6.3.3.4 Inlet valve configurations

So far we have used six port valves for mounting the sample loop and performing the injection. However, several other configurations can provide some rather special additional capabilities in the injection of samples. For example, when using a ten port valve, one sample can be injected onto two separate analytical systems using two sample loops as shown in

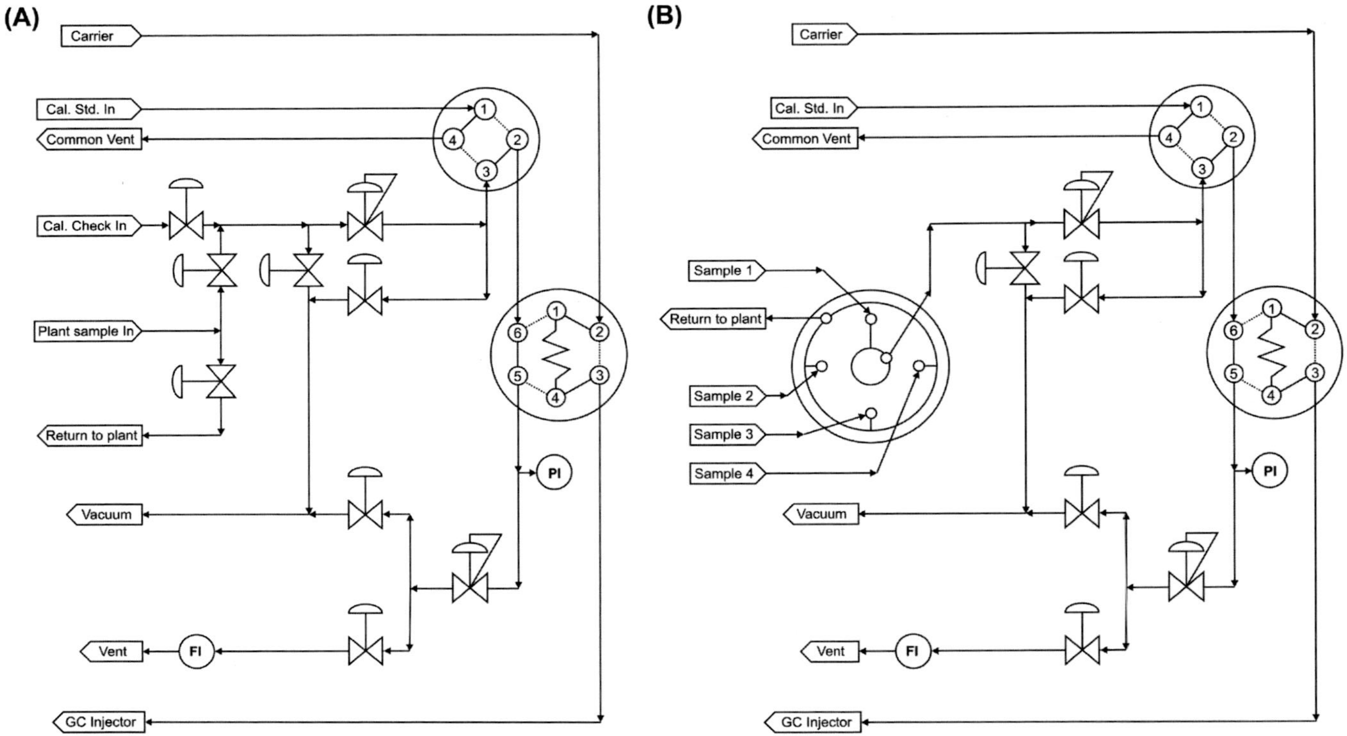

Fig. 6.18 A diagram of an at-line sample inlet using separate valves for a single plant stream (A) or a stream selector valve for multiple streams (B).

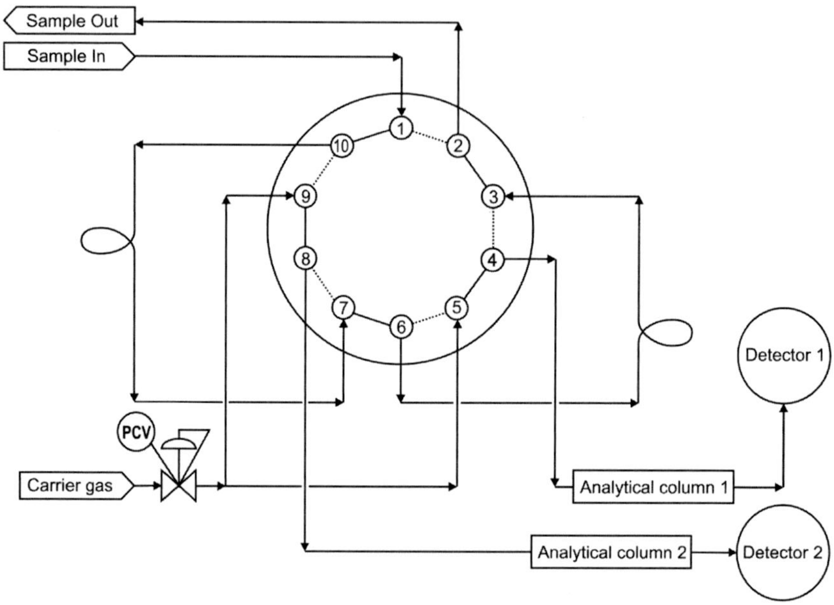

Fig. 6.19 Plumbing diagram of a ten port sample injection valve for injection of the same sample on two different analytical systems.

Fig. 6.19. This might be required when complete separation cannot be obtained on a single column and some of the components can then be separated on one column connected to the first detector while other components are separated on a second column and passed on to a second detector. Similarly, identical columns could be used with two detectors of different selectivity. One detector could then be used to detect some compounds, for example all hydrocarbons, while the other might be specific for certain compounds such as sulphur containing compounds. The two sample loops may also be of different volumes to cover different concentration ranges.

Alternatively a 10-port valve can be plumbed with two sample loops of the same or different sizes to allow the simultaneous injection of two samples onto the same column as shown in Fig. 6.20. This is a way to inject an internal standard with each sample which is very difficult to achieve by any other means when working with gases. This configuration will require a sample inlet with two separate channels with selection valves, pressure indicators and regulating valves as described for the general purpose sample inlet.

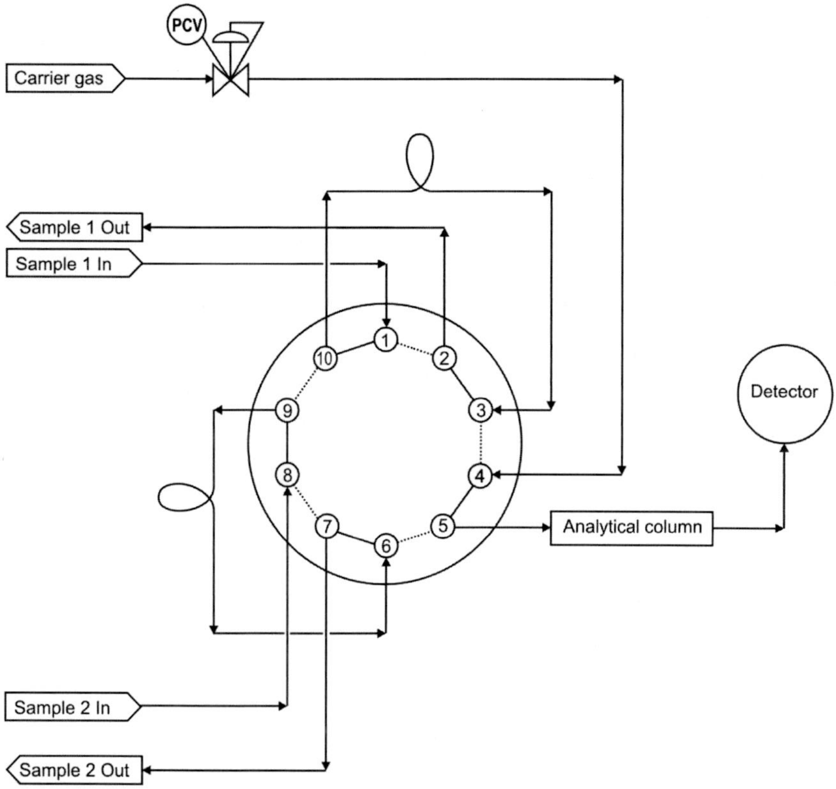

Fig. 6.20 Plumbing diagram of a ten port valve configured for the simultaneous injection of two separate samples onto the same column.

6.3.4 Liquefied gas sample injection

When the sample contains components that are not all gas at room temperature and pressure a representative sample is best obtained by injecting it at elevated pressure to ensure that a liquefied sample is injected. The sample is injected as a liquid which vaporises in the valve or on the head of the column. Since we know that a liquid sample expands up to 1000 times as it expands to gas phase these internal volumes are typically between 0.1 and 2 μL. If a liquid sample is loaded at a very high pressure it will expand a lot when the valve is switched and the pressure lowered to that of the carrier gas. This expansion of the liquid sample into a volume of gas can have the same effect as using a too large sample or injecting a gas sample at too high a pressure and may result in broad, flat-topped peaks. For samples at very high pressure in may be necessary to reduce the pressure of the sample or preferably to convert it to a gas prior to injection.

Although liquids are generally not compressible, there is some degree of compression especially with liquefied gases and the volume of the sample injected will vary in proportion to the sample inlet pressure; that is, as sample pressure increases, the sample volume will also increase. For example, a 250 kPa sample with a 0.5 μL valve provides a 0.5 μL injection, but the same valve with a 3 MPa sample may inject as much as 0.8 μL. Therefore, the inlet pressure of the sample and of the calibration standard must be regulated and be the same for both.

Natural gas may have to be sampled at pressures as high as 20 MPa and refinery gas at pressures as high as 2 MPa and in both cases some of the sample will be liquefied at these pressures. Reducing the pressure will result in cooling of the gas sample which may result in condensation of the heavier components and then neither the gas phase nor the liquid phase will be truly representative of the original sample point. In such cases a heated pressure regulator, or gasifier as shown in Fig. 6.21, can be used to reduce the pressure while maintaining all components in the gas phase. It is useful to remember that 1% gas in a sample volume is equal to 0.004% relative to the same volume of liquid and that 1% liquid in a sample volume is 250% relative to the same volume of gas.

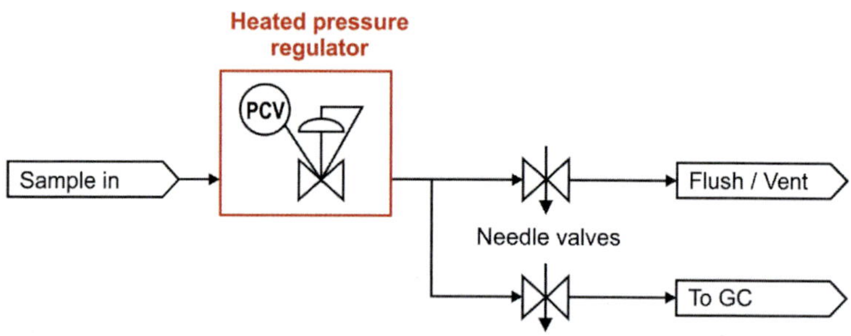

Fig. 6.21 Schematic of a heated pressure regulator or gasifier.

Another problem with mixed phase samples where part of the sample is liquefied and the rest is in the gas phase is that the relative amounts of the compounds in the sample change with repeated sampling. This is especially troublesome if the sample is relatively small and liquefies under its own vapour pressure. For example, if the headspace of a 1:1 mixture of liquefied propane and butane is analysed many times, the propane will become

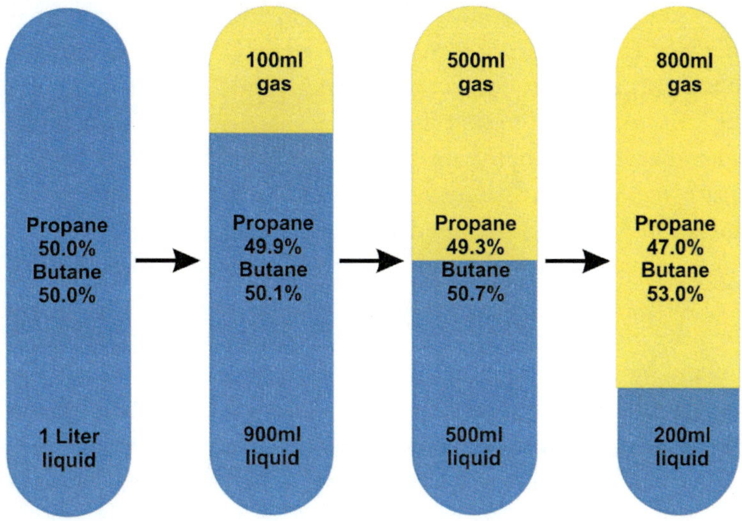

Fig. 6.22 Relative amounts of propane and butane after withdrawing different volumes of gas.

relatively less and the butane relatively more as the propane, having a lower boiling point, will occupy more of the headspace than butane at equilibrium conditions. This is graphically illustrated in Fig. 6.22.

6.4 Practical problem solving

There are many creative ways of getting gas samples into the gas inlet and it seems as unlimited as human ingenuity. Usually the creativity is sparked by a particular problem in the laboratory and a solution is found for that problem. That does not make it universally applicable and any approach to sampling and sample introduction must be proven as fit for purpose as we stated earlier.

One of the problems with Tedlar bags is that they have to be analysed individually. When many samples have to be analysed it can become a big problem since there are no automatic samplers that can be used for gases as there are for liquid samples. This usually means that samples can only be analysed during working hours and that the GC idles for most of the day and as someone in a commercial laboratory once remarked; 'we make most of our money from the instruments that work during the night'. The expensive solution is to have analysts working in shifts or to find a way to 'automate' the bag analyses.

One such solution comes from a laboratory where the analytical requirement is for normalised results of relatively high concentrations of hydrocarbon gases only and contamination with laboratory air is not a serious concern. They have a headspace analyser with autosampler and evacuate the vials. Samples are transferred from the Tedlar bags to the headspace vials using a gas-tight syringe and the vials are then analysed using the autosampler that is also fitted with a gas tight syringe. It is not really quantitative and only valid if the results are normalised, but it works for them. This technique does however allow an easy and efficient way for trends to be monitored.

A more quantitative approach involves connecting a stream selector valve to the gas inlet on the GC, using the external events in such a way that the streams can be selected using the sequence file of the GC software. Several Tedlar bags can now be connected to the stream selector valve and opened as the stream selector valve effectively closes them off except for the stream that is selected. A small diaphragm pump at the outlet of the gas inlet is now used to draw the sample from the selected bag, through the gas inlet at atmospheric pressure. The selection valve is rotated to a new sample bag position and the pump started a short time before the injection of the next sample using the external events of the GC software. In this way the process can be automated for unattended operation.

If it is possible to evacuate the sample inlet, atmospheric pressure samples in glass bulbs can be drawn into the inlet. This is certainly the preferred method, especially when using a glass bulb that is fitted with a Swagelok connection as shown in Fig. 4.4. Other methods that may work well for specific applications were discussed in Section 4.3.2.3.

To sample from a canister at or below ambient pressure it can be pressurised to above ambient pressure with carrier gas, then sampled and analysed. If absolute quantitation is required then there will be a need to know the pressure increase exactly but if results are expressed relative to an internal standard or otherwise normalised then no corrections are needed.

For the certification of the purity of a gas that liquefies under its own vapour pressure, the headspace is analysed to determine the gaseous impurities in the product gas. It often happens that the headspace analyses show that the impurities are within specification on day one, only to be out of specification the next day, or even worse, when it arrives at the customer's site. The reason for this is that the impurities are dissolved in the liquid and when the headspace is sampled, the impurities are withdrawn from the headspace only. After analysing three of more replicates the impurities in the headspace are depleted and not in equilibrium with the impurities that are still dissolved in the liquid, so it seems to be within specification. Upon standing and transport the equilibrium will re-establish itself and the product may then appear to be out of specification. Although resolving this is an engineering task rather than an analytical one is it more often than not the analyst that has to endure the abuse because of this effect. Headspace analysis is useful although only really quantitative when the composition of the gas being sampled can be related to the composition of the liquid with which it is in equilibrium.

Water is a common contaminant in many gas samples and is rather difficult to get rid of. If samples consist only of permanent gases, the moisture content can be lowered significantly by cooling the sample container and freezing the water out. Some water will however still remain as cooling the sample down sufficiently to reduce the water to negligible amounts may also cause condensable gases such as CO_2 to condense or if the water is present as a liquid significant amounts of CO_2 may dissolve in it thereby compromising the integrity of the sample. Selective membranes like Nafion will allow water to diffuse through while preventing larger molecules from doing so. While useful when sampling for larger molecules it is not useful for gases as most of them will also diffuse through these membranes.

When sampling from high pressure bulk tanks containing condensable liquid products for at-line analysis, one way of reducing the pressure without introducing leaks is to use a capillary, typically a few meters of $^1/_{16}''$ tubing connected to the tank. The small diameter results in a significant reduction of pressure. The capillary ends in a line with a larger diameter where the liquid evaporates. This causes considerable cooling of the gas which can result in condensation of water and other condensable compounds. It is therefore important that the capillary must be heated, especially at the expansion point. The gas can then be channeled to the analytical facility where the pressure is further reduced to that required by the GC.

Sampling from reactors that contain water as well as other condensable compounds entrained in the gas is particularly difficult. These so-called three-phase samples are often encountered in petrochemical plants.

One approach is to cool the stream down and collect the liquid while allowing the gas to continue on to the GC. The problem with this is that the results are never really quantitative as the liquid consists of two phases, water and organic with some gas dissolved in it while the gas still contains moisture and some of the organic compounds. The equilibrium between these phases constantly changes during the collection as the composition of the liquid changes. The best approach to this problem is to keep everything in the gas phase up to the point of injection into the GC. However, reactor streams are often at temperatures and pressures above the rating of the injection valves and reducing the pressure will result in condensation of part of the sample. Condensation occurs when the partial pressure of any compound in the gaseous mixture exceeds it vapour pressure so the solution may lie in reducing the partial pressure of the condensable compounds by diluting the sample stream. This can be done by adding a precisely measured stream of a heated inert gas followed by temperature reduction and then pressure reduction to levels that can be safely accommodated by the at-line instruments [3,4].

6.5 Volatile organics in gas

Volatile organic compounds are present in many gaseous matrices and are present in ambient air, incinerator emissions and automobile exhausts to name but a few. They are also emitted by solids such as contaminated soil and liquids such as polluted water. The volatiles can sometimes be analysed directly in the headspace of a sealed vial but often need to be concentrated to achieve the desired detection limits. There is extensive literature on the sampling, concentration and analysis of volatile organic compounds and it could easily be the subject matter for a whole book. In fact there are several books on the subject, some of which are added to the references [5–9] but there are many more. We therefore do not discuss it in detail but felt it prudent to briefly mention these techniques.

6.5.1 Static headspace

Volatiles entrained in solids or liquids can be released in various ways and this phase separation allows only the volatile components to be introduced into the analysis system. For Static Headspace analysis the sample of a liquid or solid is placed in a sealed container that may have been flushed with an inert gas. The container is heated and probably mechanically agitated until equilibrium is reached between the gas phase and the liquid or solid. The gas phase is then injected into the analysis system. Although this is a partial extraction it is highly reproducible when each sample is handled under identical extraction conditions. This requires automation and a typical headspace sampler will handle tens of samples. Applications for headspace sampling include transformer oil gas analysis, volatiles in beverages, residual solvents on solids and blood alcohol. Depending on the sensitivity and selectivity of the detector used it is possible to quantify concentrations in the low ppb range.

6.5.2 Thermal desorption

In Chapter 1 adsorption is defined as being temporary and reversible and that desorption increases with temperature. These two properties are used extensively in sampling and sample preparation for the determination of volatile organic compounds in gas matrices. We can further use these properties to trap the volatile samples and to pre-concentrate the volatiles for trace analysis. In principle a tube or trap is filled with a suitable sorbent that will selectively adsorb the types of compounds to be determined. A known volume of the gas is passed through the trap that will strip the volatile organics from the gas phase. For example, air can be sampled for a long time period to give an integrated sample that will allow trace levels to be determined. The volatiles are adsorbed onto the sorbent in the trap similar to the mechanism described for adsorption chromatography. The adsorbed volatiles are released into the GC by heating the trap.

6.5.3 Dynamic headspace

Dynamic Headspace is similar to static headspace except that the gas phase is constantly removed by flushing and trapped on an external trap. After a selected period of time the purging is stopped, the trap is rapidly heated to desorb the components and the flow path is switched for injection. In this way very large samples can be handled and by trapping for a long time it will allow very low concentration analyses.

6.5.4 Purge-and-trap

A further development is the Purge-and-Trap technique used to degas liquid samples such as water. An inert gas is bubbled through the liquid and all the volatile compounds released are trapped on a suitable material.

After some preset period, the trap is rapidly heated so that the compounds are desorbed and the gas flow switched to inject the sample onto the gas analysis system. Again very low concentrations can be determined.

6.5.5 Operation

Whichever way compounds have been trapped on an adsorbent they will need to be desorbed using heat before introduction into the GC system. This means that these techniques are not suitable for thermally labile compounds. Dynamic Headspace and Purge-and-Trap systems usually have built-in desorption heaters while a Thermal Desorption sampler may be required for sample traps such as sorbent tubes, breath analysers or badge monitors. Because of the high sensitivity of these systems it is virtually impossible to get a clean blank and some ghost peaks may be present. These are often due to contamination during storage. Blank runs and proper sorbent conditioning need to be part of any analytical sequence to ensure that there is no carry-over from sample to sample.

The choice of trap material is crucial in obtaining meaningful results. Some of the early traps were made of materials such as activated charcoal and Tenax. However today there are many choices that must be carefully chosen according to the required application. It is important that there is no breakthrough of the very volatile and high concentration components during sampling while at the same time the less volatile and low concentrations must be adsorbed.

There is always the possibility that thermal desorption injection tends to broaden the bandwidth of sample injected into the GC. Some systems use a cryogenic trap to reduce the bandwidth but this will not work with gases as their boiling points are too low. Care must be taken to ensure that the injection bandwidth be minimised by rapid heating and possibly using a small split on the injector.

6.5.6 Splitless injection

Although we said earlier that splitless injection is not used in gas analysis, it is used when analysing volatile organic compounds. With the splitless injection technique the split valve is kept closed during injection, typically for less than a minute, depending on the length of the column. Most of the sample is therefore transferred with the result that the sample will be more concentrated onto the column compared to a split injection. However, the peaks will tend to tail and to counter this, the split valve is opened shortly after injection to eliminate the tailing by cleaning out the remainder of the sample that may still be in the injector.

When a solvent is used the splitless injection technique can be used with the so-called solvent effect. The sample is introduced in splitless mode with the column oven well below the boiling point of the solvent. The solvent will then condense at the injector end of the column as a small liquid zone in which the volatile organics will redissolve creating a very narrow injection band. The column oven is then programmed normally and the solvent will elute first followed by the compounds to be determined.

6.5.7 Programmed temperature vaporisation

The programmable temperature vaporisation injector is available from most manufacturers and takes the splitless effect a step further by introducing liquid CO_2 or liquid N_2 as a coolant to the injector. We already mentioned that concentration can be done using a cooled sample loop and the PTV allows for a similar concentration to be achieved by sub-ambient cooling of the injector.

We have successfully used the PTV as a desorption oven for desorbing volatiles that were adsorbed onto a sorbent packed inside an injection liner with silanised glass wool plugs preventing the sorbent from falling out. While this poor man's thermal desorber is somewhat cumbersome and cannot be automated it works adequately if a proper thermal desorber is not available.

6.5.8 Stacked injection

While the volatile organics can be concentrated using adsorbents and introduced using thermal desorption, it is a rather time consuming way to do it. In an application note from Agilent contributed by Luong, Mork, Sieben and Winniford of the Dow Chemical Company [10], the stacked injection technique is described as an alternative that works well.

A column and oven temperature is chosen that will strongly retain the volatile organics but that will have almost no retention for the matrix. The sample is injected many times using a standard gas sample valve while keeping the column at the initial temperature. At this temperature the matrix is eluted while the analytes to be determined are effectively focused at the beginning of the column. After several injections, the column is heated and the analytes that have been concentrated by a factor equal to the number of injections, are eluted (Fig. 6.23).

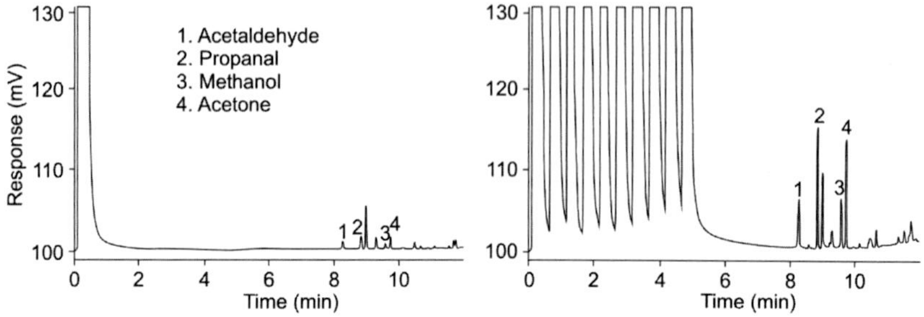

Column: Agilent Lowox, 0.53 mm x 10 m fused silica PLOT (Part no. Cp8587). Temperature: 50°C hold 5 min, increased to 270°C at 30°C/min. Carrier Gas: He, 41 cm/s, 15 kPa. Injector: Valve, 1 and stack of 10 injections, T = 150°C. Detector: FID, T = 300°C. Concentration Range: 10 - 50 ppm oxygenates in pentane.

Fig. 6.23 Example of a stacked injection used to concentrate low levels of analytes [10]. ©Agilent Technologies, Inc. 2010, 2011. *(Reproduced with permission, courtesy of Agilent Technologies, Inc.)*

6.6 Sample excess

One aspect of gas analysis that is often neglected is the disposal of excess sample. We have indicated in the discussion on the sample inlet that ultimately very little sample is necessary for analysis by GC. Consider that injection volumes seldom exceed 2 mL and therefore less than 10 mL is typically required for triplicate analyses. If the sample inlet is made using 1/16″ tubing and therefore has a very small internal volume, there is seldom any need for samples larger than 50–100 mL at the chosen analysis pressure. There is a belief that taking larger samples at high pressure is somehow more representative but this is not really true. Whether a 1 L or a 10 L sample is taken from a bulk storage tank of say 50 ML capacity, the sample that is analysed in the end seldom exceeds 2 mL and the volume of the sample taken at the tank is therefore not significant provided it was taken properly.

Large samples at high pressure create all kinds of problems in that the pressure must be reduced for analysis which in turn increases the chances of contamination by ambient air and condensation of some components. Getting rid of these samples after withdrawing the analysis sample creates problems of its own. For many years sample excess was vented into fume hoods or similar ventilation systems where it was mixed with large volume of air before being released into the atmosphere. The slogan 'dilution is the solution to pollution' was often used to justify this. It is however irresponsible and laboratories should plan for the proper and safe disposal of excess samples as well as the effluent from cleaning canisters and other sample containers.

Before accepting any samples for analysis the laboratory should carefully consider its ability to dispose of excess samples as well as what mitigating procedures will be used in cases of accidental sample leakage or spillage, not only during analysis but also during sampling, transport and storage. For corrosive, reactive and hazardous gases appropriate wet or dry scrubbers can be used and the safe disposal of the scrubber material or solutions should be included in the planning of the laboratory. Combustible gases are commonly burnt in a flare but that also creates pollution of its own. If the laboratory is not equipped to dispose of excess sample, the sample containers should be returned to the submitters for disposal of the sample excess and cleaning of the containers. This sometimes creates a problem of its own as the people who take the samples are often not sufficiently diligent in ensuring the cleanliness of the sampling vessels prior to taking the next sample. The responsible approach is to limit the size of samples to that which is absolutely necessary to perform the analysis and obtain reliable results.

In the laboratory it is a good idea to install a common vent line with a continuous flow of plant nitrogen or ambient air through it into which all vent lines from sample inlets, split vents and vacuum pump exhaust can exit. If plant nitrogen is not available a small pump can provide air flow to move all effluents to the end of the line where a cryogenic or chemical trap can be used to remove any potentially hazardous compounds from the vent stream.

References

[1] Restek Corporation, SilcoCan® Canister with RAVE™ Valve Instruction Sheet, 2014.

[2] Hamilton Company, Inc, Syringe Selection Guide. Complete Guide to Selecting the Right Hamilton Microliter, Gastight, and Speciality Syringes for Your Application, vols. Lit. No. L20082 Rev. B - 05/2017.

[3] W. Farthing, L. Felix, T. Snyder, Method and Apparatus for Maintaining Multi-Component Sample Gas Constituents in Vapor Phase During Sample Extraction and Cooling, February 12, 2008. USA Patent No: US 7,328,629 B2.

[4] L. Felix, W. Farthing, J. Irvin, T. Snyder, Apparatus and Method for Maintaining Multi-Component Sample Gas Constituents in Vapor Phase during Sample Extraction and Cooling, May 11, 2010. USA Patent No: US 7,712,384 B2.

[5] H. Bloemen, J. B, Chemistry and Analysis of Volatile Organic Compounds in the Environment, Springer Science & Business Media, Berlin/Heidelberg, Germany, 2012.

[6] R. Koppmann, Volatile Organic Compounds in the Atmosphere, Blackwell Publishing Ltd, John Wiley and Sons., 2007.

[7] R. Cumeras, X. Correig, Volatile Organic Compound Analysis in Biomedical Diagnosis Applications, Apple Academic Press, New York, 2018.

[8] L. Nollet, Chromatographic Analysis of the Environment, third ed., CRC Press, 2005.
[9] I. Horvath, J. de Jongste, Exhaled Biomarkers, European Respiratory Society Monographs, U.K, 2010.
[10] J. Luong, C. Mork, L. Sieben, B. Winniford, Oxygenates. Analysis of Trace Oxygenates in Ethylene, in: Application Note: Energy & Fuels, vol. A01612, Agilent Technologies, Inc, 2011, pp. 1–2.

CHAPTER 7

GC columns for gas analysis

Contents

A practical guide to gas analysis by gas chromatography
ISBN 978-0-12-818888-0
https://doi.org/10.1016/B978-0-12-818888-0.00007-3

The purpose of this section is to provide an overview of major column parameters that must be considered in the separation of gases. The column is the heart of the separation process and the entire chromatographic system is designed to be compatible with the column and to enable optimal separation to take place inside the column. Compared to the millions of semi-volatile compounds there are relatively few gases and this often lures analysts into thinking that gas analysis by gas chromatography must be relatively easy to do. In some cases this is true, but there are separations that are particularly difficult to achieve, for example analysing residual oxygen in argon to name but one. Separating the 14 major components in ambient air listed in Table 1.3, requires at least three different columns, two different carrier gases, different injection loop sizes and split ratio's as well as different detectors, which complicates the analysis considerably.

For many centuries up to about 1667 air was considered an element and oxygen was not known but became part of the so-called 'phlogiston' theory in which it was believed that in burning or oxidation 'phlogiston' was lost with air being an inert carrier. Daniel Rutherford is credited with the discovery of nitrogen in 1772 [1] although several other scientists in independent laboratories were simultaneously studying the gas left behind after combustion. This marked the beginning of quantitative analysis for the air separation industry. Oxygen was ultimately recognized as the agent that caused an increase in the mass of metal when oxidized and Antoine-Laurent Lavoisier showed that combustion requires oxygen by weighing closed vessels [2]. Today the measurement of gases is much less cumbersome provided the correct column system is used for the gases to be separately determined.

All that is needed to make a column is some tubing, a stationary phase and possibly a support but to prepare a column that will provide the desired separation will require much more. First of all, the materials that are best suited for the required separation of the given sample must be selected.

Then consideration must be given to the compatibility of all the sample components with the stationary phase and if the sample contains compounds that are not compatible with the column, ways must be found to get rid of the components that may damage the column or extend the run time excessively. Although few chromatographers are equipped to produce GC columns from scratch it is necessary for all chromatography users to know exactly what columns to specify when purchasing commercial columns for a particular analysis.

7.1 Column types

There are two main types of columns based on the way in which the stationary phase is held within the column. If the stationary phase is a powder, either as is, or coated with a liquid stationary phase and the column is filled with it, it is referred to as a packed column. If the stationary phase is coated or bonded to the inside surface of the column wall it is known as an open tubular column. Packed columns require tubing with a relatively large internal volume to accommodate the packing and are generally shorter and wider than open tubular columns which are generally longer and narrower and therefore also known as capillary columns.

7.1.1 Packed columns

In the early days of gas chromatography all columns were packed and tubing with diameters of a $^1/_4''$ and even $^1/_2''$ was commonly used. As described in Chapter 1, Golay and others realised that longer thinner columns would give better separation and columns with such large internal diameters are not used anymore. Most commercially available packed columns are provided in deactivated stainless steel tubing with an outer diameter of $^1/_8''$ and inner diameter of 2.2 mm.

Very thin or capillary columns are much more efficient for the separation of most compounds, even gases, and provide separation at lower temperatures and with shorter analysis times than packed columns with similar stationary phases as can be seen in Fig. 7.1. Packed columns have however continued to be used in gas analysis long after the majority of these have been replaced by capillary columns in almost all other fields of gas chromatography. One reason for this has been the unavailability of suitable phases but especially because of the poor loadability of capillary columns.

Whereas capillary columns are generally more efficient, the amount of sample that can be injected before overloading occurs is rather small.

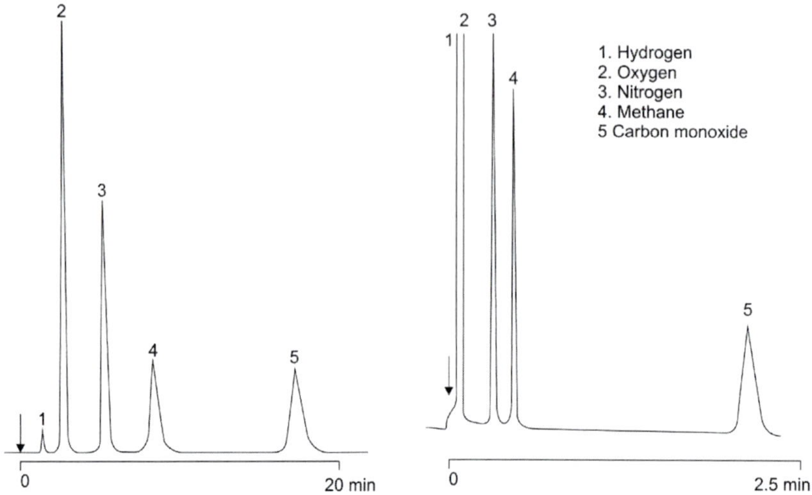

Columns: (LEFT) Molecular Sieve 5A, 60/80 mesh, 12' x 1/8" tubing. Oven temperature: 70°C. Helium carrier gas at 20 mL/min. (RIGHT) Molecular Sieve 5A, 10 m x 0.53 mm, df = 50 µm.
Oven temperature: 25°C. Hydrogen carrier gas at 80 kPa, 25 mL/min.

Fig. 7.1 A comparison between separation of some gases using packed and capillary columns showing the improved peak shape and shorter analysis time when using the latter.

On packed columns which contain larger amounts of stationary phase much larger volumes can be injected without overloading the column and this is especially advantageous when analysing trace contaminants in high purity gases. Since the packed column is filled with particles, as shown in Fig. 7.2, and the particles can be coated with a liquid phase, the capacity is larger than other types of columns as it contains more stationary phase.

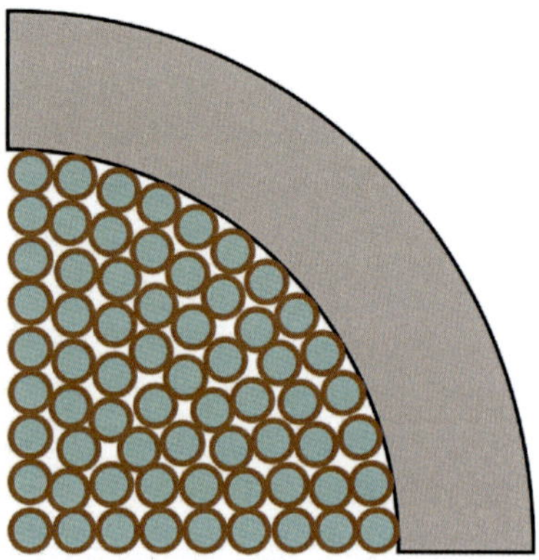

Packed column Ø2.2 mm

Fig. 7.2 Partial cross section of a packed column.

Since the powdered stationary phase inside a packed column is loose, silanised glass wool plugs are generally used at both ends of the columns to keep the packing material from falling out when handling the column.

Due to their large internal volumes, the carrier gas volumetric flow through a packed column is usually around 15–30 mL/min to achieve acceptable linear velocities which makes it less economical than capillary columns. When using hydrogen as carrier gas with capillary columns the flow through the column is much lower and therefore the risk of hydrogen build-up and explosion, should a leak occur, is also low. Using hydrogen with packed columns increases the risk of explosion as leaks, if they occur, can be much larger. However, as we indicated earlier, most modern gas chromatographs will sense the flow change and shut the instrument down.

7.1.2 Micro-packed columns

Micro-packed columns have an internal diameter of 0.75 or 1 mm and are typically made from 1⁄16″ deactivated metal tubing. Obviously the particle sizes of the packing material for these columns must be much finer than those used for conventional packed columns, Fig. 7.3. The column ends are closed using some deactivated device such as braided wire to keep the packing intact under high pressures, but still allow uninterrupted flow of carrier gas. Micro-packed columns require a higher head pressure than 2.2 mm internal diameter packed columns due mainly to the finer packing

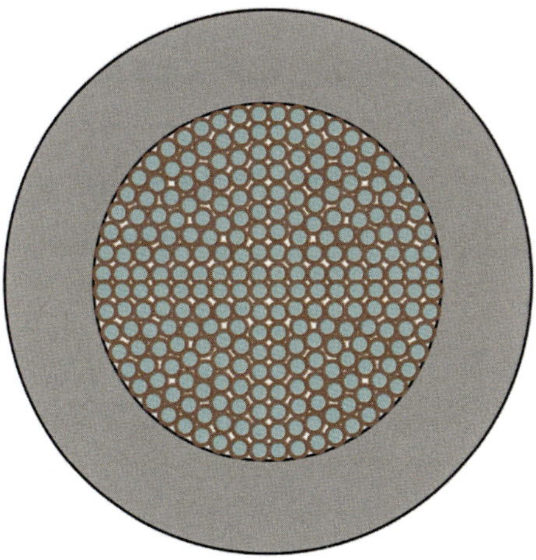

Fig. 7.3 Cross section of micro-packed column.

material and to some extent the smaller diameter. For example, for a flow of 20 mL/min at 25 °C a 2 m micro-packed column will require about 400 kPa helium. For a 2 m × 2.2 mm i.d. packed column the required head pressure for helium will be approximately half of this value.

Micro-packed columns offer a good compromise between separating power and capacity or loadability and for this reason are becoming more and more popular in gas analysis. It has been shown that micro-packed columns can handle a sample which is as much as forty times larger than that which can be loaded onto wall-coated open tubular columns under comparable conditions [3].

7.1.3 Capillary

Capillary columns have an internal diameter of between 0.1 and 0.53 mm and can be made from various materials of which the most common is probably fused silica followed by deactivated metal. Capillary columns are all open tubular columns meaning the stationary phase is coated or attached to the inner surfaces of the tubing leaving the center open making them a lot more permeable than micro-packed columns.

Capillary columns did not gain widespread use until the late 1970's due to several reasons such as their small sample capacity, difficulties in coating the column, the fragility of thin glass columns and the problems associated with sample introduction and connection to the detector. In 1979 fused silica columns that were drawn from specially purified silica that contains metal oxides appeared. These columns have much thinner walls than the glass capillary columns and are given mechanical strength by an outside protective polyimide coating. They are flexible, can be bent into coils and have the advantages of physical strength, flexibility and low reactivity which accounts for their current popularity.

There are mainly three different types depending on the way in which the stationary phase is distributed on the column wall.

7.1.3.1 SCOT

Support Coated Open Tubular columns were developed to try and replace packed columns with the same coated phases, but with a smaller internal diameter that also has low flow resistance. Solid supports such as diatomaceous earth, similar to that used in packed or micro-packed columns, and onto which the stationary phase has been adsorbed, are fixed to the inner walls of the column. These columns are generally quite fragile. Vibration inside the oven caused by the fan can cause the stationary phase layer to 'cave in' blocking the column and leaving uncoated columns walls which becomes active sites. Most applications have moved to either PLOT or WCOT columns and SCOT columns are only used in some specialised applications.

7.1.3.2 PLOT

Porous Layer Open Tubular (PLOT) columns are very popular for the separation of gases and very volatile analytes at, or above, ambient temperatures. While many of the volatile compounds can be analysed using liquid phase columns, it often needs very thick phase columns and sub-ambient cooling of the oven. Even then complete resolution may prove to be elusive. The small, porous particles that are stuck to the inner wall of the capillary tubing in gas-solid PLOT columns separate compounds by the differences in their adsorption properties meaning that differentiation between compounds are largely due to the different sizes and shapes of analyte molecules. These particles that form the stationary phase are responsible for the separation of the compounds opposed to SCOT columns where the stationary phase is coated onto inert particles.

Porous Layer Open Tubular columns typically have internal diameters of between 0.25 and 0.53 mm. The stationary phase is based on an adsorbent or a porous polymer resulting in columns that are more robust than the SCOT columns. These columns have a relatively thick (3 μm on a 0.25 mm column up to 50 μm on a 0.53 mm column) layer of the small, porous particles coated onto the inner wall of the column as shown in Fig. 7.4.

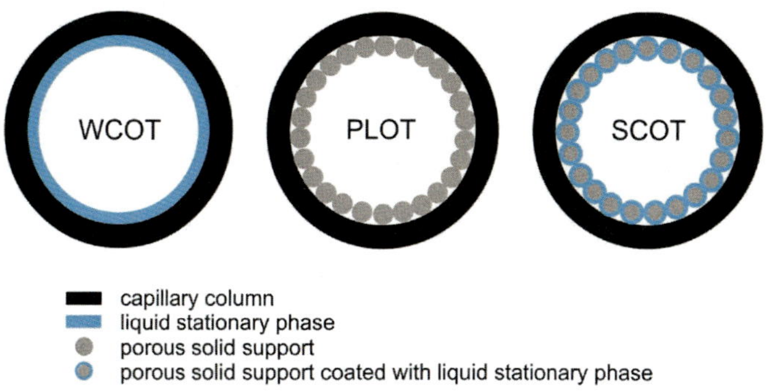

Fig. 7.4 Cross sections of a WCOT, PLOT and SCOT columns showing the stationary phase coated on the inner wall of the column. The stationary phase is either a liquid, or particles or coated particles.

In all columns the porous layer of particles are held in place by a chemical binder. With about 10^{12} particles in a PLOT column it is hardly surprising that some particles can be lost due to mechanical disturbances such as valve switching or column oven vibrations. For this reason most columns were supplied with a particle trap to be fitted ahead of a valve or detector as shown in Fig. 7.5. From time to time it is necessary to blow through or to replace the particle trap. The symptom that blocking is starting to occur is an increase in retention time or an increased column head pressure.

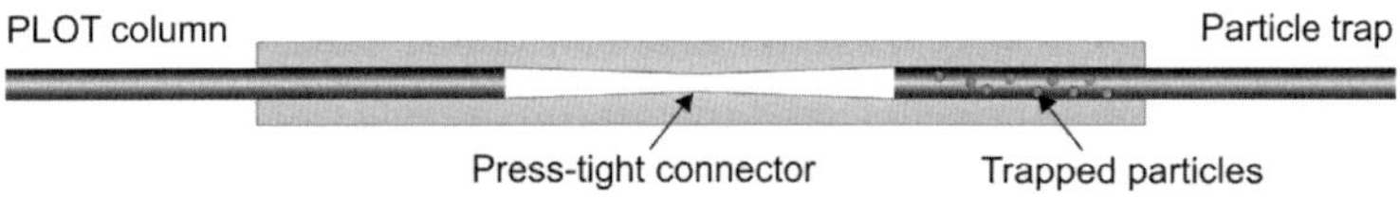

Fig. 7.5 Column connector with a PLOT column connected to a particle trap.

The later generation of highly stable bonded PLOT columns eliminates particle loss and the need for particle traps. However, it is still prudent to connect a meter or two of a liquid phase column (liquids trap particles) between the PLOT column and the detector, especially when connecting the column to a mass spectrometer. The high vacuum of the MS results in very fast flow rates in the last section of the column and this may dislodge some particles that will end up contaminating the ion source of the MS. With modern PLOT columns there is also a reduction in column bleed and bonded PLOT columns have significantly higher maximum operating temperatures than columns containing equivalent non-bonded phases, for example a Porabond Q column has a maximum temperature that is 70 °C higher than a Poraplot Q column.

7.1.3.3 WCOT

Wall Coated Open Tubular columns have internal diameters between 0.1 and 0.53 mm and are the most widely used of all column types in chromatography in general but not in gas analysis. The stationary phase in WCOT columns is a liquid film coated on the deactivated wall of the column, Fig. 7.6. The most common stationary phases are polysiloxanes and polyethylene glycols that have low vapour pressure and can therefore withstand high temperatures without 'boiling' out of the column. These columns are extremely efficient and more robust than most of the other columns but unfortunately have a rather limited application in gas analysis.

The principal mechanism of separation in WCOT columns is solvation with polar compounds dissolving more readily in polar stationary phases and non-polar compounds more readily in non-polar stationary phases.

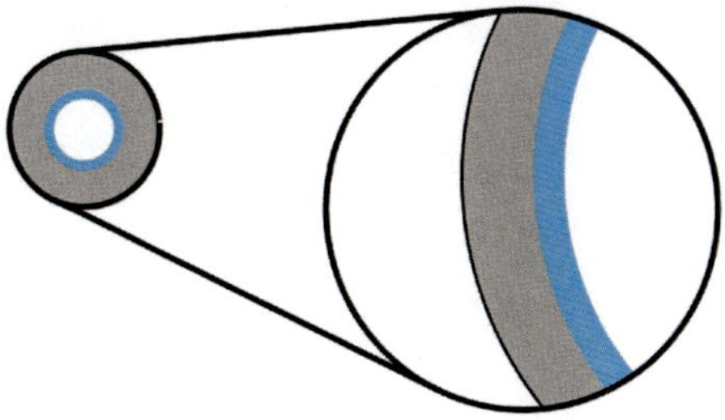

Fig. 7.6 Cross section of a WCOT column showing the layer of liquid stationary phase coated on the inside of the column wall.

Compounds in vapour form will dissolve in cold liquids but escape at higher temperatures. This means that compounds are separated on WCOT columns due to a combination of their boiling points and polarity. Generally speaking the boiling point increases as the degree of oxidation increases making it more polar, for example ethane has a boiling point of −89 °C while ethanol has a boiling point of 78 °C.

Generally the lower the boiling point of a compound the lower its solubility will be in a liquid at a set temperature and pressure and therefore the less time it would spend in a liquid stationary phase. Given the low boiling points of gases this accounts for the limited usefulness of WCOT columns for gas analysis. It is often used for very volatile compounds in gas matrices but requires thick stationary phases and low temperatures to obtain enough retention of the volatile compounds to obtain separation. The film thickness of these columns is also much thinner than for PLOT columns which further exacerbates this problem.

7.2 Dimensions

Columns for GC have always been a hotchpotch of metric and imperial dimensions. We may still talk of a 1.2 m × $^1/_8$″ stainless steel column packed with Molecular sieve 5A, 60−80 mesh but this makes more sense to most practicing chromatographers than a 1.2 m × 2.2 mm i.d. stainless steel column packed with Molecular sieve 5A 150−180 μm. Fortunately PLOT and WCOT columns are more consistent where we have for example a 30 m × 0.53 mm i.d. column with a film thickness of 25 μm.

The section on column selection and optimisation at the end of this chapter provides more detailed information on the selection of ideal column dimensions. Some typical column dimensions are given below but most column manufacturers are willing to prepare custom made columns for special applications.

7.2.1 Column length

In terms of separating efficiency column length is the least significant parameter. However, if we need to do a marginal separation such as the separation of argon and oxygen, a longer column will be needed. Packed column lengths are typically between 3 and 10 feet (1–3 m) while capillary columns are typically between 5 and 50 m and micropacked columns typically between 1 and 2 m. Longer columns of all types are available but as the column length increases so does the difficulty in producing an efficient column and very long columns are often made by joining two columns with a zero dead volume connector. The maximum practical column length may be determined by the maximum column head pressure that can be used.

7.2.2 Internal diameter

Most packed columns are manufactured from $^1/_8''$ tubing with an internal diameter of 2.2 mm. As a guideline one can expect to use between 1 and 2 g of packing material per meter length of column. Micropacked columns are typically made with either $^1/_{16}''$ or 1.5 mm tubing having 0.75 or 1 mm internal diameters while capillary columns have internal diameters from 0.1 to 0.53 mm.

It follows from the theory that the smaller the internal diameter, the smaller the HETP can be and therefore the more theoretical plates per meter of column. However, it is also more difficult to push the same amount of gas through a smaller diameter column and clearly loadability of sample is much reduced. When multi-column systems are required to effect the desired separation, it is important to match the diameters of columns that are connected in series to ensure optimum column efficiency. However, optimum efficiency is sometimes deliberately sacrificed in favour of other considerations; a typical example would be the two dimensional GCxGC applications.

7.2.3 Film thickness

Where column length can be regarded as the least significant parameter in column specification, film thickness is arguably the most significant parameter as was shown from the van Deemter equation in Chapter 1.

Very few support coated columns are used in gas analysis but occasionally one uses a column such as 5% Krytox on HayeSep Q. The concentration of the liquid determines the film thickness of the film in this type of packed or micro-packed column. If the layer is too thin the support particles may not be covered sufficiently leading to separation where the support takes part in the separation in addition to the coating. Such dual separation mechanisms would be very difficult to reproduce.

In the case of PLOT and WCOT columns film thickness is a direct measure of the film and is defined in the manufacture of that column. Film thickness for PLOT columns are typically between 4 and 50 μm and for WCOT columns between 0.1 and 5 μm. Both the loading and the separations are highly dependent on film thickness.

7.2.4 Supports

If the column uses a support, the particle size of the support is generally expressed in terms of screen openings since screens or sieves are normally used to prepare them. The typical particle sizes used in GC are: 60/80 mesh or 250−177 μm; 80/100 mesh or 177−149 μm and 100/120 mesh or 149−125 μm.

The designation 60/80 mesh means that the particles have passed through a 60 mesh screen (−60) and will not pass through the 80 mesh screen (+80). This means that the particles are between 250 and 177 μm in size. The column efficiency improves with decreasing particle size and the narrower the range of mesh size the better for good efficiency. The 80/100 mesh is probably the most popular size, but 100/120 mesh is used for more efficient columns provided that the system can supply sufficient carrier gas pressure. Generally therefore, one would use the smallest particle size available as long as the application can handle the pressure drop. It is also important to use a narrow range of particle size to prepare an efficient column hence 100/120 mesh and not something like 60/120 mesh.

7.3 Materials

One of the most important specifications for GC columns is the material of manufacture. Typically they are made of stainless steel, fused silica or glass which can be a silicate glass such as alkali-borosilicate glass, borosilicate glass or alumina silicate glass. As described in Section 7.2, the length of the columns is relatively unimportant while the inner diameter and the film thickness are more important. It is also important to consider the operating pressure required for the application and the nature of the compounds to be separated on the column.

7.3.1 Glass

Glass was the material of choice for early packed and capillary columns because of its inertness and the relative ease by which it can be drawn into capillary columns. It is however very fragile and breaks easily. After installing a glass column, one of the most frustrating sounds for chromatographers was the soft 'click' sound coming from the oven as soon as it starts to heat up. This meant the column broke either at the injector or detector end due to the expansion of the connecting nut and ferrule. Glass capillaries were more susceptible to this than packed columns and when they broke it meant that the column end had to be straightened and deactivated before the column could be reinstalled. It is therefore not surprising that chromatographers were looking for more robust materials. The answer came in the form of fused silica.

Glass columns are still used for some specific applications where high inertness is required, such as for the analysis of active analytes. Commercial glass columns undergo high temperature silanisation to ensure inertness and glass capillaries are made with fused silica ends to facilitate installation.

7.3.2 Fused silica

Fused silica is glass consisting of pure, non-crystalline silica and contains none of the additives used in other glasses to lower the melting point. Fused silica has a very high melting point but is extremely brittle. Fused silica columns are therefore coated on the outside with polyimide to impart mechanical strength which makes the columns very robust. It is safe to say that the majority of columns used in gas chromatography are fused silica. The outer coating limits the maximum column temperature and polyimide coated fused silica columns cannot be used at temperatures above 300 °C, not that such high temperatures are ever needed in gas analysis. For high temperature columns, the fused silica columns were coated with metal, such as aluminum. These tend to go very brittle with extended exposure to high temperatures so that advances in deactivation of metals has led to highly inert metallic columns that have largely replaced the metal clad fused silica columns.

7.3.3 Stainless and deactivated stainless steel

Packed columns have traditionally been made using $^1/_8$″ stainless steel tubing although in the early years of gas chromatography larger diameters were used. Smaller diameter tubing is used for micro-packed and open tubular columns. Stainless steel is much more robust than glass or fused silica and also more economical and is used for less demanding applications where the inertness of glass is not required.

For more demanding applications where inertness is required deactivated stainless steel is used. These are available under trade names such as Silcosteel™, Ultimetal™ or Sulfinert™ and are produced using proprietary deactivation methods by different manufacturers. In essence the deactivation involves coating the inside of the tubing with several layers that deactivate the surface and allow coating of the stationary phase as illustrated in Fig. 7.7.

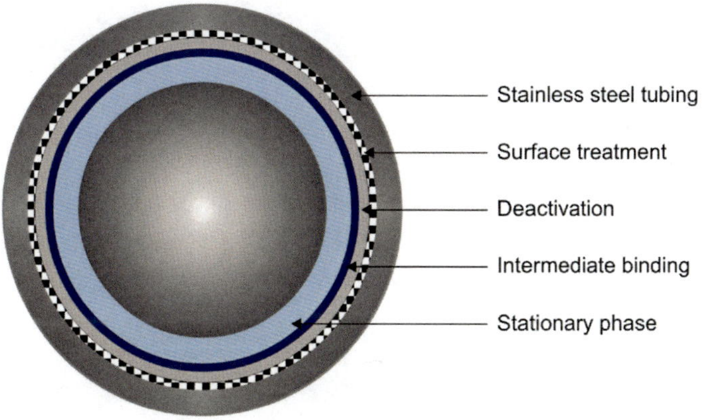

Fig. 7.7 Cross section of a typical deactivated stainless steel column showing the different layers applied for deactivation of the surface and coating it with stationary phase.

SilcoSmooth™ and Ultimetal™ tubing have the inertness of glass but the strength and flexibility of stainless steel. The use of deactivated stainless steel has demonstrated significant improvements in peak shape especially for analytes like carbon monoxide at low concentrations. If the analysis involves low concentration sulphur compounds Sulfinert™-treated tubing offers the best results and is required for analyses at ppb levels.

7.3.4 Other column materials

For specialised applications other tubing materials may be mandatory. Although copper is a general purpose tubing and easy to work with it is not recommended for columns. If the analysis involves reactive compounds the fragile and inflexible glass columns are sometimes used or if the application uses low temperatures Teflon columns can be used. PTFE columns offer the flexibility of metal with inertness approaching that of glass but are permeable to most gases making it unsuitable for trace analyses. Hastelloy® tubing is a nickel-chromium alloy with excellent inertness. It is used for highly corrosive or oxidising compounds and gases while nickel tubing is often used for analyses of caustic or oxidising compounds or gases.

7.4 Column handling

7.4.1 Cutting

It is sometimes necessary to cut columns, usually to get rid of contamination in the first part of the column. Often adequate separation can be obtained on shorter columns and if time is of the essence columns are made shorter to accommodate this requirement.

To cut a stainless steel packed column a tube cutter should be used to cut through the tubing while holding the end to be cut off pointing upwards. This is to prevent the stationary phase from falling out of the part of the column that will still be used. The cut is made with a tubing cutter suitable for the column diameter and when the top part comes off, some of the stationary phase in that part of the column will fall out but care should be taken not to lose the stationary phase in the rest of the column. Once the cut is made, deburr the cut and then carefully empty the first 1−2 cm of the column by gently tapping the end to shake out a small amount of the stationary phase. Then plug the column with a plug of silanised glass wool to prevent the rest of the stationary phase from falling out when handling the column. In columns used for gases that will react with the glass wool, Teflon wool can be used to plug the end but keep in mind that it will not withstand high temperatures. Protective gloves must be worn when handling glass wool.

Fused silica columns are usually cut with a ceramic wafer. There are variations to the technique and we are sure that every experienced chromatographer has their own way of doing it. What works for us is to hold the column in one hand with the column lying flat on your index finger and then slide the edge of the wafer across the column and at a right angle to the column to score through the polyimide layer. The ceramic wafer will not cut your finger and minimal force is required. Then gently bend the top to be cut off away from the scored mark and the column will snap off. Inspect the cut with a magnifier to make sure the cut is clean and that the polyimide did not tear away from the cut. If the cut is not perfect, redo the cut; losing a centimeter or two from a column is not serious but a bad cut can compromise the peak shapes and separation. Cutting columns this way becomes progressively harder with increasing column diameter and for the beginner it is useful to practice on an old 0.53 mm column to master the technique. Cutting of the column must be done after the column has been pushed through its ferrule to ensure that no ferrule material has entered the column.

Metal capillary tubing is usually cut by scoring the wall with a needle file, a ceramic wafer or with a special metal column cutter that is available from some suppliers. With all cutters, score the tubing all the way around the column, making sure not to cut through the column wall. Wipe away all filings then bend the tubing away from the score mark. As the score opens, bend the tubing back towards the score and repeat this action until it snaps in two. As with the fused silica columns, the column end should be inspected with a small magnifier to see the integrity of the cut. The tubing must still be round, the cut at 90 degrees to the length of the tubing and the bore not constricted. If not the procedure must be repeated.

Glass columns are cut with a diamond glass cutter. For glass packed columns the glass is scored all the way around the column and then snapped off while glass capillaries can be cut by scoring one side of the column and then snap it off. It is generally not recommended to cut glass columns as that exposes an unsilanised area of the glass.

When cutting columns it is important to wear safety glasses as fine silica, metal or glass shards can fly away when the column is snapped off and can get into your eyes. Cut ends of glass columns are sharp and will cut through skin if sufficient care is not taken while thin fused silica capillary column ends can pierce skin. While these injuries may not necessarily require medical treatment it is not pleasant and best avoided.

7.4.2 Coiling

In all cases it is not advisable to try and coil any column too tightly. In the pneumatic section it was shown that the tubing has a finite minimum coiling diameter. In general, tightly coiling a column will cause some of the packing to be crushed and particles will no longer be of the same size. GC ovens are normally quite large but it may be necessary to fit a tightly coiled column into an auxiliary oven or process GC oven. In this case it is advisable to custom order the column of the required coiling diameter.

7.4.3 Installing

When fitting columns into the instrument, the very first assumption is that the column oven as well as the injector and detector have been cooled before any work commences. Like minor cuts, blisters on fingertips are not pleasant.

If the system is equipped with an injector, the optimum insertion distance must be obtained from the instrument documentation. Similarly, all detectors have an optimal insertion distance and these too must be found in the instrument's manual. When one works with only one instrument or instruments from the same manufacturer these distances are quickly memorised. However, in larger laboratories with different instruments from different manufacturers and using different injectors and detectors it is useful to put a label on each with the correct insertion distances. Another useful tip is to mark these distances on the front of the oven but behind the door where it would be inconspicuous. A permanent marker or lines scratched with a sharp tool works well. This makes it very easy to install columns without having to find a ruler every time.

Packed and micro-packed columns are supplied with the appropriate fittings and column plugs so that they are simply attached to the connectors in the column oven. Make sure that the column does not touch any part of the oven wall but hangs freely in the air.

To install a capillary or PLOT column, hold the column end facing downwards and slide the nut over the column end first and then the ferrule with the correct size hole for the column. Cut the first 1–2 cm of column off in case some of the ferrule material might have scraped off and ended up in the column. Measure the correct distance from the column end and make a mark on the column using a permanent marker. Carefully insert the column into the injector or detector and tighten the nut until finger tight. Slide the column until the mark is in line with the bottom of the nut and then tighten the nut until the column does not move any more. Check for leaks using a leak detector and if leaking tighten the nut a $^1/_4$ turn or less and recheck. Overtightening a column nut can crack the ferrule of even extrude part of it through the hole in the nut and can cause leaks. In most cases the column nuts will require tightening another $^1/_8$ turn after one or two heating and cooling cycles of the oven after which it should remain tight.

Instead of using a permanent marker some chromatographers use an old septum or white typist correction fluid to mark the distance from the column end. Both have the advantage that the ferrule will not slide down the column further than the mark and the septum will also prevent the column nut from sliding down the column.

When installing metal or metal clad columns into a FID it is important to switch off the electronics to avoid electrical shock. Some manufacturers use a grounded jet but the majority has a dc voltage of a few hundred volts on the flame jet. Typically one would turn off the electrometer, insert the column until it touches the lower side of the flame jet, then withdraw about 1–3 mm (or whatever the manufacturer specifies) before tightening the column nut.

When all the connections are made, hang the column on the rack that is present in the GC ovens of all manufacturers. Always make sure that the column hangs freely in the GC oven and that it does not touch the sides of the oven. The oven is equipped with a fan that will cause vibration which can destroy PLOT and SCOT columns. If any part of the column touches the side wall of the oven, the vibration can cause the polyimide to be abraded off from the column and it will then break at that point. Generally a column with more than two ends does not work. All is not lost though as polyimide coated capillary columns can be joined using a Press Fit connector which is a deactivated glass tube with inside tapers on both sides as shown in Fig. 7.5. The columns ends to be joined are cut and pushed into the ends of the glass tube. A little pressure using your fingers will allow the polyimide to form a seal where it touches the glass and the seal will become permanent when heated. There are also other types of column connectors but we found the Press Fits works pretty well. Some of these connectors use a spot of polyimide glue to improve the sealing and requires conditioning at elevated temperatures.

Whatever the system configuration, it is important that the system be tested for good connection of the columns. Unswept volume is the volume of any portion of a fitting which is in the flow path but with an increase in diameter compared to the primary flow orifice or any other not directly swept by the flow of gas. This is known as 'dead volume' if it is not efficiently swept. The contents of any blind volume will be exponentially diluted and if analytes flow through any unswept volume, they will tail as illustrated in Fig. 7.8.

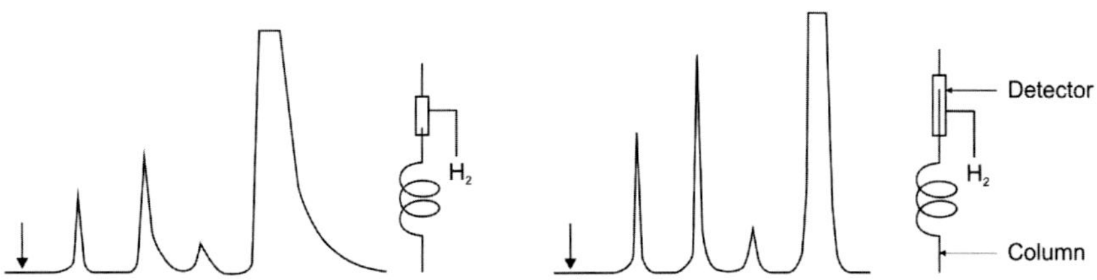

Fig. 7.8 Schematic showing the unswept volumes and its effect on the peak shapes. With the column not inserted far enough into the detector (left) and installed correctly (right).

7.4.3.1 Connecting fused silica columns directly on to valves

If the column is directly connected to a valve this will be done with the appropriate fitting and a special tool available from the valve supplier. It will be tightened as already described in Chapter 3. This connection can be done using fused silica adapters with or without a removable liner. The liner is slid over the column with the large end forward. A small amount of fused silica tubing is cut clean and square to prevent particles that may have been scraped from the liner entering the fitting. Slide the small end of the liner into the specially bored nut making sure that the liner bottoms in the counter bore of the nut. Slide the special ferrule over the liner which must be held down by the tool provided. Pull the tubing back slightly until its end is flush with the liner. The whole tube and liner assembly must retain the pressure from the fitting tool as the assembly is screwed into the fitting. The nut is screwed in until finger tight. A spanner is then used to tighten the nut in increments of about 15 degrees of rotation until the ferrule is starting to grip the column. Do not over-tighten but it will be necessary to retighten after the first high temperature run. Leak check the fitting preferable with an electronic leak detector and with hydrogen or helium as carrier gas. This installation is not without some challenges. Since two hands are required during installation it is almost impossible to do this successfully in a cramped space. Another difficult task is to keep pushing the column through the liner while tightening the nut and ensuring that the column is still level with the end of the liner. If it is too far out of the liner it will never give a leak tight seal. Conversely if it is pushed too far through the liner it could touch the valve stator and there is a very high probability that a small amount of fused silica will break off. If the incorrect valve bore is used the column could even touch the rotor. The first actuation of the valve will result in scoring the valve body which will require replacement of the valve.

There is a more robust way of connecting columns but unfortunately this increases the number of unions and cannot be avoided. However, the advantages far outweighed the disadvantages. Select appropriate metal tubing that has the same internal diameter as the column and order pre-cut 5 cm lengths. Matching the internal diameters of the column and union ensures that the column cannot go through the tubing. Connect these tubes to any port that is to be connected to a column using the standard metal fittings. Onto these tubes connect the internal connection of an internal to external union. Remove the metal ferrule from the external fitting and replace it with a one piece ferrule made from graphitised Vespel. The advantage of this is that the tubing can be bent away from very restricted areas and all column connections can be aligned for easy access. Although this is contrary to our aim of keeping joints to a minimum, it does make changing columns very easy.

7.4.4 Temperature

Each stationary phase has an upper allowable temperature limit above which the column should not be operated. Many stationary phases are polymers that consist of materials having a range of molecular weights. As the column temperature is increased, the more volatile portion of the polymer is swept out of the column by the carrier gas. The volatile products could also be formed by thermal degradation of the stationary phase especially due to oxygen in the carrier gas while the column is being used. This is called column bleed and is seen on the detector system as a rise in the baseline or as increased noise.

Stationary phases also have minimum allowable temperature limit. Most PLOT columns like Molecular sieve, alumina and the PLOT Q and U columns can be used down to −100 °C. Columns with liquid stationary phases or liquids coated on packing material have minimum temperatures where the liquid freezes at which point the analyte molecules cannot dissolve in the phase and therefore no retention is possible. Most columns are therefore supplied with a minimum and two temperatures for the upper limit, for example −60 to 220/240 °C. This means that the column may be used at any temperature above −60 °C up to a maximum of 220 °C isothermally and may be taken up to 240 °C for a short time during a temperature program.

Care must be taken to ensure that no part of the column is overheated and most applications will have part of the column in the detector and injector if fitted. For example, the maximum isothermal temperature for the column is 220 °C but the detector is set to 280 °C. This means that the last few centimeters of the stationary phase in the column will be destroyed and could easily become an active site which could cause tailing of peaks. Likewise the injector temperature must not exceed the column maximum temperature. In a case where the column has a lower temperature limit than the required detector temperature a piece of deactivated tubing must be joined to the end of the column so that this connector and tubing plays no role in the separation.

7.4.5 Conditioning the column

Commercial columns are pre-conditioned but during transport there is going to be a certain amount of air ingress. Before any column is used for the first time, it must be conditioned for a period of hours to rid it of the very volatile portion of the stationary phase and also the last traces of solvent which were used in the coating step. Packed columns often contain traces of the solvent used during packing to eliminate static.

The column must be conditioned with a flow of carrier gas and it is recommended to use a temperature program starting at a low temperature, say 30 °C for 5 min and then to slowly increase the temperature at about 1 °C/min to the maximum isothermal column temperature and then to leave it there for 3—4 h. One can start to use a column after it has been conditioned at maximum temperature for 3—4 h, but conditioning overnight may be better. The best practice is to condition the column as recommended by the column supplier. If the column is to be used with a highly sensitive detector such as PDHID we normally recommend a first bake-out of 36 h or more. There is also no need for the column to be connected to the detector during conditioning and bake-out and leaving the detector end disconnected during this procedure will prevent possible contamination of the detector. However, it is good practice to blank off the detector so that normal fuel gas flows are running without a chance of them entering the column oven.

7.5 Stationary phases

There are many different stationary phases that are used in gas analysis and it is not possible to provide an extensive list of all the possible separations and combinations that are used for the many different applications. Attempting to provide example chromatograms for all of these applications would probably require several volumes and we therefore provide only a few examples to illustrate the properties of the different phases. There are many application notes from the different suppliers and even more examples on the Internet that can be consulted for industry specific analyses.

Information on packed columns and the different coatings used for these columns are increasingly difficult to find and they were replaced with capillary columns in many laboratories. We therefore included some information on the different packing and coating materials.

7.5.1 Molecular sieves

Molecular sieve columns are used to separate many noble and permanent gases.

7.5.1.1 Aluminium silicate

Most of the early stationary phases used for gas separations were 'rocks' or Zeolites, which are some 40 minerals of aluminium silicate. These were originally mined but today are synthetised as it is cleaner and much more reproducible. Because the separation mechanism is based on molecular size these columns were named Molecular sieves. The suffix 4A, 5A or 13X refers to the pore size in Ångström (10Å = 1 nm) and the 'X' meaning 'approximately 10 Å'. Since helium has a slightly smaller molecular diameter than hydrogen a Molecular sieve column elutes helium ahead of hydrogen. In turn this will be followed by argon, oxygen, nitrogen, methane and carbon monoxide. With its slightly larger pores, the Molecular sieve 13X column provides a better peak shape for carbon monoxide because it is somewhat less retained than on a Molecular sieve with smaller pores such as 5A as shown in Fig. 7.9.

There are some anomalies. Molecules that are too large to enter the sieves are not retained and therefore elute without any retention. For example, sulphur hexafluoride is eluted before helium on Molecular sieve 4A but is strongly retained on Molecular sieve 13X. This is discussed in more detail in Section 11.10.2.

Molecular sieve columns are robust but will be altered by samples with high water vapour concentrations. The most common symptoms of this happening are the disappearance of carbon monoxide or the inability of the column to separate methane and carbon monoxide. Ultimately there will be reduced separation of oxygen and nitrogen.

Molecular sieve PLOT columns are more efficient and provide better peak shapes than the packed equivalent as is evident when comparing Figs 7.9 and 7.10. Molecular sieve PLOT columns with the correctly chosen dimensions are able to separate argon and oxygen at ambient temperatures but this separation will become less reliable as the column ages and especially if water vapour is present in the samples. It is also much easier to

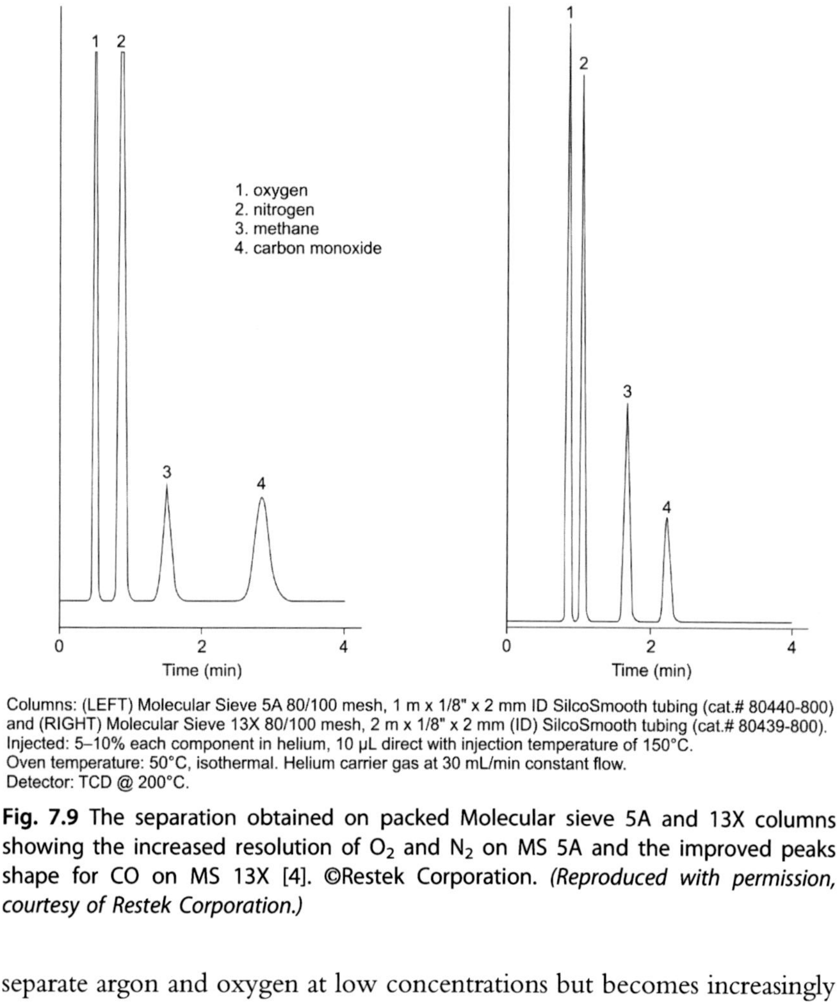

Fig. 7.9 The separation obtained on packed Molecular sieve 5A and 13X columns showing the increased resolution of O_2 and N_2 on MS 5A and the improved peaks shape for CO on MS 13X [4]. ©Restek Corporation. *(Reproduced with permission, courtesy of Restek Corporation.)*

separate argon and oxygen at low concentrations but becomes increasingly difficult with increasing concentrations, especially if one of the pair is present in much higher concentration than the other. This separation can be considerably improved by sub-ambient cooling of the column oven.

If high concentrations of water vapour are present in the samples a system using backflush-to-vent as described in Chapter 10 should be used. If this is not available it will be necessary to bake out the column after perhaps 20 injections. If the valves are mounted in the GC oven or there are columns with low maximum operating temperature, the Molecular sieve column has to be conditioned in an external oven. A new Molecular sieve column needs to be activated by heating under carrier gas flow to 350 °C or more for at least 4 h. However, some chromatographers find the

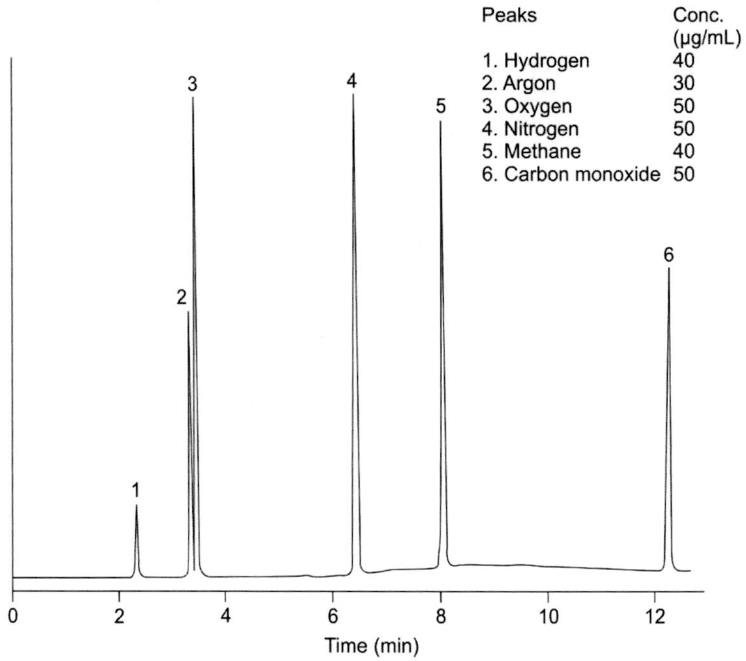

Column: Rt®- Molecular Sieve 5A, 30 m, 0.53 mm ID, 50 μm (cat.# 19723). Injected: Permanent gases, 5 μL direct with a 6-port Valco® valve at ambient and injector temperature at 200°C. Oven temperature: 27°C, hold 5 min. increase to 100°C at 10°C/min and hold for 5 min. Helium carrier gas at constant flow of 5 mL/min. Detector: Valco® Helium Ionisation Detector @ 150°C.

Fig. 7.10 When using a MS 5A PLOT column the peaks are considerably sharper and resolution improved. It is also possible to separate Ar and O_2 at ambient temperatures [5]. ©Restek Corporation. *(Reproduced with permission, courtesy of Restek Corporation.)*

best way to get a good Molecular Sieve column is to bake from 30 °C up to 300 °C at 5 °C/min and hold for 3–4 h. They believe that the extra heating causes some undesirable changes to the stationary phase which cause some tailing especially for CO. Baking to remove adsorbed water can be done at 300 °C for 2–3 h. The use of dry air as carrier for bake-out of Molecular sieve columns appears to help to recover separation efficiency.

Early workers believed that carbon dioxide was irreversibly absorbed on Molecular sieves but by accident (not turning off the recording system overnight) it was found that high retention existed. Under typical conditions for the separation of argon, oxygen, nitrogen, methane and carbon monoxide, carbon dioxide will have a retention time of several hours. A robust and routine system must not allow injection of mixtures containing

CO_2 on to Molecular sieve and temperature programming is not desirable with most of the detectors used in gas analysis. However, in practice, for the analysis of contaminants in beverage grade CO_2 on a batch basis, analysts inject this on Molecular sieve, do ten or fifteen samples and then bake the system at elevated temperature for an hour or more to elute the CO_2.

7.5.1.2 Carbon molecular sieves

As early as 1905 Ramsey utilized activated charcoal columns to separate various gases [6]. This had limited use in gas chromatography until Kaiser developed a carbon molecular sieve that had a pore structure similar to the zeolites and with a surface area of 1000 m^2/g [7]. Modern high surface area carbon molecular sieves such as Shincarbon ST have a surface area as high as 1500 m^2/g.

Currently available Carbosieves, Carboxen, Carboplot and Shincarbon polymers will separate permanent gases and C1−C3 hydrocarbons on a single column. However, all will elute oxygen and argon together with marginal separation of oxygen and nitrogen at ambient temperature. Most sales brochures show a reasonable oxygen nitrogen separation that includes several other gases but the oxygen nitrogen separation is adequate only at low concentrations. If the matrix is air many users simply integrate the whole area and allocate 20% to oxygen and 80% to nitrogen. Another drawback is that separations involving permanent and heavier gases may require temperature programming which is not desirable with several detectors. If one uses a TCD with argon carrier gas for the isothermal portion of the chromatogram and in series with a methaniser FID for carbon monoxide and carbon dioxide, a nice flat baseline can be achieved. The chromatogram shown in Fig. 7.11 shows the marginal separation between oxygen and nitrogen as well as good separation of the other 6 components using temperature programming.

7.5.2 Alumina

Scientists who use alumina as a catalyst support are often taken aback by the fact that alumina is a very important GC phase for the separation of C1−C5 hydrocarbons and their isomers. Alumina PLOT columns are well suited for the separation of C1−C10 hydrocarbons and small aromatics. Alumina has to be deactivated either by treatment with KCl or Na_2SO_4, with KCl deactivation the most common. The different deactivations slightly change the polarity of the columns which changes the elution order for some of the hydrocarbon gases.

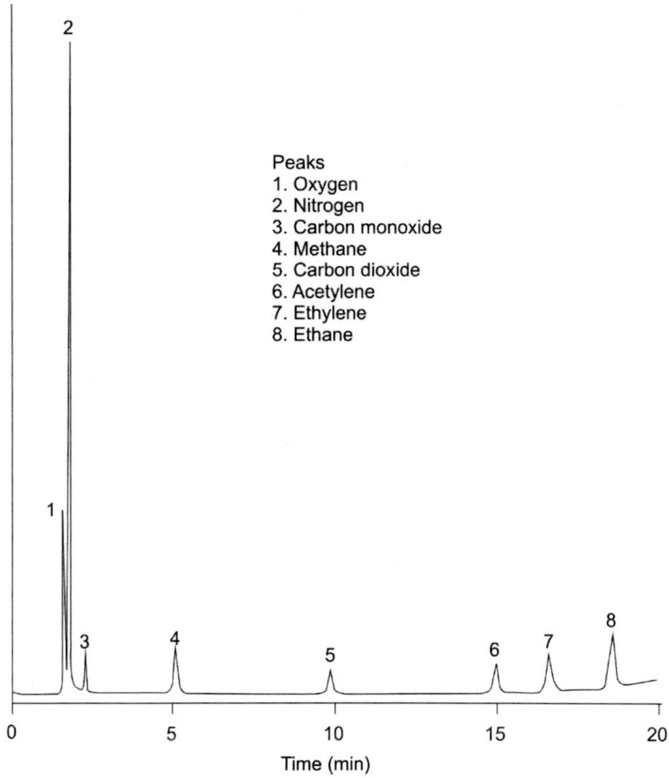

Column: ShinCarbon ST, 100/120 mesh, 2 m, 1/16 in. OD, 1.0 mm ID (cat.# 19808). Injected: Permanent gases + C1 + C2 hydrocarbons, 5 μL direct packed on-column with injection temperature of 100°C.
Oven temperature: 40°C, hold 3 min. increase to 250°C at 8°C/min and hold for 10 min. Helium carrier gas at constant flow of 10 mL/min.
Detector: Helium Ionisation Detector @ 200°C.

Fig. 7.11 Representative chromatogram showing the separation obtained on a ShinCarbon column [8]. The marginal separation between oxygen and nitrogen is only possible at relatively low concentrations and the two peaks will merge into one at higher percentage levels. ©Restek Corporation. *(Reproduced with permission, courtesy of Restek Corporation.)*

Alumina columns are somewhat sensitive to water vapour in the sample as well as in the carrier gas. Fortunately this can be baked out well within its temperature limit of 200 °C and no damage is done to the column. If the sample contains water vapour it is easy to hold the column oven at 200 °C for 10 min at the end of the run. If large amounts of water vapour are in the sample, it may be necessary to use a polar pre-column and backflush the water vapour. If the carrier gas contains only a few ppm of water vapour it would be advisable to bake the column about once every two weeks depending on the number of samples and the amount of water vapour in the samples analysed.

These PLOT columns have a relatively thin stationary phase which means that they are prone to overloading. The elution patterns of hydrocarbon gases and volatiles are very similar on both these columns and they provide separation of the all the alkane and alkene isomers up to butane and elutes the pentane-, hexane-, heptane- and octane isomers as separate groups making it well suited for the analysis of natural and refinery gas. There are subtle differences that can be exploited to provide a slightly different elution order when required as shown in Fig. 7.12. On the

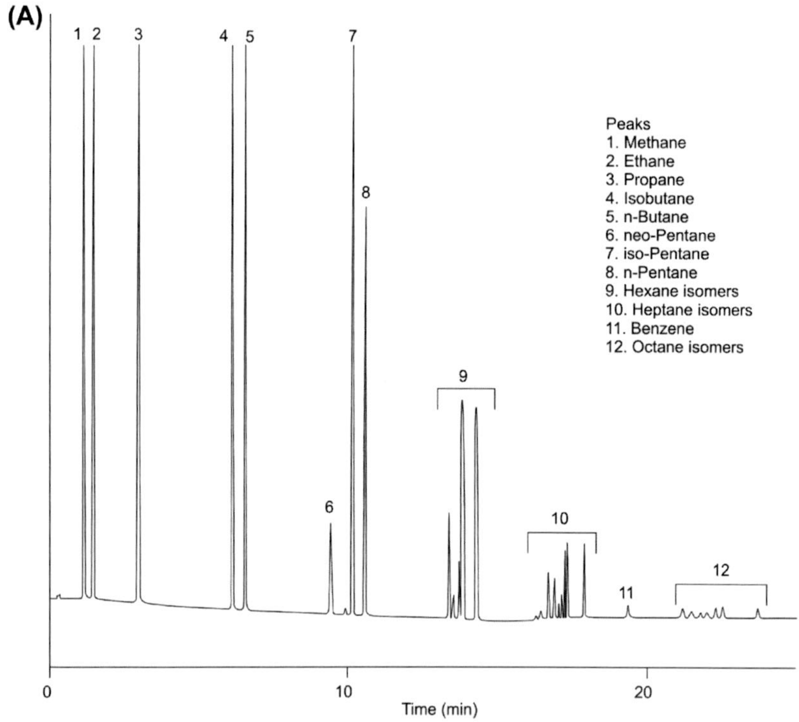

Column: Rt®-Alumina BOND/KCl, 50 m, 0.53 mm ID, 10 μm (cat.# 19756). Injected: Natural gas, 500 μL split in 2.0 mm ID Single taper liner with injection temperature of 200°C and split vent flow of 50 mL/min.
Oven temperature: 45°C, hold 1 min. increase to 200°C at 10°C/min and hold for 8.5 min. Hydrogen carrier gas at constant pressure of 55.2 kPa and average linear velocity of 45 cm/sec. @ 45°C. Detector: FID @ 200°C with nitrogen make-up. Data acquisition rate: 20 Hz. Instrument: HP5890 GC.

Fig. 7.12 Natural gas (A & B) and refinery gas (C & D) on Alumina columns deactivated with KCl (A & C) and Na_2SO_4 (B & D) respectively [9]. ©Restek Corporation. *(Reproduced with permission, courtesy of Restek Corporation.)*

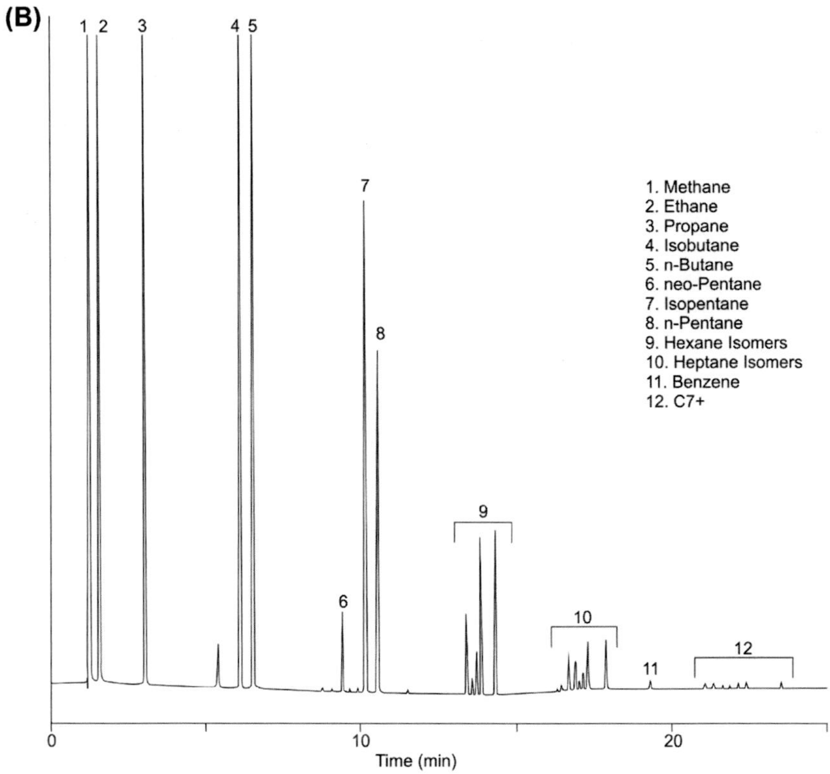

Column: Rt®-Alumina BOND/Na_2SO_4, 50 m, 0.53 mm ID, 10 µm (cat.# 19756). Injected: Natural gas, 500 µL split in 2.0 mm ID Single Gooseneck liner with injection temperature of 200°C and split vent flow of 50 mL/min. Oven temperature: 45°C, hold 1 min. increase to 200°C at 10°C/min and hold for 8.5 min. Hydrogen carrier gas at constant pressure of 55.2 kPa and average linear velocity of 45 cm/sec. @ 45°C. Detector: FID @ 200°C with nitrogen make-up. Data acquisition rate: 20 Hz. Instrument: HP5890 GC.

Fig. 7.12 Cont'd

Alumina KCl column acetylene will elute before n-butane and methyl acetylene before 1,3-butadiene. On the Alumina-Na_2SO_4 column acetylene and propadiene elute after the butanes and methyl acetylene after 1,3-butadiene. Also cyclopropane elutes before propylene making it well suited for the determination of impurities in propylene and this column provides the best separation of the butene isomers. Benzene elutes between the C7 and C8 isomer groups.

The main advantage of these columns is their ability to separate these isomers at temperatures above ambient whereas this separation on a non-polar WCOT capillary column will require sub-ambient column temperatures to achieve a similar separation. Since sub-ambient operation is not practical in a production environment these columns are very common

(C)

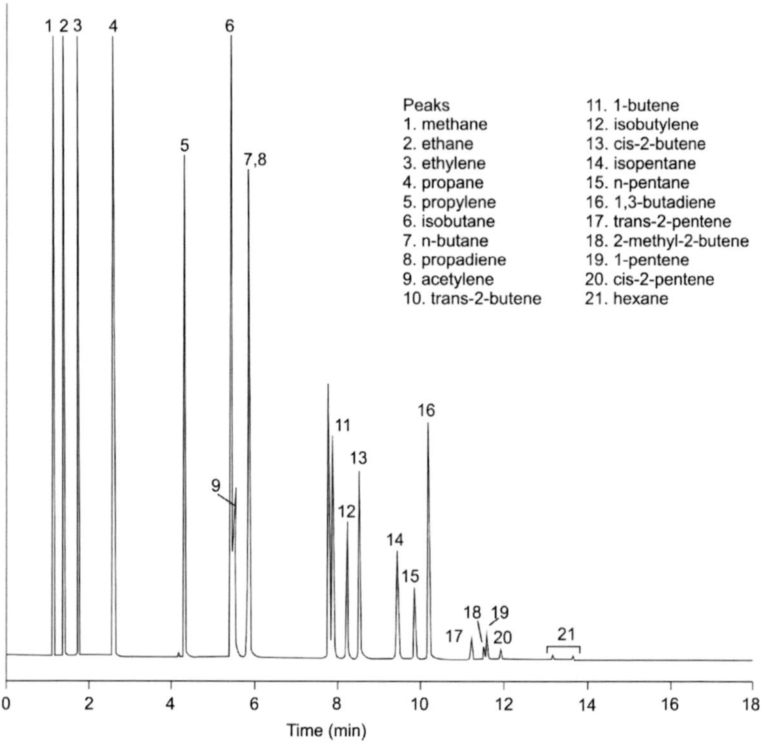

Column: Rt®-Alumina BOND (KCl), 50m, 0.53mm ID, 10μm (cat.# 19760). Injected: Refinery gas, 10 μL split in 2.0 mm ID Single taper liner with injection temperature of 200°C and split vent flow of 80 mL/min. Oven temperature: 45°C, hold 1 min. increase to 200°C at 10°C/min and hold for 3.5 min. Hydrogen carrier gas at constant pressure of 55.2 kPa and average linear velocity of 74 cm/sec. @ 45°C. Detector: FID @ 200°C with nitrogen make-up. Data acquisition rate: 20 Hz. Instrument: HP5890 GC.

Fig. 7.12 Cont'd

in petrochemical production facilities, both as laboratory instruments and at-line analysers.

Because alumina has a maximum temperature of 200 °C, the injector and detector temperatures must be below 200 °C to prevent irreversible damage to the stationary phase in the ends of the column that are inserted in the injector and detector. If higher injector or detector temperatures are required for some reason, short lengths of deactivated tubing can be joined to the column ends.

7.5.3 Porous polymer packed columns

The separation mechanism of porous polymer stationary phases is selective adsorption. Porous polymers for use in gas chromatography were developed

(D)

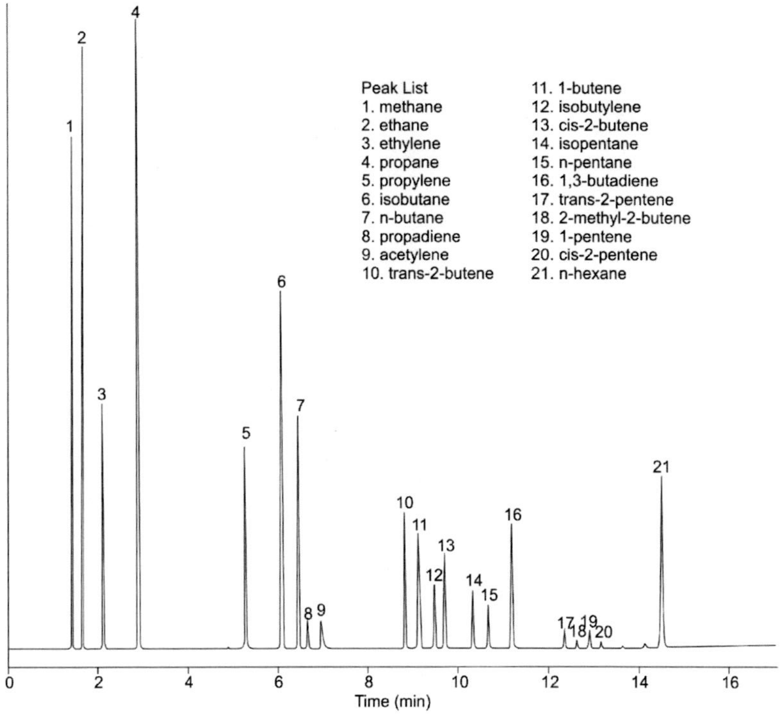

Column: Rt®-Alumina BOND Na_2SO_4, 30m, 0.53mm ID, 10 μm (cat.# 19755). Injected: Refinery gas hydrocarbons through C6, 10 μL split in 2.0 mm ID Precision™ liner w/wool with injection temperature of 200°C and split vent flow of 40 mL/min. Oven temperature: 60°C, hold 2 min. increase to 200°C at 10°C/min and hold for 1 min. Helium carrier gas at constant pressure of 34.5 kPa and average linear velocity of 37.3 cm/sec. @ 60°C. Detector: FID @ 200°C with nitrogen make-up. Data acquisition rate: 20 Hz. Instrument: HP5890 GC.

Fig. 7.12 Cont'd

in the 1960s and were sold under the names such as Porapak™ and Chromasorb™. Early polymers were plagued with problems related to batch-to-batch variations, shrinkage and inadequate clean-up. These inconsistencies lead to difficulties in reproducing separations and hence porous polymers were used only when other packings were unavailable.

Further development by Hayes Separations, Inc. in 1983 lead to the introduction of the HayeSep™ range of porous polymers which are supplied both cleaned and preconditioned for packed column applications. These treatments produce polymers that are consistent, with no shrinkage and low column bleed. Porous polymers are available in a variety of polarities with the more polar polymers giving the sharpest peaks for really polar analytes such as water vapour. Also the elution order for a mixture of compounds varies considerably. For example HayeSep Q elutes ethylene before ethane

Table 7.1 Summary of the HayeSep phases [10,11].

HayeSep A	Copolymer of high-purity divinylbenzene and ethylene glycol dimethylacrylate. Separates permanent gases at ambient temperatures. Used at higher temperatures to analyse C2 hydrocarbons, hydrogen sulphide, and water.
HayeSep B	Copolymer of divinylbenzene and polyethyleneimine. Separates C1 and C2 amines and trace levels of ammonia and water.
HayeSep C	Copolymer of divinylbenzene and acrylonitrile. For the analysis of polar molecules. Separation characteristics are similar to those of Chromosorb 104.
HayeSep D	High-purity divinylbenzene polymer. Exhibits superior separation characteristics for light gases, carbon monoxide, carbon dioxide, and acetylene (ahead of other C2 hydrocarbons). For the analysis of water and hydrogen sulphide. Three different variants are available.
HayeSep N	Copolymer of divinylbenzene and ethylene glycol dimethylacrylate. Used for the separation of all the C2's but has low temperature limit.
HayeSep P	Copolymer of divinylbenzene and styrene.
HayeSep Q	Divinylbenzene polymer. The most commonly used porous polymer.
HayeSep R	Copolymer of divinyl-benzene and N-vinyl-2-pyrollidone.
HayeSep S	Copolymer of divinyl-benzene and 4-vinyl-pyridine.
HayeSep T	Polymer of ethylene glycol dimethacrylate. Very useful for reactive components.

and acetylene which co-elute. HayeSep N separates all of these in the order ethylene, ethane and acetylene. A summary of the HayeSep phases is given in Table 7.1.

One of the advantages of using packed columns with Hayesep phases is that the stationary phase can be modified for specific purposes, for example a Hayesep DB that will have slightly different adsorption properties than either the Hayesep D or the Hayesep B.

Most column manufacturers will prepare packed columns on request and many phases are available also as micro-packed columns. A range of HayeSep PLOT columns are also routinely available.

7.5.4 Porous polymer PLOT columns

There are three commonly used columns of this type with various prefixes depending on the manufacturer or type of bonding of the phase to the column wall. The distinguishing part is the letters Q, S and U which is also the order of increasing polarity. As with the packed column phases the Q refers to a co-polymer of styrene and divinylbenzene, the S to a copolymer of divinylbenzene and vinyl pyridine while the U refers to a copolymer of divinylbenzene, ethylene glycol and dimethyl acrylate.

An outstanding feature of the porous polymers is that they are hydrophobic and not affected by water in the samples. Water may even elute with a good peak shape so that it can be quantified, obviously only if a detector is used that can detect water.

7.5.4.1 PLOT Q

This is probably the most widely used PLOT column as it is amenable to many diverse applications. Air can usually not be separated into oxygen and nitrogen on porous polymer columns and in most applications O_2, N_2 and CO will co-elute from these columns as shown in Fig. 7.13. However at sufficiently low concentrations and oven temperatures, some separation occurs between nitrogen and oxygen with the important feature that nitrogen elutes before oxygen. If separation between O_2, N_2 and CO is required in addition to heavier components, a Molecular sieve column is often used together with a PLOT Q or Alumina column. This type of application will be discussed in some more detail in Chapter 10 on multidimensional gas analysis.

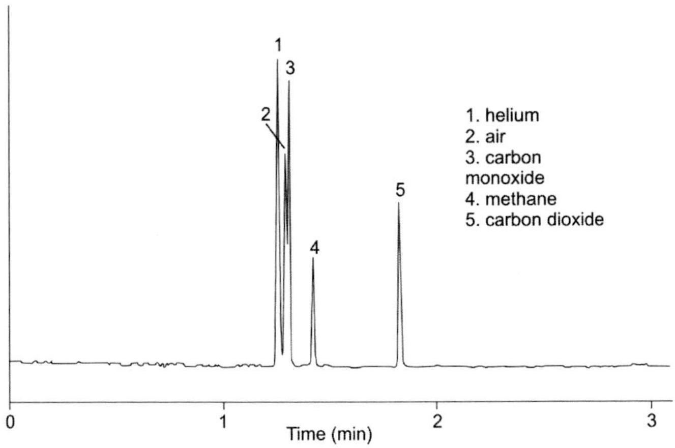

Column: Rt-QPLOT™ PLOT (cat.# 19718), 30 m, 0.32 mm ID. Injected: Permanent gases 2-5 mol%, 30 µL split 40:1 with injection temperature of 30°C.
Oven temperature: 30°C, isothermal. Hydrogen carrier gas at a linear velocity of 38 cm/sec. Detector: HP µTCD @ 200°C.

Fig. 7.13 Permanent gases on a PLOT Q column [9]. ©Restek Corporation. *(Reproduced with permission, courtesy of Restek Corporation.)*

For refinery gas analysis, the PLOT Q column provides slightly better separation for C1−C3 hydrocarbons than the alumina PLOT columns, but C4 and higher hydrocarbons are better separated with an alumina PLOT column than the PLOT Q. PLOT Q exhibits long retention times and significantly broadened peaks for hydrocarbons higher than C6 and aromatics. These are not gases any more so a liquid phase column is better suited for the heavier components.

The PLOT Q separates sulphur gases from each other and from most light hydrocarbons and is also useful to separate chloro-fluoro-carbons (CFC's) and volatile polar solvents.

7.5.4.2 PLOT S

The PLOT S columns are medium polarity columns and often used for the analysis of solvents. While these solvents are not gases per our definition, the analysis of residual solvents in manufactured products and pharmaceuticals is often done by headspace sampling in which cases the residual solvents are in the gas phase. Similarly many of these are found in ambient air of some industrial processes and it is often required to analyse them for legal compliance purposes.

7.5.4.3 PLOT QS

The polymer phases can be combined to provide a column that combines the best properties of the component phases. One such an example is the QS phase that is useful for solvents, Fig. 7.14, and refinery gas analysis, Fig. 7.15. While some separation of the C4 and C5 isomers is possible on this column, the Alumina KCl would be preferred.

7.5.4.4 PLOT U

The most polar of the porous polymer phases allows the determination of highly polar compounds such as formaldehyde, water and methanol, as in Fig. 7.16, with good peak shapes but can also be used for non-polar compounds. It is also highly inert, like the others in this family, and suitable for the analysis of H_2S, COS and mercaptans in hydrocarbon streams.

7.5.5 Silica gel

Historically silica gels or porous glasses were highly variable but newer grades are much more standardised. Chromasil, Porasil and Spherosil with varying surface area and pore size have been used where it is necessary to elute CO_2 after ethane. Because of its use as a desiccant it was not compatible with water vapour but modern bonded phase PLOT columns

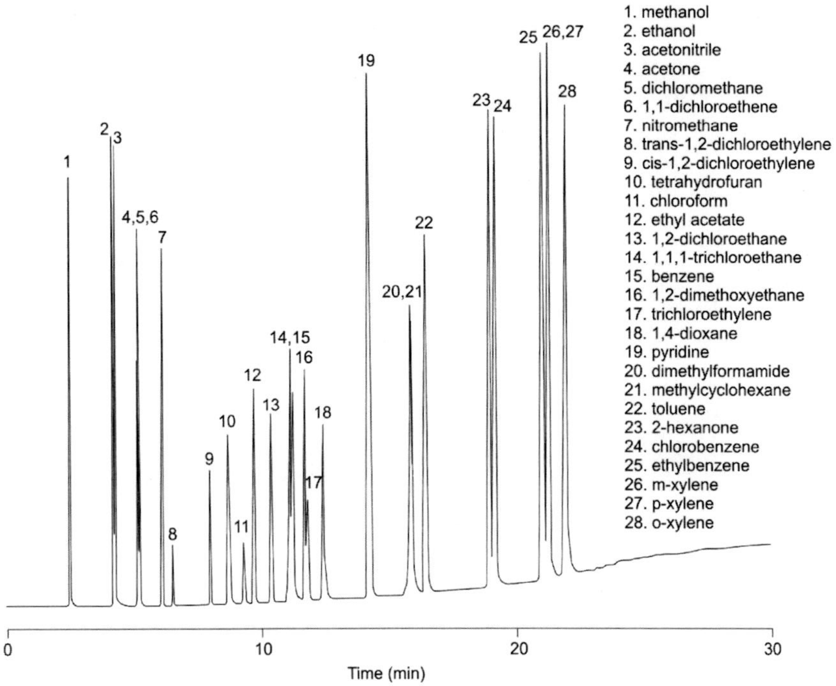

Column: Rt®-QS-BOND, 30 m, 0.53 mm ID, 20 μm (cat.# 19738). Sample: solvent mixture. Inj.: 1.0 μL, split (split vent flow 100 mL/min.), 4 mm single gooseneck liner (cat.# 20798). Inj. temp.: 200°C. Carrier gas: hydrogen, constant pressure, 4.2 psi. Linear velocity: 40 cm/sec. @ 120°C. Oven temp.: 120°C to 240°C @ 5°C/min. hold 5.0 min.). Detector.: FID @ 240°C.

Fig. 7.14 Solvent mixture on PLOT QS [9]. ©Restek Corporation. *(Reproduced with permission, courtesy of Restek Corporation.)*

such as GS-GasPro are water vapour resistant. In recent years it has become popular for sulphur gas analysis but is quite versatile and also suitable for the separation of light hydrocarbons and halocarbons as in Fig. 7.18. In the example in Fig. 7.17 butane will elute just before carbon disulphide.

7.5.6 Solid phases

There are many other solid phases that are used for specialised gas analysis applications and many are still used where the phases listed above cannot provide the desired separation. A few are listed here, but there are more; some of which are often also used as solid phases for extracting volatiles from air in environmental sampling.

7.5.6.1 Kel-F

Kel-F is a chlorofluoro-homopolymer formed from chlorotrifluoroethylene (CTFE) as the monomer, similar to Teflon or polytetrafluoroethene (PTFE) that is formed from tetrafluoroethylene as monomer. The name is a

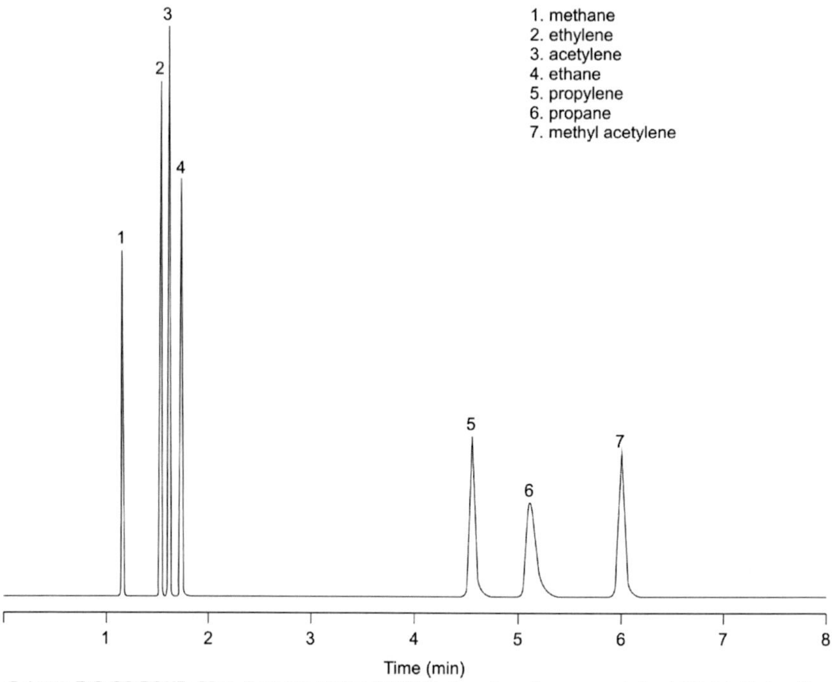

Column: Rt®-QS-BOND, 30 m, 0.53 mm ID (cat.# 19738). Sample: refinery gas mix (mole%). Inj.: 2 μL split (split ratio 10:1), 4 mm single gooseneck liner (cat.# 20798). Inj. temp.: 240°C. Carrier gas: hydrogen, constant flow. Flow rate: 3.6 mL/min. @ 60°C. Oven temp.: 60°C (hold 15 min.) to 240°C @ 25°C/min. (hold 2 min.). Detector.: FID @ 240°C.

Fig. 7.15 Refinery gas mixture on PLOT QS [9]. ©Restek Corporation. *(Reproduced with permission, courtesy of Restek Corporation.)*

combination of the name from the company M W Kellogg that commercialised it in the 1950s and the word fluoropolymer. Poly-CTFE is also known as Kel-F 81 while other fluoropolymers have different numbers like the copolymer poly(chlorotrifluoroethylene-*co*-vinylidene fluoride) that is known as Kel-F 800.

Kel-F has high tensile strength, is non-flammable and resistant to heat up to 175 °C which means it cannot be used in high temperature applications. Its low thermal expansion coefficient, good chemical resistance, zero-moisture absorption and non-wetting properties make it useful in GC. It is resistant to most chemicals and oxidizing agents due to its high fluorine content making it one of the few phases that can withstand attack by very aggressive gases such as fluorine and chlorine. It is resistant to oxidation due to the absence of hydrogen atoms but may swell somewhat in halocarbon and aromatic solvents as well as some ethers and esters.

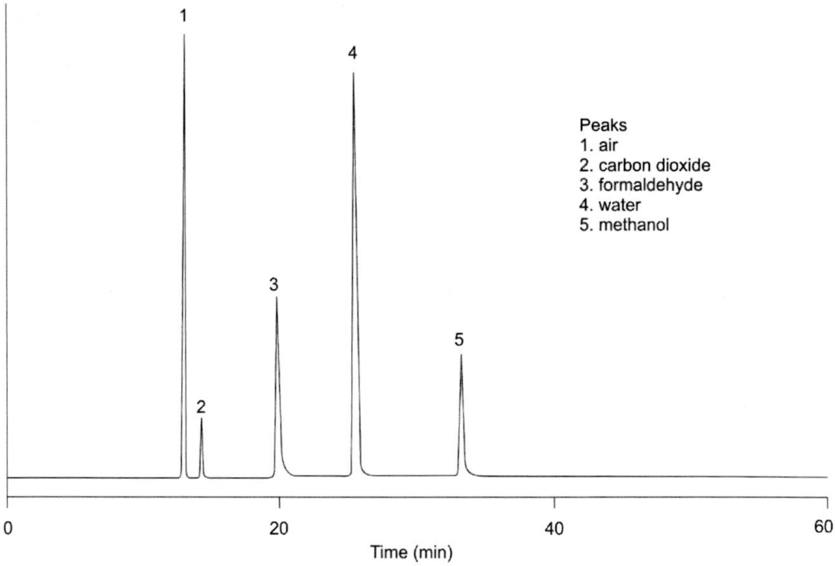

Column: Rt®-U-BOND, 30 m, 0.53 mm ID, 20.0 µm (cat.# 19750). Injected: Formaldehyde (manual headspace), 10 µL split ratio 1:10 in 2 mm split Precision™ liner w/wool with injection temperature of 200°C and split vent flow of 40 mL/min. Oven temperature: 100°C, hold 1 min. increase to 150°C at 25°C/min and hold for 3 min. Helium carrier gas at constant pressure of 53.1 kPa and average linear velocity of 39 cm/sec. @ 100°C. Detector: TCD @ 200°C. Instrument: Agilent 5890 GC.

Fig. 7.16 Formaldehyde on a PLOT-U column that also allows the determination of other polar compounds such as water and methanol [9]. ©Restek Corporation. *(Reproduced with permission, courtesy of Restek Corporation.)*

7.5.6.2 Durapak®

Durapak® was developed by Waters Associates, Inc. in the early days of chromatography and was probably the first bonded phase to be commercially available. At that time it was claimed to have no vapour pressure and therefore no bleeding during temperature programming. It has been largely replaced by columns from other manufacturers using n-octane bound to a silica gel surface. These columns are often used in process chromatographs for the separation of C1 to C5 hydrocarbons as they are able to separate the C4 hydrocarbons including 1-butane and isobutylene which is one of the more difficult separations to obtain reproducibly with other stationary phases.

7.5.6.3 Carbopack B

Carbopack is a graphitised carbon black and can be non-porous or porous. It has a very pure surface with excellent adsorption properties due to the process of graphitisation. As an adsorbent it is somewhat weaker than the

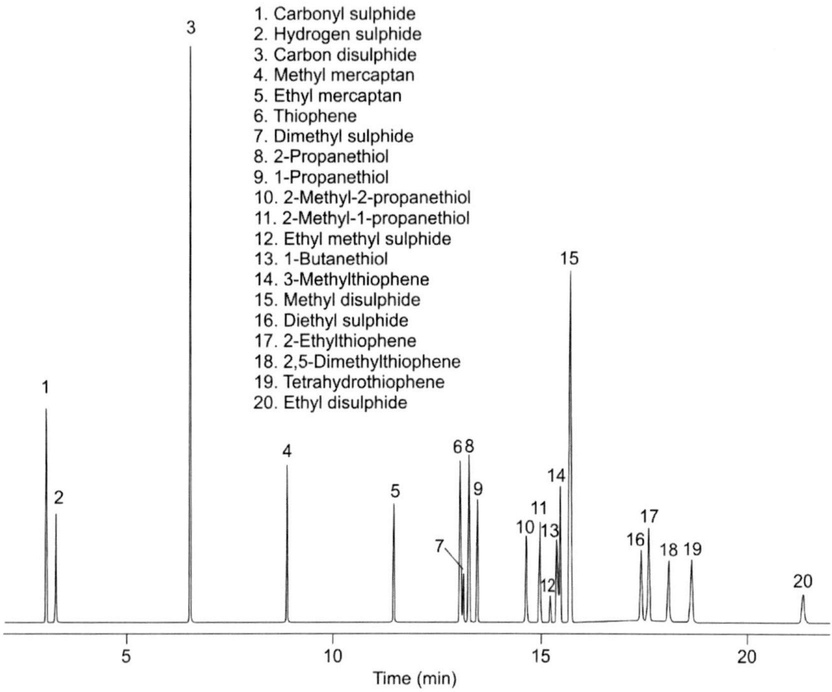

Column: GS-GasPro, 30 m x 0.32 mm I.D., J&W P/N: 113-4332. Carrier: Helium at 10 psig (16 cm/sec measured off COS at 80°C). Oven: 80°C for 2.5 min, 80°C to 260°C at 15°/min, 260°C for 8 min.
Injector: Split 1:10, 200°C, 1000 µL of sulfur gas standard, 3 ppmV each component.
Detector: PFPD (OI Analytical), 270°C.
J&W wishes to thank Air Toxics, Ltd. (Folsom, CA) for providing the standard mixture shown in this chromatogram.

Fig. 7.17 Separation of sulphur gases on a GS-GasPro column. ©Agilent Technologies, Inc. *(Reproduced with Permission, Courtesy of Agilent Technologies, Inc.)*

carbon molecular sieves. It is commonly used as an absorbent to trap larger semi-volatile organic molecules in environmental sampling from as low as the C4 hydrocarbons to larger molecules such as the polychlorinated biphenyls (PCBs).

It is a popular stationary phase support in many packed GC columns because they are synthesised under controlled conditions and may offer better consistency between lots than the diatomaceous Chromosorb® adsorbents mined from the ground.

7.5.7 Liquid phases

In gas liquid chromatography, many stationary phases were coated on inert support material. For example, a column could be specified as 5% n-octane

1. Chlorotrifluoromethane (13) (peak not shown)
2. Trifluoromethane (23)
3. Bromotrifluoromethane (13B1)
4. Chloropentafluoroethane (115)
5. Pentafluoroethane (125)
6. 1,1,1-Trifluoroethane (143a)
7. Dichlorodifluoromethane (12)
8. Chlorodifluoromethane (22)
9. 1,1,1,2-Tetrafluoroethane (134a)
10. Chloromethane (40)
11. 1,1,2,2-Tetrafluoroethane (134)
12. Bromochlorodifluoromethane (12B1)
13. 1,1-Difluoroethane (152a)
14. 1,2-Dichloro-1,1,2,2-tetrafluoroethane (114)
15. 2-Chloro-1,1,1,2-tetrafluoroethane (124)
16. 1-Chloro-1,1-difluoroethane (142b)
17. Dichlorofluoromethane (21)
18. Trichlorofluoromethane (11)
19. Chloroethane (160)
20. Dichloromethane
21. 1,1-Dichloro-1-fluoroethane (141b)
22. 2,2-Dichloro-1,1,1-trifluoroethane (123)
23. 1,1,2-Trichloro-1,2,2-trifluoroethane (113)
24. 1,2-Dibromo-1,1,2,2-tetrafluoroethane (114B2)

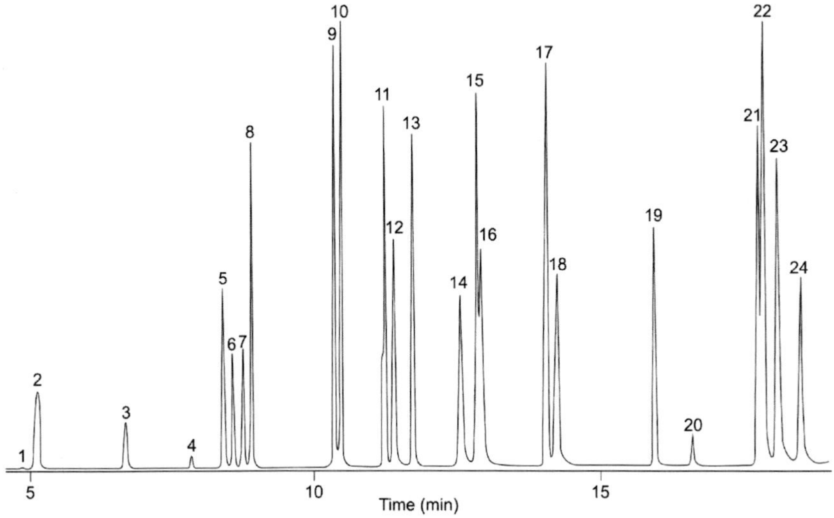

Column: GS-GasPro, 60 m x 0.32 mm I.D., J&W P/N: 113-4362. Carrier: Helium at 35 cm/sec, constant velocity. Oven: 40°C for 2 minutes, 40-120 at 10°C/min, 120°C for 3 min, 120-200 at 10°C/min.
Injector: 250°C, Splitless, 0.20 min purge activation time, 1.0 μL of 100 ppm mixture of Accustandard M-REF & M-REF-X in methanol. Detector: Shimadzu QP5050A MSD, 280°C, full scan 45-180 amu.

Fig. 7.18 Halogenated gases on a GS-GasPro column. ©Agilent Technologies, Inc. *(Reproduced with permission, courtesy of Agilent Technologies, Inc.)*

coated on Chromosorb W-HP 80–100 mesh. The support in this case has very little influence on the separation as the separation mechanism is partitioning. This means that the analytes dissolve into and out of the stationary phase (n-octane). The better an analyte dissolves, the longer will be the retention. In gas analysis, there are a few special cases that do require partition separations but these are not sufficient to warrant a detailed discussion. Many references can be found such as 'A practical guide to the care maintenance and troubleshooting of capillary gas chromatographic systems' by Dean Rood [12]. The majority of these support coated columns can be replaced by alternative phases today.

7.5.7.1 PDMS

Most of the capillary columns used today for the analysis of liquids utilise polydimethlysiloxanes (PDMS) as major components in the non-polar and intermediate polarity columns. While these phases are used for most of the analysis where the sample is in liquid form its application in the field of gas analysis is rather limited. The non-polar WCOT columns are however present in virtually all refinery and natural gas analysers for the separation of hydrocarbons with five or more carbon atoms. Since these hydrocarbons do not elute from the traditional gas analysis columns in a reasonable time, this group of compounds is usually back flushed from the gas columns onto WCOT columns connected to flame ionisation detectors. Analysis done on non-polar PDMS columns are very close to being true boiling point separations More polar WCOT columns are often used for the analysis of volatile amines, acids and aldehydes, especially ethyl amine, formic acid, acetic acid, formaldehyde, acetaldehyde, and other polar volatile liquids that may be present as vapours in gas streams. Some of these are discussed in some more detail in Chapter 11 on selected applications. These PDMS columns come under a wide variety of trade names many of which are equivalent and therefore interchangeable.

7.5.7.2 Krytox™

This refers to a group of colourless polymeric fluoroether synthetic lubricants that were developed by scientists at DuPont and designed as a liquid or grease form of PTFE. Usually formed from polyhexafluoropropylene oxide it has a general chemical formula $F-(CF(CF_3)-CF_2-O)_n-CF_2CF_3$, with the n in the formula between 10 and 60. Other names that are used to refer to this group of compounds include perfluoropolyether (PFPE), perfluoroalkylether (PFAE) and perfluoropolyalkylether (PFPAE) or CAS number 60164-51-4. Krytox is insoluble in water, acids, bases, and most organic solvents, is nonflammable, nonvolatile and thermally stable over a temperature range −75 to 350 °C. It is often used in nuclear and aerospace industries because of its high resistance to ionising radiation. Above 260 °C it may emit vapours that can be irritating and damaging to lungs upon inhalation. When used as a coating in a packed GC column it makes small selectivity changes for certain analytes such as NF_3.

7.6 Phase ratio

The phase ratio is a very useful parameter in selecting the best column dimensions for a specific application. It is calculated by dividing the column

Table 7.2 Phase ratios of columns with different dimensions.

Column diameter		Film thickness (μm)							
(mm)	(μm)	0.10	0.25	0.50	1.00	1.50	3.00	4.00	5.00
0.10	100	250	100	50	25	—	—	—	—
0.25	250	625	250	125	63	42	21	16	13
0.32	320	800	320	160	80	53	27	20	16
0.53	530	1325	530	265	133	88	44	33	27

diameter by four times the film thickness, both measured in micron making it a unitless number. Phase ratios for columns with different dimensions are given in Table 7.2. Columns having lower phase ratios are better for low boiling and low molecular weight compounds as they provide more retention. Columns with a phase ratio of less than 100 are therefore preferred for gas analyses. Conversely, high boiling, high molecular weight compounds will require columns with a phase ratio of 1000 or more. As a general rule one would select the thinnest column with the thickest film (lowest phase ratio) for gas analysis.

7.7 Column selection and care

Selecting the right column for the job is a daunting task but a systematic approach can narrow the choices and save time by reducing trial and error. There are many parameters that affect the efficiency of separations but several of these are fixed by the column dimensions. From the theory section we know that there must be an optimum average linear velocity that is best for a particular choice of carrier gas. Unless there is a critical separation, we often run slightly faster than this optimum in order to reduce the run time without sacrificing too much column efficiency. Application notes, chromatography blogs and your local column supplier can all provide valuable inputs into choosing the best column for the sample to be analysed. Selected applications are discussed in Chapter 11. As we said before, the sample dictates the entire chromatographic analysis. A summary containing the elution order for some common permanent gases on the commonly used stationary phases is given in Table 7.3.

Since carrier gas and detector compatibility is vital, take into account the carrier gas to be used and the relationship between head pressure, column length, linear velocity, flow rate and column efficiency. Understand the influence of column length, diameter and film thickness on column

Table 7.3 Elution order of some permanent gases and volatiles on different stationary phases.

Molecular sieve	PLOT Q	PLOT U	Carboxen	Alumina	Shincarbon	Silica
He	He, H_2	He, H_2	He, H_2	He, H_2, N_2, Ar, O_2, CO	He, H_2	He, H_2, N_2, Ar, O_2, CO
H_2	N_2, Ar, O_2, CO	N_2, Ar, O_2, CO	Ar, O_2		Ar, O_2	
Ne			N_2	CH_4	N_2	CH_4
Ar	CH_4	CO_2	CO	C_2H_6	CO	C_2H_6
O_2	CO_2	CH_2O	CH_4	C_2H_4	CH_4	C_2H_4
N_2	C_2H_6	H_2O	CO_2	C3's	CO_2	C_2H_2
CH_4	H_2S	CH_3OH	C_2H_2	C4's	C_2H_2	COS
CO	H_2O		C_2H_4	C_2H_2	C_2H_4	C3's
	C3's		C_2H_6	C5's	C_2H_6	H_2S
	C4's		H_2O	C6's +	C3's	C4's
	C5's		C3's		C4's	C5's
	C6's +		C4's		C5's	
			C5's			

efficiency and analyte retention as discussed in the chapter on chromatography theory. Be conscious of the differences between thin- and thick stationary phase columns and how the film thickness affects the column capacity, bleed and inertness. Consider the minimum and maximum allowable column temperatures and the column cost.

When working with reactive gases it is important to know whether the column can tolerate all the components of the samples, even those components that are not being determined. For example high concentrations of moisture in the gas may seriously compromise the efficiency of the column and even moderate amounts of oxygen can be detrimental to some stationary phases, especially at elevated temperatures.

In certain stationary phases very low concentrations of some gases, such as oxygen, cannot be determined. For example, if ultrapure helium is used with some Hayesep phases it will desorb traces of oxygen that were previously adsorbed in the column. When a standard containing trace amounts of oxygen is injected, part or all of the oxygen will bind at these sites and will therefore not be detected. To prevent this, the carrier gas can be purposely contaminated with a trace amount of oxygen to ensure that the active sites remain saturated. This is generally not advisable as some detectors as well as some stationary phases may be damaged by the continuous oxidation caused by elevated levels of oxygen. Also it is difficult to ensure the batch to batch reproducibility of this blended carrier gas.

Modern columns are very robust and can last for many years provided they are cared for properly. Since most gas analyses takes place at moderate temperatures it is seldom necessary to operate the column at or close to its maximum allowable temperature. Columns should never be exposed to temperatures above the maximum allowable operating temperature. When several columns are present in the same oven it is wise to set the oven temperature limit of the instrument to the maximum allowable temperature of the column with the lowest maximum. If there are any valves in the same oven, ensure that they too are able to handle the temperatures that the columns can. Columns should never be heated without a flow of carrier gas through them as that will cause irreversible damage to the column. Very rapid heating of the column oven should also be avoided as far as possible as the thermal shock, especially with short, rapid runs will shorten the column lifetime. It is recommended to always use slow temperature programs even when baking and fill the column with an inert gas before sealing the ends of all columns for storage. Also store the columns in an environment without thermal extremes and free of vibration.

Ensure that fused silica columns do not touch each other in multi-column configurations and that the columns are hanging freely without touching the sides of the column oven. The smallest break in the polyimide coating of the column will cause it to break at that point. Not only will it mean that the column should be joined at the break or replaced, but if hydrogen is used as carrier and the column breaks close to the detector, the built-in safety measure of the instrument might not activate resulting in an oven full of a potentially explosive gas mixture waiting for a spark from the fan or heat from the detector to ignite it.

Last but not least, whenever possible do not overload the column and always use the best quality carrier gas.

References

[1] M. Weeks, Daniel Rutherford and the discovery of nitrogen, Journal of Chemical Education 11 (2) (1934) 101.

[2] American Chemical Society, Academie des Sciences de L'institut de France, Societe Francaise de Chimie, The Chemical Revolution. An International Historic Chemical Landmark, American Chemical Society, Paris, France, June 8, 1999.

[3] V. Pretorius, T. Smuts, Sample capacity in open tubular and micro-packed columns for GC, Journal of High Resolution Chromatography 2 (7) (July 1979) 444–446.

[4] Restek Corporation, Petrochemical Applications. Molecular Sieve 5A and 13X. Packed GC Columns for Permanent Gas Separations, 2007 vol. Lit. Cat.# 59538A.

[5] Restek Corporation, Restek's PLOT Column Family - The Benchmark for Performance, 2015 vols Lit. Cat.# PCSS1163F-UNV.

[6] S. Ahuja, Chapter 9. Gas chromatography, in: Chromatography and Separation Science, vol. 4, Academic Press (Elsevier Science, USA), San Diego, CA, 2003, p. 134.

[7] R. Kaiser, Carbon molecular sieve, Chromatographia 3 (1) (1970) 38–40.

[8] Restek Corporation, ShinCarbon ST Micropacked GC Columns. Above-Ambient Analyses of Permanent Gases and Light Hydrocarbons, 2011. No. Lit. Cat.# PCTS1472-UNV.

[9] Restek Corporation, Restek Searchable Chromatogram Library, Restek Corporation, Bellefonte, PA, USA., 2017 [Online].

[10] VICI AG International, HayeSep® Porous Polymers, VICI AG International, 2018 [Online]. Available: https://www.vici.com/hayesep/hayesep.php.

[11] VICI AG International, ValcoPLOT® Capillary Columns from VICI Metronics, VICI AG International, 2018 [Online]. Available: https://www.vici.com/columns/vp.php.

[12] D. Rood, A Practical Guide to the Care, Maintenance, and Troubleshooting of Capillary Gas Chromatographic Systems, third revised ed., Wiley-VCH, New York, 1999.

CHAPTER 8

GC detectors

Contents

A practical guide to gas analysis by gas chromatography
ISBN 978-0-12-818888-0
https://doi.org/10.1016/B978-0-12-818888-0.00008-5

8.1 General properties of detectors

We are all used to using detectors on a daily basis such as listening to sounds, seeing light and smelling your food. All of these detectors have limits of detection in terms of range, sensitivity and selectivity. For example, humans can only hear frequencies up to 15 kHz. We can smell rotten eggs (H_2S) at ppm and sub-ppm levels but cannot smell 79% nitrogen in the air that we breathe. Depending on the analysis required we must distinguish between selectivity and sensitivity. Some detectors give response for all compounds other than the carrier gas while others may be selective for certain classes of compound. Even though the response of some detectors is universal, in other words it will give a response to all compounds other than the carrier gas, the response is not equal for all the compounds. Some might be detected better than others and the detector will therefore be more sensitive towards the better detected compounds and less sensitive towards the more poorly detected compounds as shown in Fig. 8.1. Sensitivity is determined as the ratio between the concentration and the response and can therefore be quantified. However, there is no such thing as absolute selectivity so that if we are analysing for 1 ppm analyte in a pure matrix the detector will give a response of some sort for the matrix. Even if the detector is

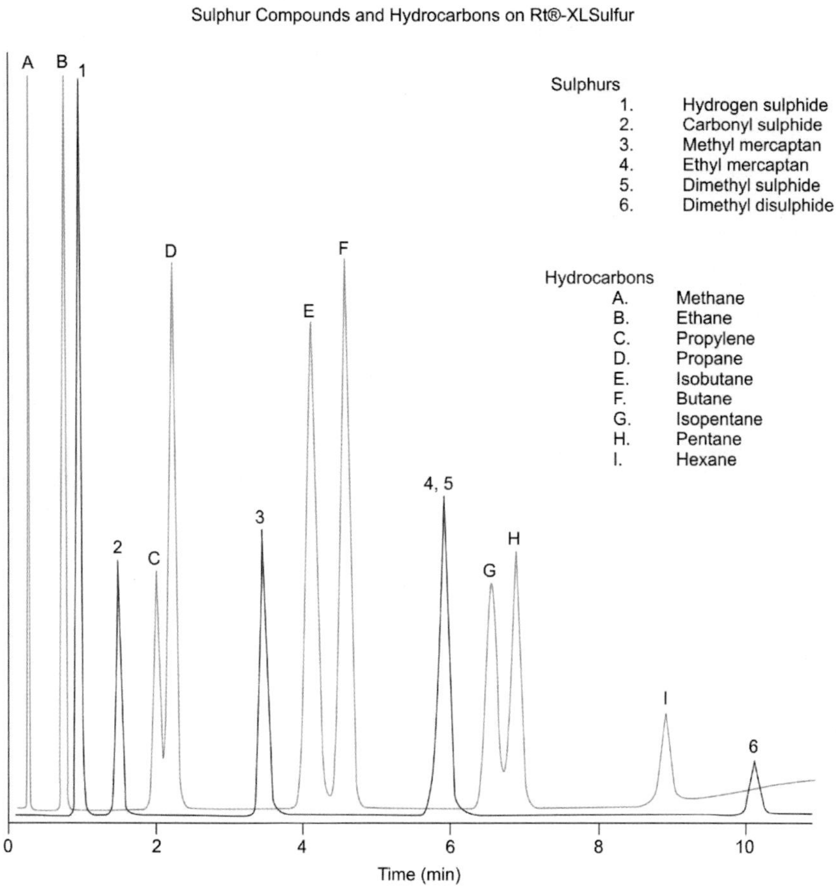

Column: Rt®-XLSulfur,1 m, 0.95 mm OD, 0.75 mm ID (cat.# 19806)
Sample: Conc.: 50 ppb each. Injection: packed not on-column
Oven Temp: 60°C to 230°C at15°C/min. Carrier Gas: He, constant flow
Flow Rate: 9 mL/min. Detector: SCD/FID
Acknowledgement: Sulphur standards courtesy of DCG Partnership 1Ltd., Pearland,TX.

Fig. 8.1 A sample of a mixture of hydrocarbon and sulphur compounds separated on the same column but using different detectors showing the difference in selectivity between the two detectors. On the FID, the sulphur compounds cannot be seen at all while the sulphur compounds can be detected using a SCD which has almost no response toward the aliphatic hydrocarbons [1]. ©Restek Corporation. *(Reproduced with permission, courtesy of Restek Corporation.)*

selective by a ratio of 10^5:1 for that compound relative to the matrix, the matrix response will still be ten times larger than the signal from the 1 ppm analyte. This confirms why separation is necessary irrespective of the detector's selectivity.

The first function that is vital in the performance of any detector is that column resolution is not compromised due to the volume or speed of the detector. Especially with capillary columns, the analyte peaks eluting from the column are often less than 1 s wide at half height. If we assume a carrier gas flow of 1 mL/min this means that the detector volume should be less than 16.7 μL. Most detectors do not meet this requirement and have a much larger 'dead' volume. Another potential problem is that many detectors have hot metal surfaces that can cause catalytic decomposition of the analyte. The residence time of each eluted analyte in the detector must therefore be minimised. Assuming that the column is correctly positioned in the detector, all of these problems can be resolved by sweeping the end of the column with an additional flow of gas. Most detectors are designed to work with a total carrier gas flow of 30−40 mL/min so the difference between carrier gas flow and this is called make-up gas. Not only does it stabilise the detector's operation, it minimises the effective detector volume. However there is one downside. Make-up gas dilutes the column effluent which reduces peak sizes on all concentration dependent detectors.

Certain detectors require a specific gas to function correctly for example a pulsed discharge helium ionisation detector requires helium to operate. Other detectors cannot tolerate certain carrier gases for example oxygen on the hot wire of a thermal conductivity detector. Temperature can play an important role in determining the performance of some detectors. They should be operated at a temperature high enough to prevent any condensation of any product eluting from the column. In gas analysis this is easily achieved by heating the detector to 120 °C.

Several important criteria must be considered when choosing a detector for the analysis required. Response is the magnitude of the signal generated by the detector relative to the amount of a specific compound being detected. Linearity or linear dynamic range is the range where the response is directly proportional to the amount of a compound. In gas chromatography it is only acceptable to quantify results in the linear range unless the detector responds exponentially. In this case the square root of the response versus the amount curve must be linear. The limit of quantitation is the lowest concentration of a compound that can be detected at a certain level of confidence. Some users talk about the minimum detectable peak as being a value of 3−5 times noise level. This is usually a rather optimistic value and in practice a signal to noise ratio of 10:1 is required to give a reliable and measurable peak. Other considerations include reliability, stability, ease of operation, robustness as well as the original and operating cost of the detector.

Two distinct types of detectors are used in GC namely mass dependent or concentration dependent. A mass dependent detector responds to mass per unit time entering the detector, not mass per unit volume, which results in a response that is directly proportional to each ng of the compound entering the detector and almost independent of flow rate. The area of a peak remains virtually constant whether the peak width varies or if a make-up or diluent gas is added at the column exit. This is typical for the flame ionisation detector (FID) and flame photometric detectors (FPD and PFPD). Mass dependent detectors are destructive and the compounds eluting from the column are chemically modified or destroyed in the detector. However only a proportion of analyte molecules are destroyed and vents from these detectors must be disposed of correctly especially when analysing toxic materials.

Concentration dependent detectors are non-destructive and can therefore be used in series with other detectors. However, it is only makes sense to use a less sensitive detector ahead of a more sensitive or selective detector. Typical concentration dependent detectors are thermal conductivity (TCD), photoionization (PID), electron capture (ECD) and pulsed discharge helium ionisation (PDHID) detectors. The area and height of a peak will depend on the flow rate through the column as well as by the dilution caused by the addition of make-up gas. If the retention time of any peak is changed for example, by changing the temperature program, it is necessary to recalibrate all components. The detector responds to local concentration differences in the detector and measures some physical property such as conductivity, rather than a chemical or physical change of the analytes. The outlets of these detectors must be correctly vented to ensure the integrity of the laboratory atmosphere.

There are many detectors available that are selective towards different groups of compounds but most of these are seldom used for gas analyses and when they are, it is usually for a very specific and special application. The most common detector for gas analysis is probably the thermal conductivity detector due to its universal response although it is not as sensitive as the other detectors. The pulsed discharge helium ionisation detector provides increased sensitivity for gas analyses and is also a universal detector. For hydrocarbon gases the flame ionisation detector is used mostly due to its selective response towards hydrocarbons. The TCD and FID are often used in combination for the determination of most of the permanent gases. We will therefore focus our attention on these detectors and will also mention other detectors for the sake of completeness. Some detectors such as the

TCD and PID are used without prior GC separation, especially in industrial stream monitoring and we did not include those applications in the following discussion. Other detectors that would also fit this category and may or may not employ prior separation by GC such as MS and FTIR are however discussed briefly at the end of the chapter.

8.2 Thermal conductivity detector (TCD)

One of the first detectors used in gas chromatography was the katharometer or hot wire detector which has evolved as the thermal conductivity detector (TCD) [2]. Although this is a relatively insensitive detector it has survived because of its universal or general response to all compounds other than the carrier gas and is used especially for gas analysis.

The TCD is one of the most universal detectors available. Depending on the compound, the TCD responds with a detection range of 0.001−100% (10−1,000,000 ppm). The classical TCD consists of four in-line filaments housed in a metal detector block. The TCD detector block is installed in its own thermostatically controlled oven for stability. Since the four TCD filaments can be damaged or destroyed if energized in the absence of carrier gas flow, some form of filament protection is provided in all TCD-equipped GCs.

8.2.1 Principle of operation

The TCD detector measures the difference in thermal conductivity and specific heat between the reference carrier gas flow and the eluting carrier gas plus analyte flow called the eluate. Every compound possesses some degree of thermal conductivity, and may therefore be measured with a TCD detector. Due to its high thermal conductivity and safety, helium carrier is most often used. However, other gases such as nitrogen, argon, or hydrogen may be used.

The Wheatstone bridge circuit in the TCD as in Fig. 8.2 uses four general-purpose tungsten-rhenium filaments for sample analysis. Two of the filaments are exposed to the sample-laden carrier gas flow and provide the actual chromatographic signal. The other two filaments are provided with clean carrier flow, enabling them to be used as a baseline reference signal. When the eluate from the column flows over the two sample stream filaments, the bridge current is unbalanced with respect to the reference signal. This deflection is translated into a signal which is digitised by the data system for analysis.

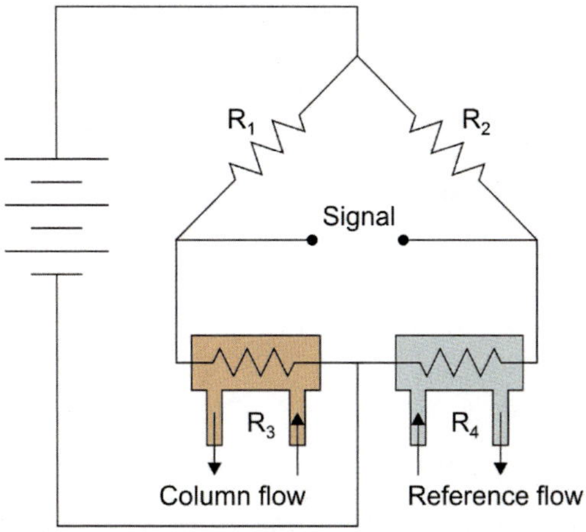

Fig. 8.2 Schematic of the cross section of a thermal conductivity detector showing the reference and measuring channels connected into a Wheatstone bridge circuit.

Agilent uses a TCD system that modulates the flow to the single filament, single cell detector at about 10 Hz so that the reference and sample flows are alternately compared. The difference in the sample and reference is measured and produces the signal. This design allows for a smaller volume cell and, since this is a concentration dependent detector, will improve sensitivity.

Hydrogen and helium have similar thermal conductivity which is almost ten times higher than the majority of permanent gases as shown in Table 8.1. This means that a certain concentration of say carbon dioxide

Table 8.1 Thermal conductivities of some common gases [3].

Gas	Thermal conductivity in [a]W/(mk) @ 293 K	Gas	Thermal conductivity in [a]W/(mk) @ 293 K
Hydrogen	0.17064	Carbon monoxide	0.02446
Helium	0.15016	Oxygen	0.02222
Deuterium	0.13806	Ammonia	0.02200
Freon 12	0.07300	Argon	0.01795
Neon	0.04794	Carbon dioxide	0.01653
Methane	0.03542	Water vapour	0.01600
Air	0.02587	Krypton	0.00946
Nitrogen	0.02475	Xenon	0.00568

[a]1 W/(min K) = 0.58984 kcal/(hrC).

will give a peak area nearly ten times larger with hydrogen carrier gas as compared to nitrogen carrier gas.

Fig. 8.3 shows a (very old) TCD that has been cut in half to show the reference and measuring channels of approximately 200 μL each. In line with miniaturisation trends in chromatography there has been substantial reduction in the size of TCD's in recent years. An extreme example would be the Chrompack micro-TCD shown on the right in Fig. 8.3, that has dimensions of 1 × 1 cm and an internal volume of 20 nL.

Fig. 8.3 A cut away of a TCD showing the two channels (left) and a micro TCD (right).

8.2.2 Operation

Since the signal for a peak on a TCD is essentially a small change in temperature of the measuring filament, thermal stability of the detector housing is essential. It is common practice to use one temperature controlled oven inside another and to leave the TCD oven temperature on at all times and only switch off the filaments for standby mode. The large thermal mass of the detector requires considerable time to reach equilibrium. Since the TCD is a universal detector all impurities in the carrier and make-up gas must be minimised. Clearly the purer the gas the lower the background and therefore lower concentrations can be determined. Since it operates with a heated filament, the oxygen content is important for reducing oxidation thereby increasing the filament lifetime.

8.2.3 Precautions

As a general rule, switch off filament current before opening the column oven door so that any testing or maintenance can be undertaken without the chance of damage to the detector. Never attempt to operate a TCD without first ensuring that there is an adequate flow of carrier and reference gas. Some systems use the incoming carrier gas as the reference flow and the column effluent as the gas supply for the measuring side. In a leak tight system these two flows will obviously be the same. Most TCD's have a limited zero range which implies that the same gas must be used as carrier and reference gas. There are some special TCD's that allow two different gases to be used simultaneously so that the hydrogen channel with argon carrier gas and the other channel with helium carrier gas are in one detector.

8.2.4 Make-up

Because of the large volume of the classical TCD systems, low carrier flows will lead to loss of peak shape. This problem can be overcome by adding a flow of gas at the column exit to sweep the analytes into the detector. The amount of make-up gas must be limited since this will dilute the analytes in this concentration dependent detector.

8.2.5 Detector temperature

The detector is normally held at a temperature that is hot enough to ensure that there is no chance of condensation especially of water vapour. On the other hand sensitivity reduces with increasing temperature which means that 120 °C is a good default temperature setting for gas analysis.

8.2.6 Troubleshooting

There is no routine maintenance for TCD's. To confirm that it is working on the data handling system, one can very briefly disrupt the flow by placing a finger on the gas outlet of the detector. This will give a small sharp peak which will probably be both positive and negative.

The most common symptom of a damaged TCD is difficulty in zeroing the signal. Some systems have a percentage readout of zero signal but on many systems this is not available so the actual signal level must be monitored.

The TCD usually provides no surprises with the exception of hydrogen when using helium carrier gas. As shown in Table 8.1, the thermal conductivities of hydrogen and helium are similar, so hydrogen will give a poor response when helium is used as a carrier gas. At low concentrations the hydrogen peak would be positive and at increased concentration the peak would adopt a characteristic skewed 'M' shape and at even higher concentrations would show a negative peak followed by a positive peak. None of these peaks can provide any useful quantitative information. In an interesting Agilent application note [4], scientists at the 'Energieonderzoek Centrum Nederland (ECN)' used helium spiked with 8.5% hydrogen as a carrier gas in a Micro-GC to eliminate this effect. With this carrier gas a negative but quantifiable peak was obtained for hydrogen without compromising the detection of the other permanent gases although low concentrations of hydrogen cannot be determined this way. This meant that all the permanent gases could be determined using the same carrier gas whereas conventional systems would require two channels using different carrier gases to determine hydrogen as well as the heavier permanent gases.

8.3 Pulsed discharge helium ionisation detector (PDHID)

Several versions of ionisation detectors have been offered for high sensitivity universal gas detection but all suffered reliability problems. Argon and helium ionisation detectors both used radioisotopes for the energy source and were difficult to use routinely.

The PDHID is one of the most universal non-destructive and sensitive detectors used in GC and can be used in several modes of operation. It was developed from studies done at the University of Houston, Houston, Texas by Prof Wentworth and his co-workers in the early 1990s [5]. For gas analysis it is particularly important in the helium photoionisation mode where the response to organic and inorganic compounds is linear over a wide range. Response to gases is always positive and it is capable of detecting ppb concentrations.

The discharge gas can be doped with noble gases which reduces the energy in the discharge zone and consequently changes selectivity for the analysis of aromatics or aliphatics. A non-radioactive electron capture mode is also available for the detection of halogen compounds such as pesticides. However, these applications are of limited application for gas analysis and will not be further discussed. We will concentrate on this detector in the helium photoionisation mode as this is such an important detector for trace gas analysis.

A miniaturised version of the PDHID has been introduced which uses only about 20% of the amount of ionisation gas. The performance remains similar to the standard unit which will make this ideal for use in portable GC analysis.

8.3.1 Principle of operation

The discharge gas must be ultra-pure and generally requires 'on demand' purification using heated Zirconium getter systems that produce very dry helium with less than 1 ppm (v/v) total impurities. Clearly extreme care is required to design a system that is capable of achieving and maintaining these standards. For example, a joint sealed with PTFE plumber's tape will not be tight enough to effectively use this detector.

A helium flow of approximately 30–35 mL/min enters the discharge zone of the detector and is ionised by a pulsed high voltage discharge between the discharge electrodes, schematically shown in Fig. 8.4. This helium continuum has a wavelength between 70 and 90 nm and is the primary source of energy for the ionisation of analytes eluting from the column. Since helium has an ionisation potential of 17.7 eV, any compound with a lower ionisation energy will be ionised as shown in Table 8.2. The only analyte that has a higher ionisation energy is neon which will therefore not give a response.

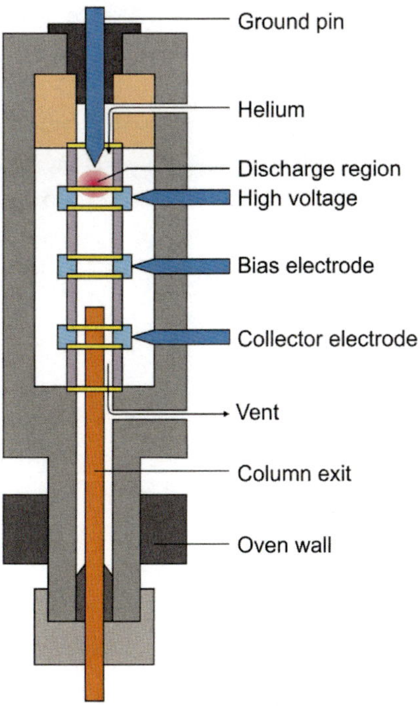

Fig. 8.4 Schematic of a cross section of the PDHID showing the helium ionisation gas, the discharge electrodes and discharge zone.

Table 8.2 Ionisation energy of gases and volatiles often encountered in gas analysis [6].

	Compound	Ionisation energy (eV)		Compound	Ionisation energy (eV)
Permanent gases	Helium	17.7	Alkanes	Methane	12.6
	Nitrogen	15.6		Ethane	11.6
	Water	12.6		Propane	11.2
	Oxygen	12.1		Butane	10.8
	Argon	15.8		Pentane	10.6
	Krypton	10.6	Alcohols	Methanol	10.8
	Xenon	9.6		Ethanol	10.4
Inorganics	HF	16.2		1-Propanol	10.2
	Ar	15.8		2-Butanol	10.0
	F_2,BF_3	15.6	Acids	Formic	11.6
	Kr,CO	14.1		Acetic	10.6
	CO_2	13.9		Propionic	10.4
	NF_3	13.1		Butyric	10.2
	H_2O,ClF	12.7	Aldehydes	Formaldehyde	10.9
	COS	11.3		Acetaldehyde	10.4
	NO	9.4		Propionaldehyde	10.1
				Butyraldehyde	9.9

Although the PDHID can use a standard FID electrometer, its background signal and noise is about two orders of magnitude greater than the FID. The background signal is typically specified to be less than 1 nA (1000 pA). Since this is a universal detector, column bleed contributes a large amount of background signal and for this reason temperature programming should be avoided if possible. It is vitally important that only well-conditioned columns be attached to the detector. For packed and micro-packed columns this detector might require conditioning for a relatively long time.

The reason for using ultra-pure helium is demonstrated in Fig. 8.5 where a sample of helium 5.0 was introduced into a system using the same helium after purification as discharge and carrier gas.

8.3.2 Precautions

Since the PDHID has a universal response to all gases except helium and neon, it is not possible to use anything other than high purity helium as a carrier gas. The PDHID should never be turned on without ionisation

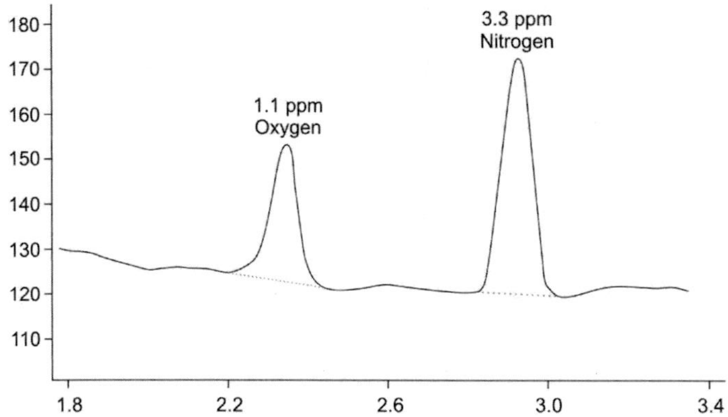

Fig. 8.5 Chromatogram of commercial helium 5.0 analysed on a custom built gas chromatograph. The same helium is used as carrier and ionisation gas after purification.

gas as this may cause damage to the electrodes. It is also important to ensure that only high purity helium is used as ionisation gas. When impure gas is used, the helium plasma would have a blue colour while pure gas would result in a pinkish colour of the plasma. This can usually be seen through a window on the side of the detector and if the detector is mounted in such a way that the window is not easily visible, a dentist's mirror can come in handy. After seeing this colour a few times, plasma colour becomes a very good diagnostic even if it is not very scientific.

Always remove the column from the detector when columns need to be baked out to prevent accumulation of high boiling compounds, decomposition products and column bleed in the detector. It is also advisable to always use a particle trap between PLOT columns and the detector to prevent particles that may become dislodged from entering the detector and giving baselines as in Fig. 8.6.

8.3.3 Make-up

When used with packed columns additional make-up gas is usually not necessary but, as with many other detectors, make-up gas can be used with capillary columns. The amount of make-up gas should be kept as low as possible since the PDHID is concentration dependent. It is important to

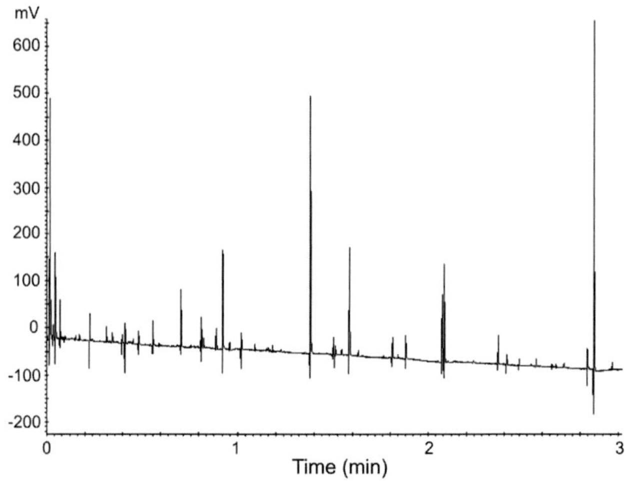

Fig. 8.6 Example of a chromatogram baseline showing the effect of particles from the column entering the detector causing sharp spikes.

ensure that the make-up gas effectively sweeps the analytes into the detector without significant dilution. The make-up gas must also be high purity helium.

8.3.4 Detector temperature

The temperature of the detector should be high enough to ensure that no condensation of any of the analytes can occur in the detector but keep in mind that the detector is temperature sensitive and is normally used for isothermal analysis to prevent excessive changes in the baseline.

8.3.5 Column installation

There are different adapters for using capillary and packed columns which allow the addition of make-up gas. It is important to insert the column the correct distance into the detector. For the VICI-PDHID this is 11.4 cm for capillary columns and 8.9 cm for the packed column adapter. When using metal columns, the column end should not be inserted into the PDHID as it would cause a short circuit. Metal columns should be connected to a fused silica insert with a Quickfit connector so that this is inserted into the detector.

8.3.6 Troubleshooting

The most common problem with the PDHID is a high background signal. This can be due to a dirty detector, impure ionisation gas, impure carrier gas or column bleed. Firstly, disconnect the column and observe the background current. If it improves to within the specified 1 nA within a few minutes, the problem is not the ionisation gas or detector. It is not necessary to plug the detector input line to do this quick check but it should be plugged if a long bake-out is done. If the background signal does not reduce sufficiently within a few minutes, the detector can be baked out and should this still not resolve the problem, the purity of the ionisation gas may be compromised. A typical symptom would be negative peaks seen with analysis of low levels of oxygen and nitrogen. This happens when there is more N_2 and O_2 in the discharge and carrier gas streams than in the sample. This could be due to leaks, a bad cylinder of helium or indicates that the purifiers are spent.

An unstable background signal is most commonly due to poor conditioning of the column and associated plumbing. This can be confirmed by heating the column oven about 10 °C and observing the baseline change. If the background is stable but no analyte peaks can be seen then the problem most probably arises from the column. Firstly check if there is actually flow from the column and then ensure that the column is inserted the correct distance into the detector and that the correct insert is used. Then ensure that there are no blockages, for example in the injection valve or sample loop that prevents the sample from being injected.

8.4 Flame ionisation detector (FID)

Historically, the unavailability of helium in remote countries for use with TCD's was responsible for the urgent need for a better GC detector and it is not surprising that this was done in these countries. A hydrogen flame was used in the Heat of Combustion Detector developed by Scott in 1955 [7]. The FID was invented in 1957 by Harley and Pretorius [8] in South Africa, and separately by McWilliam and Dewar in Australia and New Zealand [9].

8.4.1 Principle of operation

The FID employs hydrogen as the fuel gas, which is mixed with the carrier gas then passed through a narrow jet situated inside a cylindrical electrode and burnt in a controlled air stream that bypasses the flame jet; a schematic of an FID is shown in Fig. 8.7. The flow of hydrogen is set to about 10% of the flow of air. A direct current (DC) potential of a few hundred volts is applied between the jet and the collector electrode. This is shown with a grounded jet but many manufacturers bias the jet with this voltage and care must be exercised when inserting metal columns into the FID. When ignited, a certain background signal will be seen due to impurities in these three gases as well as column bleed. The size of this background signal will vary according to the application, but in a clean system should be of the order of 10 pA. Many modern instruments automatically re-ignite the FID as soon as the background signal has dropped below a pre-set value. When a combustible carbon containing analyte is burnt in the jet, the electron-ion pairs that are formed are biased toward the collector which feeds the analogue signal to the amplifier. This signal must be converted into a digital signal that can be handled by the data system. Most manufacturers do this conversion as close to the detector as possible.

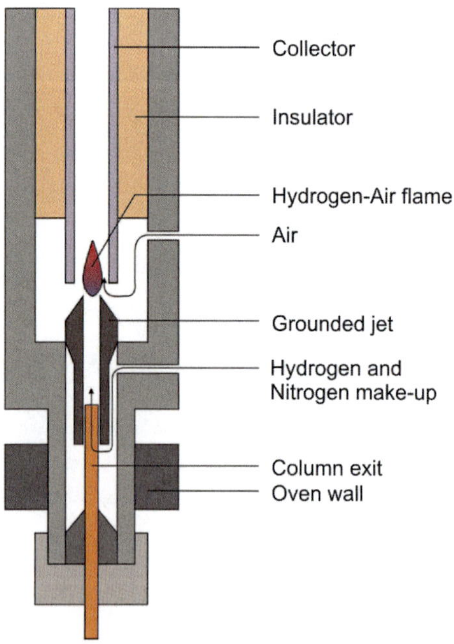

Fig. 8.7 A schematic of the cross section of an FID showing the different parts and entry points for the sample gas, the hydrogen fuel gas, nitrogen make-up gas and air.

The FID responds to mass per unit time entering the detector, not mass per unit volume, consequently the response is almost independent of flow rate. Although the column eluate is mixed with the hydrogen prior to entering the flame jet this dilution effect has no impact on the sensitivity and also has a sweeping effect that effectively reduces the apparent volume of the detector.

Although the FID ionises only a relatively small percentage of molecules entering the detector it is still regarded as a sensitive hydrocarbon detector. With the exception of methane any compound with a single carbon will give poor response on a FID, for example CO, CO_2, methanoic acid (formic acid), HCN, COS and CS_2. Although the actual chemical ionisation reactions can be quite complex [10] it suffices to say that if the compound burns in air then it will give a good response on the FID. Conversely a compound that does not burn in air will give little or no signal for example, CO_2 will give almost no response. Furthermore, the more oxidised a compound is the poorer its response will be on an FID.

The FID behaves essentially as a carbon counter whose response decreases with increasing polarity. For example, a calibration mixture that contains 1000 ppm (v/v) of methane, ethane, propane and butane will give peak areas in the ratio of 1:2:3:4. The peak area ratio for alkenes such as ethene, propene and butene will also be 2:3:4 but the peak areas will differ slightly from the alkanes. This is also true for aromatics and polycyclic aromatics. However, as soon as oxygen is present in the hydrocarbon, the ratios of responses for the number of carbons remains constant but the actual responses drop significantly. For example ethanol (C_2H_6O) will have a peak area response of about half of that of ethane (C_2H_6) although its response will still be double that of methanol. Likewise there will be similar sets of values for ketones, aldehydes, ethers and acids.

The effective carbon number is a method of calculating relative responses based on the number of oxygen molecules attached to the compound. For example, a C6 alcohol gives a response of 0.46 relative to the C6 alkane and C6 aromatic gives a response of 1.07 relative to this alkane [11]. Although calculation can never be as good as actual calibration this provides a useful method of obtaining surprisingly good relative response factors when standards are not available but the class of compound is known.

The response factors calculated by the data system during calibration are therefore highly diagnostic and must be critically evaluated before continuing with quantitative work.

During the 60 years that FID's have been used there have been several attempts at making the FID more sensitive to oxygenated compounds using a catalyst under names such as OFID. Activated Research Company produces a post column reactor system Polyarc® which converts oxygenated compounds to methane so that the FID sensitivity for all classes of compounds are equal [12]. This allows compounds such as carbon monoxide, carbon dioxide, formic acid and formaldehyde to be detected with virtually the same sensitivity as methane on the FID.

8.4.2 Operation

The FID must be heated enough to prevent water formation; this is discussed in more detail in the section on detector temperature below. Since the flame is colourless and will not be seen in daylight conditions, the easiest way to confirm ignition is to hold a cold shiny object above the detector for a few seconds and observing the dew formation. A cold glass surface such as a beaker can also be used. Most instruments have some form of readout that will show a background signal when the flame is ignited. The FID can be used with any non-hydrocarbon carrier gas but will require small adjustments in flows to compensate for this. Carrier gas must be filtered to remove hydrocarbon impurities to reduce background signals. Make-up gas must also be hydrocarbon free. If the FID is used to measure trace hydrocarbons even the purity of the air is important and zero air should be used with appropriate in-line filters.

The FID has a linear dynamic range of about 10^7 which is often way above the range of the electronics and data system. Some manufacturers use automatic range switching to handle this shortcoming, otherwise the range must be adjusted to allow peaks of interest to be on scale. This would typically be peaks having a height of less than 1000 mV on most data systems. Heights greater than this will give flat top peaks whose areas are not representative of their true size.

8.4.3 Precautions

If the ignitor does not work, the flame may still be ignited using a cigarette lighter. Matches should be avoided because of the sulphur content of the combustible material. Turbulent air flow from air conditioners can sometimes cause flickering of the flame resulting in a noisy baseline especially in systems where the top covers are removed to allow the installation of valve actuators. In systems such as these the FID body is often open and since it is hot, it can cause burns when touched. Lastly, the FID flame is usually left on and care should be taken to have no combustible materials, especially gases, in the immediate vicinity of the FID. Where an FID is used in a production environment, for example in petrochemical processing, it should be contained in a flame-proof shelter.

8.4.4 Make-up

The FID needs make-up gas for optimum performance. To improve sensitivity and to stabilise the flame it is common practice to add a make-up gas for optimum performance. This should be a dense gas that will make up the total flow of gas from a capillary column to typical packed column flows. Sensitivity will increase with increased air flow until a point is reached where ultrasonic noise will appear on the FID. As shown in Fig. 8.8, the correct combination of nitrogen, make-up and hydrogen flows are required for optimum sensitivity. Typical operating flow rates would be 1 mL/min helium through the column, 30–40 mL/min nitrogen make-up gas, 40 mL/min hydrogen and 400 mL/min air. For ignition it is common to drop the air flow by half and then increase it after the flame is burning.

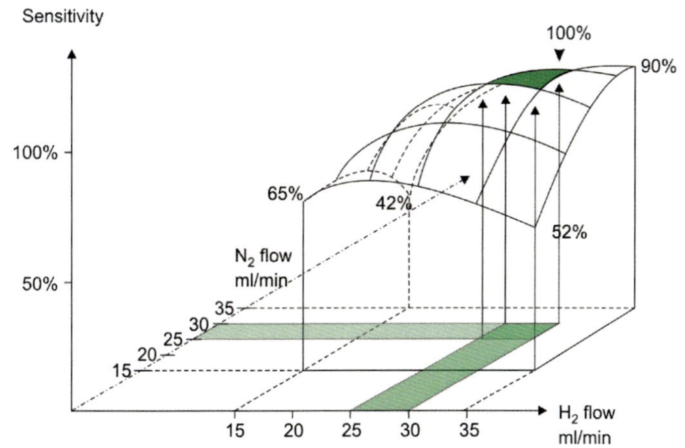

Fig. 8.8 A graph showing the effects of hydrogen and nitrogen flows on the detector sensitivity from which it is clear that there is an optimum set point for the hydrogen and nitrogen flow rate ratio.

In practice, most manufacturers blend the hydrogen and make-up gas in the flow control module and then supply this blend through a single tube to the detector.

8.4.5 Detector temperature

The FID does not change performance significantly with temperature but other practical issues need to be addressed. The detector should not be set higher than the maximum allowable operating temperature of the column material that is inserted into the detector. Normally the detector is set about 20 °C above the maximum column oven temperature used. Because of water formation at low temperature, it is important that a FID is not operated below 120 °C. Some manufacturers have an interlock that prevents flame ignition below some pre-set temperature such as 150 °C. For gas analysis it is unlikely that a temperature above 250 °C would be required. Most FID's are capable of a maximum temperature of 450 °C which is required for very high temperature analysis. Should the situation arise where the column maximum temperature is lower than the minimum FID temperature, a small length of deactivated tubing must be used at the end of the column.

8.4.6 Troubleshooting

The FID requires very little routine maintenance but periodic cleaning of the flame jet may be required. The symptom that some blockage may be occurring is that the flame may extinguish during a run. Often the FID will run satisfactorily if the make-up gas is set to a lower value, but remember that the responses for compounds may require recalibration. This allows the flame jet to be cleaned at a more convenient time. Although it has been said before it is important to remember not to switch the FID on at temperatures below 120 °C and never switch it on without gas flow.

Some common problems that may be encountered include:

Failure to ignite. This is probably the most common problem and is nearly always due to the incorrect fuel gas to air ratio. For example, after a long shutdown, air will diffuse into the hydrogen line resulting in insufficient hydrogen reaching the flame for ignition. After purging for some time the flame should ignite if the flow controllers are set correctly and delivering the correct flows.

Flame-out. In valve systems this can result from a valve switch following which the relative flows are temporarily disturbed or become unstable to such an extent that the flame goes out. It is usually remedied by slightly changing the pressures or flows in such a way that the flame keeps burning. Alternatively some form of buffering can be employed such as an extra length of column or restrictor.

Poor or elevated baselines. This usually relates to column bleed, for example, using a column that has not been properly conditioned. Another reason for this could be hydrocarbon impurities in the gases and this was discussed earlier. Baseline signals tend to reduce with time and sufficient warm-up times must be allowed particularly when using high detector temperatures.

8.5 The methaniser FID

The methaniser FID is a standard FID with a post-column reactor mounted between the column outlet or if a non-destructive detector is connected in series, the detector outlet, and the FID. Methanisers are available commercially from most GC suppliers but can also be made in-house as in Fig. 8.9. Similar reactors that are built into the detector will be described later in the sections dealing with detectors such as the SCD but the methaniser is discussed separately as it is not an integral part of the FID but complementary to it and finds wide application in gas analysis.

The methaniser uses a hot catalyst with excess hydrogen to convert CO and CO_2 to methane for the determination of low concentrations on FID. It will also convert oxygen containing hydrocarbon gases such as formaldehyde. Typically a nickel-zirconium catalyst at a temperature in excess of 350 °C is used. The catalyst material must all be at this temperature otherwise peak shapes will be badly affected. Some methanisers are packed in U-tubes and the catalyst must only be in the very hot section of this tube. The hydrogen flow is not critical and can be 15 mL/min up to the total

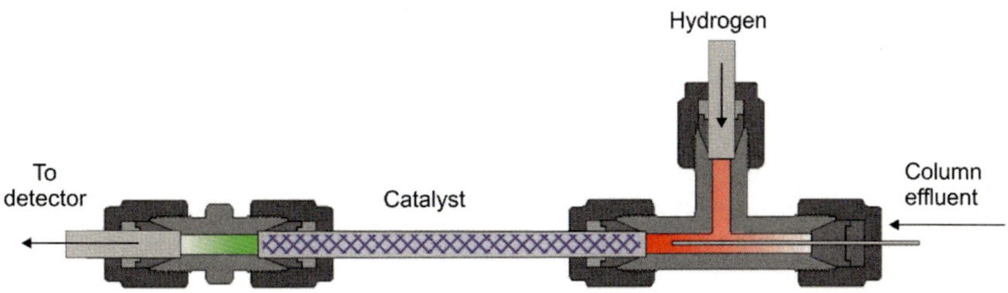

Fig. 8.9 Schematic diagram of the methaniser.

FID hydrogen flow of 40 mL/min. Since the CO and CO_2 are separated from the methane in time they are detected as methane. The most common application is to use a methaniser after a TCD and before the FID. When argon carrier gas is used for hydrogen determination the TCD sensitivity for CO_2 is poor but using a methaniser in series will give excellent sensitivity on the FID. The sensitivity of the methaniser is several orders better than the TCD for traces of CO and CO_2 even when helium or hydrogen carrier gas is used. The reactions that take place are: $CO_2 + 2H_2 \rightarrow CH_4 + O_2$ and $2CO + 4H_2 \rightarrow 2CH_4 + O_2$.

8.5.1 Operation

A new catalyst needs to be conditioned at elevated temperature before its efficiency is stable. Typically a nickel catalyst will be run with correct gas flows for 15–20 min at 400 °C and then cooled to about 370 °C and efficiency calculated as shown below. Other catalysts may require different temperatures to work efficiently. Once installed and configured, there is not really any specific actions required to operate the methaniser. However, since it contains a catalyst, periodic checking of its efficiency is necessary. It is also important to maintain accurate temperature control and ensure sufficient hydrogen flow.

8.5.2 Precautions

Large concentrations of oxygen will give some response through the methaniser but this does not damage the catalyst. However, care must be taken to protect the methaniser from any sulphur gases and high concentrations of unsaturated hydrocarbons. A four port valve can be plumbed to bypass the methaniser by venting the unwanted components to waste.

8.5.3 Make-up

Since hydrogen is added as reactant gas this also serves partly as make-up gas. However the FID may still need additional make-up gas which must not be pushed back through the methaniser. Care should also be taken to ensure that the additional make-up gas does not add excessive backpressure on the TCD. If there is a choice of flame jet diameters always use one with a larger orifice as this will minimise the back pressure effect on the TCD outlet. Gas quality considerations are the same as for the FID except that CO and CO_2 impurities in carrier gas or hydrogen would elevate the FID background signal.

8.5.4 Detector temperature

The FID temperature does not have to be adjusted because of the methaniser. One manufacturer fitted the catalyst inside the FID so in that case the detector temperature is actually the catalyst temperature and must be set accordingly high.

8.5.5 Troubleshooting

Efficiency must be calculated by monitoring the areas given for known concentrations of CO and CO_2 versus the area for a known concentration of methane.

For example, the standard contains 1000 ppm (v/v) each of CO, CO_2 and methane and the areas recorded using the methaniser for the three gases are 192,354, 193,678 and 199,764 pA.s respectively. The methaniser efficiency for CO is 192,354/199,764*100% which is 96% and for CO_2 this will be 193,678/199,764*100% which is 97%. Although this can be as high as 99%, it is usually acceptable to use the methaniser at any efficiency above 80%.

Ultimately there are only three parameters required for the optimum operation of the methaniser namely temperature, hydrogen excess and the catalyst. If the first two are correct, replace the catalyst and condition it as described in Section 8.5.1.

Methanisers do require replacement and the cooling, reheating and conditioning of the catalyst can be quite time consuming. A recent development by Activated Research Company allows the proprietary catalyst to be deposited in the flame jet using 3D printing technology [13–15]. This means that a quick change of flame jet is all that is needed to replace the methaniser. At this stage this 'Jetanizer' is only available for Agilent and Shimadzu GC's. The Jetanizer temperature is the detector temperature and must be maintained at 400 °C.

8.6 Flame photometric detector (FPD)

Originally developed in 1966 by Brody and Chaney [16], the FPD has a lot of similarity to the FID with the main difference being that the luminescence of the flame is monitored by a horizontally mounted filter photometer. For improved optical emission, the flame uses higher hydrogen flows, is hotter and larger than the flame used in FID. The whole chamber is made light-tight to lower the background on the photomultiplier tube.

Compounds entering the flame decompose and form energetic radicals that lose their energy by emitting light that can be detected by the photomultiplier. In later developments the FPD's were constructed with two combustion regions to separate the reaction (decomposition) and emission regions.

Sulphur compounds are broken down in the flame to form the $S^{\cdot}_2$ radical which then emits light in the UV range. Since every two S atoms produces one $S^{\cdot}_2$ radical, the FPD exhibits a quadratic response towards sulphur. To selectively detect sulphur compounds a bandpass filter of about 390 nm is mounted between the flame and the photomultiplier. At this wavelength some hydrocarbon emission also occurs which unfortunately reduces the sulphur selectivity and sensitivity. There are several emission wavelengths for sulphur in the UV region but all of these are close to hydrocarbon emission wavelengths. Phosphorus compounds break down in the flame to the $HPO^{\cdot}$ radical which emits light in the visible range and gives a linear response on the FPD. In some systems the filters can be switched for rapid switching of the two modes of operation while others require mechanical disassembly and reassembly to exchange the filters.

8.6.1 Operation

FPD's use a hydrogen rich flame. Dual flame systems have two air flow controls and one hydrogen flow controller. While not a problem in gas analysis, when samples dissolved in a solvent are introduced, the lower flame is often extinguished when the solvent elutes but is automatically reignited provided the flows are correct. Sensitivity and selectivity are influenced by flows of the fuel gases and electronic flow control is highly desirable.

8.6.2 Precautions

The response of the FPD can be significantly reduced by partial or full co-elution of the sulphur or phosphorus compound with any other compound. This is due to a 'quenching' effect caused by the closeness of the wavelengths of sulphur and hydrocarbons and is worst when hydrocarbons co-elute with the sulphur containing compounds, especially when trying to analyse trace levels of sulphur compounds in a hydrocarbon matrix. Co-elution of a large hydrocarbon peak with a sulphur containing compound will cause distortion or suppression of the sulphur response. It is often difficult to determine whether the lack of response is due to a low concentration of the analyte in the sample or due to quenching when using only a single detector. Quenching can be confirmed by comparing the FPD chromatogram with a FID chromatogram recorded at the same time. The quenching was very serious with the original single flame FPD and early users plotted twice the log of response vs the log of the sulphur concentration although it was known that square root of the response must be linear. The quenching was somewhat reduced by using dual flame FPD's. However quenching remains a serious problem with all FPD's and because of this interference, they do not provide sufficient selectivity for the analysis of trace sulphur compounds in a hydrocarbon matrix. Since there may be exposed hot metal parts inside these detectors the system must be primed as shown in Fig. 11.8.

8.6.3 Make-up

Since the FPD is a mass dependent detector the use of make-up gas does not reduce the signal by dilution. Furthermore, since the compounds are destroyed in the detector and the signal measured originates from the $S^{\cdot}_2$ radical it implies that the integrated peaks can be calibrated with virtually any sulphur containing compound. The same calibration can then be used for different sulphur containing compounds without having to calibrate for each, keeping in mind that if there are two or more sulphur atoms present in a molecule, as in CS_2, the response must be adjusted accordingly. Since the wavelengths of hydrocarbon emissions are close to sulphur emissions the carrier and make-up gases must have hydrocarbon filters installed.

8.6.4 Detector temperature

Most FPD's must be used at a relatively moderate maximum temperature due to the limited temperature that the filter photometer can accommodate. At higher temperature the sulphur sensitivity is reduced and the noise from the photomultiplier increases with increasing temperature. On the other hand, the temperature must be maintained at a high enough level to prevent condensation of the sample components on the optical window as this will reduce sensitivity due to decreased light transmission through the window.

8.7 Pulsed flame photometric detector (PFPD)

The pulsed flame photometric detector was pioneered by Prof. Aviv Amirav [17] of the University of Tel Aviv in Israel and offers distinct advantages over the FPD. It can detect 28 elements including sulphur, phosphorous, nitrogen, arsenic and many rare earths which are unlikely to be of interest in gas analysis. Our focus will, therefore be only on sulphur and nitrogen gases for which different optical filters are used. Like the FPD it has a quadratic response towards sulphur but the response to nitrogen is linear.

The pulsing of the detector has four distinct steps which is due to a combustible gas flow rate and flame source that cannot sustain a continuous flame operation. An air and hydrogen mixture enters the combustion chamber and after mixing with the column eluate moves upwards through the inside of the quartz combustion tube. An additional flow of the air and hydrogen mixture is added and sweeps the outside of the combustion tube before entering the ignition chamber. A continuously heated glow coil in the ignitor chamber, ignites the mixture and the flame front moves downwards into the combustion chamber where the sample molecules are decomposed forming radicals as in the FPD. When the flame front reaches the bottom of the combustion chamber it extinguishes itself, for lack of fuel, with an audible 'pop' whereafter the same sequence is repeated. These steps are illustrated in Fig. 8.10. The optimum pulsing rate is 3–4 times per second.

Following combustion, the excited species in the sample undergo further reactions that result in optical emission. The emission from the flame itself, the background emission, is very short-lived, only about three

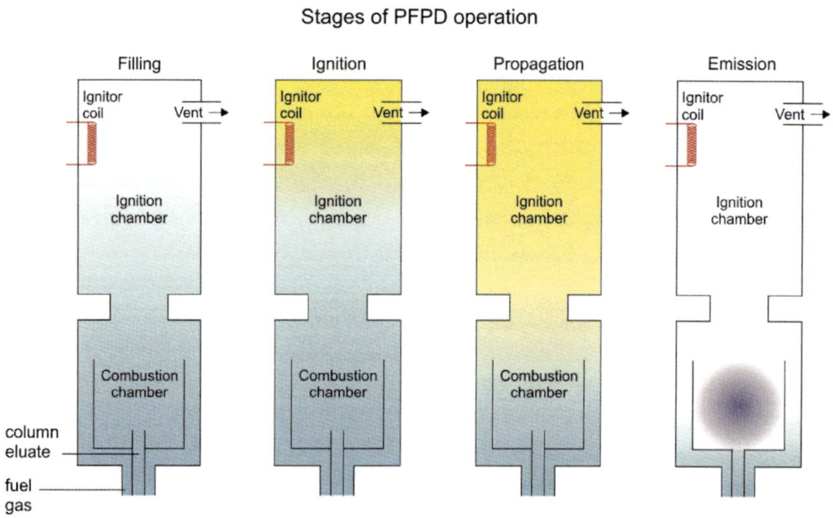

Fig. 8.10 A graphical illustration of the four stages of the PFPD operation.

to four milliseconds, while emission from sulphur, nitrogen, phosphorus and other elements that can be detected in this way, continues for a considerably longer time. Since the detector measures the signal after a few milliseconds it does not detect the emission from the flame background, which is mostly due to the combustion of carbon. By the time the signal is measured, the emission is mostly due to the species being measured and the selectivity is further enhanced by using a specific wavelength optical filter as with the FPD. The PFPD therefore differs from the FPD in that it uses a timed function to gate the signal from the detector which significantly improves the selectivity, especially for sulphur over hydrocarbons. Some suppliers offer PFPD's that can simultaneously measure the background as well as the analyte signal which gives a FID-like non-selective chromatogram in addition to the selective chromatogram of compounds containing the specific elements that are detected. In this way one can observe any co-elution of hydrocarbons that may cause quenching of the sulphur signal.

8.7.1 Operation

Since high flows of helium or hydrogen will cool the detector significantly, nitrogen is preferred as carrier gas. When using helium, the maximum column flow should be limited to 5 mL/min. Since the fuel gas acts as make-up, no additional make-up gas is required with the PFPD. It is important to accurately control the flow of all the fuel gases for optimum performance but once properly set up the PFPD is stable, reliable and user friendly.

An optional oscilloscope application is available for this detector that allows the signal to be observed in real time. This is extremely valuable in optimising selectivity and also gives an indication of the hydrocarbon response which can be observed if quenching is suspected.

8.7.2 Precautions

Compared to the FPD, the PFPD has much less quenching due to the co-elution of hydrocarbons with sulphur compounds but its response may be significantly reduced when large quantities of hydrocarbons are present. As with the FPD, the presence of quenching can be established by comparison with a simultaneous FID chromatogram.

Since this detector pulses at a frequency of about 4 Hz it will not give reliable peak size information if the peak widths are less than 0.5 s.

8.8 Electron capture detector (ECD)

The Electron Capture Detector (ECD) which is the most sensitive detector used in chromatography, is selective for electronegative compounds, especially halogens. The sensitivity to some of these compounds can be as low as the parts per trillion (ppt) range or in other units, femtogram per second. It contains a radioactive Nickel 63 source, a β-emitter of 5–10 mCi (185–370 × 10^6 Bq) in a sealed container. Because of this, its purchase, possession and disposal is regulated by authorities dealing with radioactive elements. The half-life of ^{63}Ni is about 90 years at ambient temperature and there is little chance that the source will need to be replaced during the lifetime of the GC system unless it is severely contaminated by non-volatile samples or destroyed by using excessive temperatures.

Since ^{63}Ni is a source of electrons, they ionise the carrier gas in this detector. Using a negative bias pulse results in a constant flow of electrons between the signal collector and the power electrode as shown in Fig. 8.11. This base frequency of the ECD is monitored on an ongoing basis and this value is very diagnostic of the cleanliness of the detector and system. When any analyte that needs an outer shell electron enters this zone, it captures an electron from the ionised carrier gas. To maintain the required

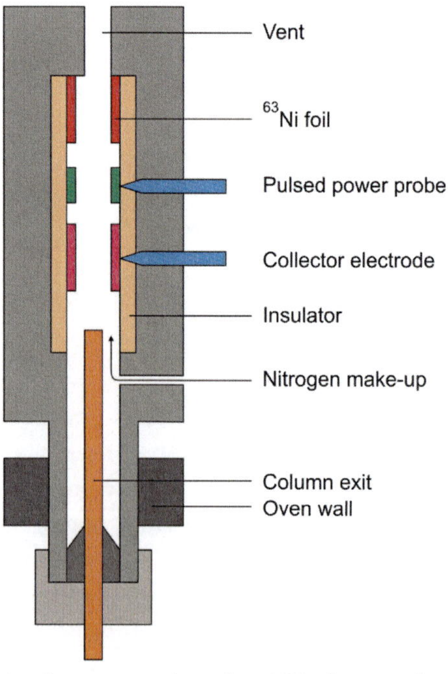

Fig. 8.11 Schematic of a cross section of an ECD showing the component parts.

electron current, the frequency of the pulse is increased to compensate and provides the signal; hence the name pulse modulated ECD.

8.8.1 Operation

The ECD detector requires carrier gas of nitrogen or an argon-methane mixture (known in industry as P5 or P10) to operate. Hydrogen can be used as carrier gas but will result in an elevated background signal which can be acceptable provided the flow does not exceed 2 mL/min and the make-up gas is nitrogen or argon-methane.

Certain column materials, such as cyano-phases, are not compatible with ECD's since these materials are electrophilic and will produce very high background signals due to column bleed. This will also be highly temperature dependent and will seriously reduce the linear operating range of the detector although this is less of a problem in gas analyses where moderate temperatures are usually employed.

Because of its high sensitivity, the purity of the carrier and make-up gas is very important. Not only must the carrier have high purity, it must also be stripped of all electrophilic impurities hence it is common practice to have in-line filters for moisture is series and ahead of an oxygen trap. The ECD has a rather small linear dynamic range of no more than 10^4 so care must be taken to ensure that all standards and samples are within this range. The detector is very selective for halogen containing gases, such as CFC's and also responds to peroxides, nitro groups and oxygen containing compounds, but not with the same sensitivity. It is relatively insensitive to hydrocarbons, amines, and alcohols.

8.8.2 Precautions

The ECD should never be operated without gas flow, in fact it should not be heated without gas flow. If the gas supply has to be interrupted for any reason, the ECD should be cooled down to less than 100 °C.

Dirty samples are the cause of many ECD problems and prevention is better than cure so attention should be given to sample clean-up to improve the reliability and lifetime of the ECD. Contaminants from gaseous samples can usually be removed by baking the detector for some time after the analysis of a batch of samples. More severe contamination can be removed by trying to reduce the contaminants under a flow about 20 mL/min of hydrogen at elevated temperature (350 °C) for a period of a few hours. After such a hydrogen clean-up, the detector may require several hours to stabilise and regain its sensitivity and selectivity.

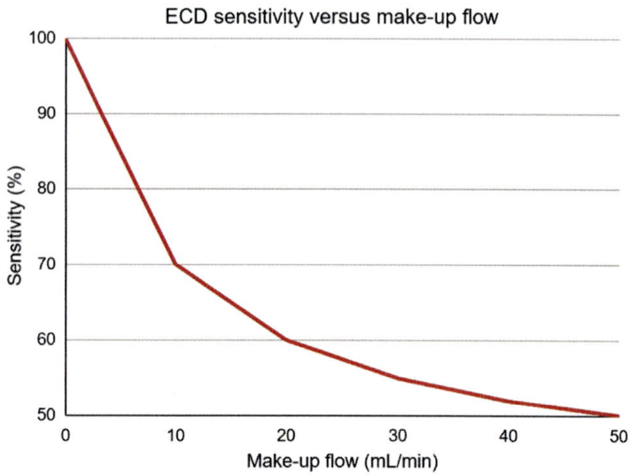

Fig. 8.12 The decrease in sensitivity with increased make-up flow of the ECD showing that good sensitivity is obtained with make-up flows of less than 10 mL/min.

8.8.3 Make-up

The ECD is a concentration dependent detector so for maximum sensitivity the amount of make-up gas must be kept low. The linear dynamic range of this detector is rather limited but the range can be extended upwards somewhat by increasing the flow of make-up gas and thereby diluting the column effluent as illustrated in Fig. 8.12.

8.8.4 Detector temperature

The detector is operated at a temperature about 50 °C above the maximum column operating temperature to ensure that sample cannot condense in the detector. In practice this may not always be achievable since the detector temperature may well be used to improve selectivity for the compounds of interest. For certain compounds the selectivity of the ECD can be highly temperature dependent. The only way to confirm this is to change temperature by 20 °C and observing any change in selectivity for the standard sample. Although the ECD has a maximum temperature rating of 400 °C rapid loss of ^{63}Ni will take place at these temperatures and should only be used for short periods during bake-out.

8.8.5 Troubleshooting

The background signal or base frequency must be monitored regularly as it is a measure of carrier gas purity which in turn measures the quality of the moisture and oxygen purifiers. When changing in-line filters, the gas flow will be interrupted so it is necessary to cool the detector before doing this maintenance. A typical symptom of a dirty ECD is the occurrence of a negative dip in the baseline after every peak.

When gas purity has been rectified and the detector has been properly cleaned there still can be situations where the sensitivity for electron capturing compounds appear exceptionally good, but noise increases and selectivity worsens. This will be an indication that the detector is contaminated and the radioactive foil needs replacement. This will probably never be necessary if the ECD is used only for gas analysis.

8.9 Atomic emission detector (AED)

The atomic emission detector is a multi-element detector that can measure twenty three different elements including carbon, hydrogen, oxygen, lead, manganese, fluorine, and silicon in addition to sulphur and nitrogen.

The eluate from the GC column is mixed with the reagent gas containing high purity helium and oxygen and fed into a quartz tube where it is combusted forming a plasma. The characteristic emission wavelengths of the different elements are focussed onto the entrance slit of a spectrophotometer with a rotating grating [18]. This allows the light spectrum to be collected on a fixed position of a photodiode array (PDA) as shown in Fig. 8.13, allowing the simultaneous measurement of a number of elements per peak eluting from the GC column.

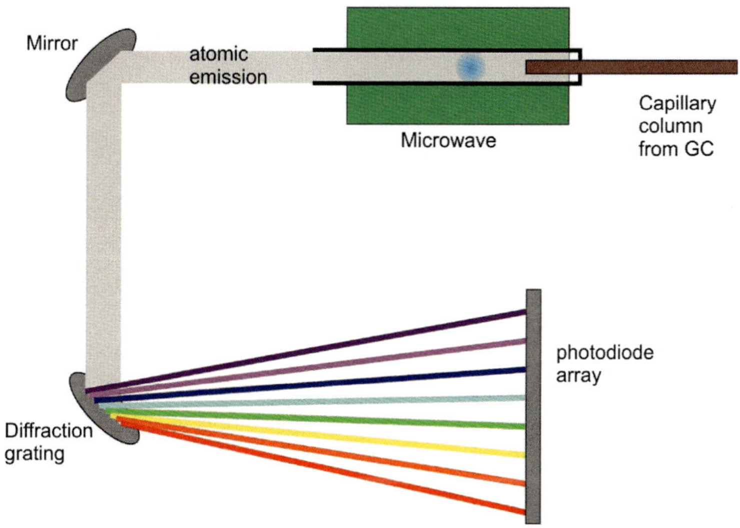

Fig. 8.13 Schematic of the optical path of the AED.

This allows a single instrument to be used for a wide range of analyses although separate injections may be required for the detection of some species. The reason for this is that the instrument setup and composition of the reagent gases must be varied to obtain optimum conditions for the detection of the different elements. For example, to detect carbon, sulphur and nitrogen in a sample, hydrogen and oxygen will be used as reagent gases. Carbon can then be detected at 179 nm and sulphur at 181 nm. For nitrogen detection, methane should be added as an additional reagent gas and the emission measured at 388 nm.

For optimum selectivity the flow of the reagent gases should be optimised as well as the amount injected. Samples containing a large excess of hydrocarbons may quench the plasma or lead to coking in the quartz tube. As with the FPD and PFPD, any compound containing the atoms of interest can be used for calibration as it is element specific and not compound specific although interference from other elements present in the same sample can occur. The quality of all gases used with the AED is critical and where possible on-demand purifiers must be used. Since oxygen can be detected by AED it is imperative that the whole system is designed for very low oxygen ingress from ambient air as has been discussed in Chapter 2.

While the multi-element capability of the AED makes it attractive, its use in gas analyses is rather limited mainly because its sensitivity towards sulphur gases is no better than the PFPD and SCD. It is expensive to purchase and to operate, service sensitive and not easy to operate routinely.

8.10 Sulphur chemiluminescence detector (SCD)

The SCD was invented by R. Benner and D. Stedman [19] and has a linear, equimolar response to sulphur compounds without significant matrix interference. Following further development by R.E. Sievers, R.S. Hutte and J.W. Birks [20], this detector became arguably the best detector, in terms of selectivity, for sulphur containing compounds and has found widespread application in various gas chromatography and supercritical fluid chromatography (SFC) applications. It is best suited for the analysis of complex mixtures and its use in gas analysis has so far been limited. The SCD sensitivity to sulphur is comparable to that of the PFPD and exhibits no quenching. It is somewhat less user-friendly, more expensive and may require more maintenance as it operates at reduced pressure.

In principle, the column eluate is combusted at a very high temperature, typically higher than 1800 °C to form sulphur monoxide (SO). This reacts

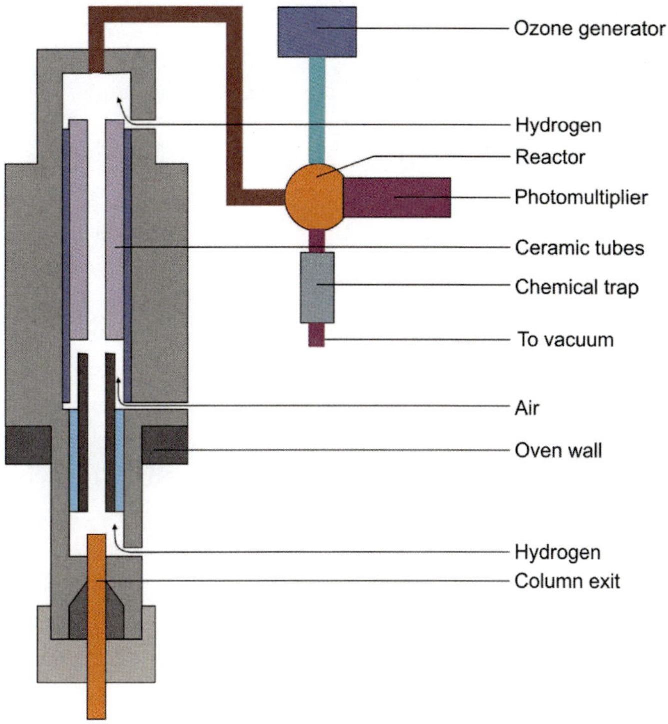

Fig. 8.14 Schematic diagram of a SCD showing the gas feeds to the combustion zone.

with ozone (O_3) and the signal of this chemiluminescence reaction gives rise to the signal that is measured. The emitted light is detected and amplified by a photomultiplier tube after passing through an optical filter. A schematic of the SCD is given in Fig. 8.14. Apart from high purity carrier gas the hydrogen and ozone must also be high purity.

8.11 Nitrogen chemiluminescence detector (NCD)

The NCD is very similar to the SCD and gives a linear and equimolar response to nitrogen containing compounds except nitrogen. It has therefore, also found widespread use in the analysis of complex mixtures but not in many gas analyses despite the fact that it will respond to most NOx species that can pass through a GC column such as ammonia, hydrazine and hydrogen cyanide.

Similar to the SCD, the column eluate enters into a dual plasma burner where the nitrogen containing compounds are converted to nitric oxide in the hydrogen and oxygen plasma. Potential interferences are prevented by using a catalyst that prevents the formation of secondary nitrogen species or destroys the same prior to detection. In a reaction with ozone, the nitric oxide becomes electronically excited and emits infrared light in the region of 600–3200 nm when it relaxes to the ground state. The emitted light is measured and is directly proportional to the amount of nitrogen in the compound.

8.12 Photoionisation detector (PID)

As the name implies, the photoionisation detector (PID) uses a UV lamp to ionise the molecules exiting the column [21] and a schematic diagram of a PID is shown in Fig. 8.15. The lamp is mounted in a thermostatted, low-volume (typically 100 μL), flow-through cell and emits radiation at the pre-set energy of the lamp used. Any molecule that has an ionisation potential below the energy of the lamp will therefore be ionised and detected, such as aromatic compounds and compounds with carbon double bonds. The ions that are produced by the UV light are accelerated by a polarizing voltage (250 V) onto a collector and the resulting current flow is measured and amplified through an electrometer.

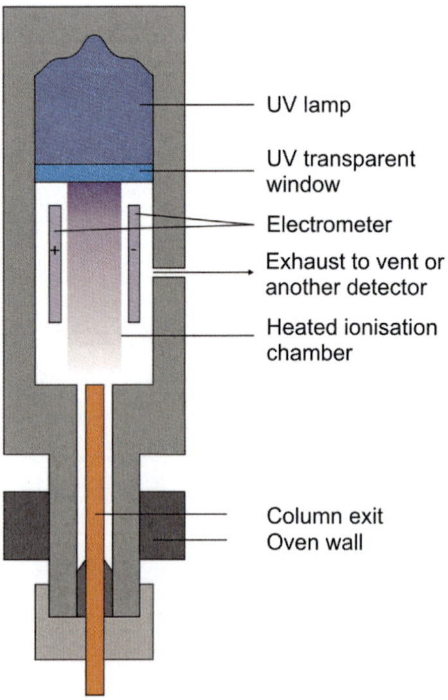

Fig. 8.15 Schematic diagram of a PID.

It is a non-destructive detector and can be used in series with other detectors, such as FID, similar to the TCD. It features in many EPA methods due to its selectivity and sensitivity and is capable of detection limits in the ppb range. The PID is more sensitive that the FID for ethylene and offers a 10:1 sensitivity advantage over the FID for the determination of BTEX components while also responding to many inorganic molecules like arsine, phosphine and hydrogen sulphide which requires the use of a high energy lamp.

The PID requires no fuel gases, is easily portable and can even be used with air as a carrier gas in the so-called gasless operation, for stream monitoring applications where no columns are used and the entire sample stream is directed through the detector.

The PID is easily maintained, has a broad linear range and uses the same electrometer as an FID. A sweep gas like hydrogen is often used to keep the sample from contacting the optical window to retain sensitivity and reduce maintenance. The lamps can be easily removed for cleaning when necessary and constitutes almost all of the routine maintenance that the PID requires. On the downside it has a maximum operating temperature of 270 °C and even lower when using high energy lamps. Common problems include low sensitivity due to energy loss, poor baselines, long warm-up times, flow instability, column bleed and like most detectors it should not be switched on without gas flow. The UV lamp is a consumable and should not be left switched on when not in use for extended periods. This is usually done by programming a stand-by method at the end of a sample sequence where the lamp is switched off. Provided the detector temperature is kept constant, warm-up times are short.

8.13 Mass spectrometer (MS)

In the field of Chromatography, most users are familiar with the coupling between Gas Chromatography and Mass Spectrometry known as GC–MS. Here we are looking at a Mass Spectrometer used this way as well as a MS used directly for gas analysis. This has the advantage of monitoring products changing more rapidly than the GC cycle time.

In classical GC–MS the sample is injected into the GC and the separated analytes are continuously fed to the mass spectrometer which is scanned giving a 'fingerprint' for the identification of unknown components. In gas analysis the need for identification of unknowns is seldom required because of the limited number of compounds that fit the definition of gases. While GC–MS can be quantified it is not as robust as a stand-alone detector.

The role of dedicated gas monitoring is the real application for MS in gas analysis. Some of these systems use a magnetic sector MS with a number of movable Faraday cup detectors for continuous monitoring certain masses of interest. Other systems use a quadrupole or time-of-flight MS with selected ion monitoring that allows any number of significant ions to be continuously monitored.

8.13.1 Principle of operation

Mass Spectrometry requires that molecules be ionised so that they can be separated according to their mass-to-charge (m/z) ratio and detected. Electron impact ionisation is one mechanism used to convert gas molecules to ions with a positive charge because the molecule loses an outer shell electron. The electrons are emitted by a filament mounted outside the ion source and heated to a high temperature. A negatively biased shield behind the filament directs electrons towards the ion source and they have an energy of typically 20—70 eV. Some of these electrons collide with the molecules inside the ion source which causes some of the molecules to become ionised. Depending on the structure of the molecule, the temperature and the amount of available energy, some fragments will form. All ionic species are electrostatically focussed into an ion beam. All of this takes place inside a high vacuum system. Also the ion source and mass analyser will be heated to prevent condensation of sample compounds and to reduce the adhesion of water vapour to the metal surfaces.

In chemical ionisation (CI) a reagent gas is introduced at low pressure into the ion source and gets ionised by the electron beam. These reagent ions then react with the analyte molecules and ionise them by charge transfer after which they are analysed as for EI. This is a softer ionisation mode and generally results in less fragmentation and can also be used to form negative ions adding another dimension to the analysis. Many different reagent gases can be used and the most common are probably methane, isobutene and ammonia. In selected ion flow tube mass spectrometry (SIFT-MS) several reagent gas ions are generated from dry or moist air using a microwave plasma [22]. One or more reagent gas ions are preselected before reacting with the sample substantially increasing the versatility of the technique and making it particularly useful for the direct analysis of VOC's in gaseous matrices. The major components of air such as nitrogen, oxygen and argon are not ionised and therefore these gases cannot be measured using this technique. Several alternative ionisation techniques can be used for different types of sample but will not be discussed as they are not commonly used for the analysis of gases.

There are also several ways of separating the ions into their respective mass or more correctly mass-to-charge ratio. Here we will only look at magnet sector, quadrupole and time-of-flight systems for which schematic diagrams are shown in Figs 8.16–8.18 respectively.

In a magnetic sector MS all ions leave the ion source with the same kinetic energy and when this ion beam enters a magnetic field, lighter ions will be more deflected than heavier ions. Multiple Faraday cup detectors are used to allow continuous monitoring of relatively high concentrations of certain gases. These collectors could typically be set to monitor oxygen, nitrogen and carbon dioxide at masses 32, 28 and 44 respectively. Some interference will take place if carbon monoxide or ethylene is present since they both have a nominal mass of 28 and will influence the nitrogen reading. In high resolution instruments, the separate determination of carbon monoxide (*m/z* 27.9949), nitrogen (*m/z* 28.0062) and ethylene (*m/z* 28.0312) is possible due to the small differences in their molecular masses. Also when carbon dioxide is ionised a certain amount of fragment ion at mass 28 will also be present. The helium mass spectrometer leak detector is a single collector magnetic sector system that is set to monitor only *m/z* 4.

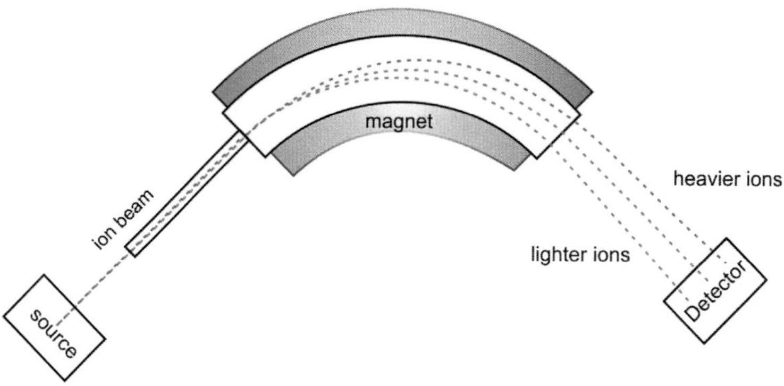

Fig. 8.16 Schematic diagram of a magnetic sector mass spectrometer.

The quadrupole consists of four accurately machined rods which act as a mass filter. A combination of direct current and high frequency voltages is applied to opposite pairs of rods and at any one particular setting only those ions with a specific mass-to-charge ratio will pass through the rod system. All other ions will collide with the rods, lose their charge and be pumped away by the vacuum system. Detection is usually done with an electron multiplier which gives a signal that is about 10^5 larger than a Faraday cup signal for the same amount of compound.

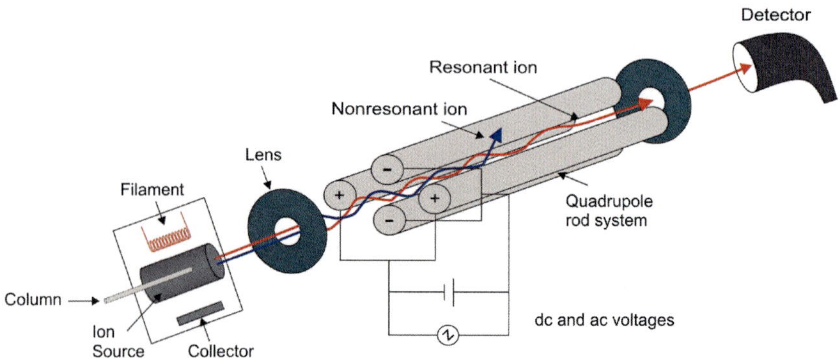

Fig. 8.17 Schematic diagram of a quadrupole mass filter.

The time-of-flight mass spectrometer consists of a long evacuated flight tube with the ion source at one end and the detector at the other. After ionisation the ions are accelerated using an electric field. Ions with a lower mass will gain more velocity than higher mass ions of the same charge and the differences in velocity means that low mass ions reach the detector before higher mass ions. The small differences in flight time are used to determine the masses of the molecules. Clearly the longer the flight tube the better the separation between masses and most TOF systems use an electrostatic reflector at the end of the tube to reverse the direction of the ion beam back towards the detector which is in this case situated at the same end as the ion source and therefore doubling the flight path. The so-called INFITOF by JEOL has a flight path in a figure-of-eight that allows the ions to be cycled many times, effectively increasing the flight path to the point where high resolution is obtained allowing separate detection of carbon monoxide, nitrogen and ethylene [23].

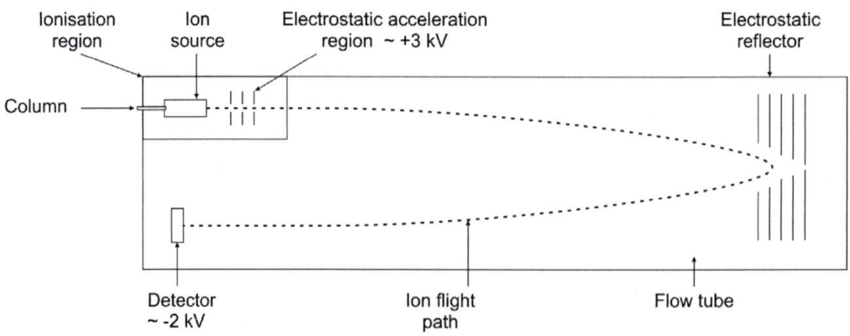

Fig. 8.18 Schematic of a time of flight mass spectrometer.

8.13.2 Operation

Sample introduction for an instrument used for on-line operation would have some form of restriction to allow only a very small amount of sample to continuously enter the ion source. This is necessary to prevent damaging the filament and to prevent the vacuum system from overloading. This can be a precision leak valve, a short narrow bore capillary or a critical orifice. Since the vacuum system can only typically handle flows of 1–2 mL/min, no make-up gas is used and in this case there is no carrier gas.

The vacuum system must be maintained in good working order and the system should never be used when significant leaks are present as the oxygen in air will rapidly oxidise the filament and the constant high background will shorten the lifetime of the electron multiplier detector. Luckily the MS is a good leak detector, for example by setting the MS to scan m/z 40, a flow of argon onto all connections will quickly show where the leaks are. The ion source and mass analyser are heated to prevent condensation of sample on any of these surfaces. Typically 120 °C should be sufficient provided no high boiling compounds are in the sample.

Given the many different ionisation, analyser and detector configurations for mass spectrometers, the instrument choice depends largely on the type of samples to be analysed and the information needed from the analysis [24,25].

8.14 Differential mobility detector (DMD)

Differential Mobility Detection is an advanced form of Ion Mobility Spectroscopy (IMS) [26]. Instead of an evacuated flight tube as used in mass spectrometry, it uses a drift tube with a transport gas at atmospheric pressure and thermally stable conditions. DMD uses additional sets of RF modulation voltages which allows more selective performance and which is tuneable for the compounds of interest and the matrix.

Electrons generated by a ^{63}Ni radioactive source are used to ionise the column eluates and these ions are pushed through a drift tube by means of a high velocity transport gas, typically zero-air or nitrogen. The drift tube consists of two parallel plates, approximately 500 μm apart as shown in Fig. 8.19. A RF modulated electric field of 1.3 MHz is applied across these plates by setting a fixed voltage and a scanning voltage (or compensation voltage). The net applied field alternates between high and low electric field strength and interacts with the ions to isolate ions of interest whose trajectory will pass straight through the filter. Those ions that are not tuned

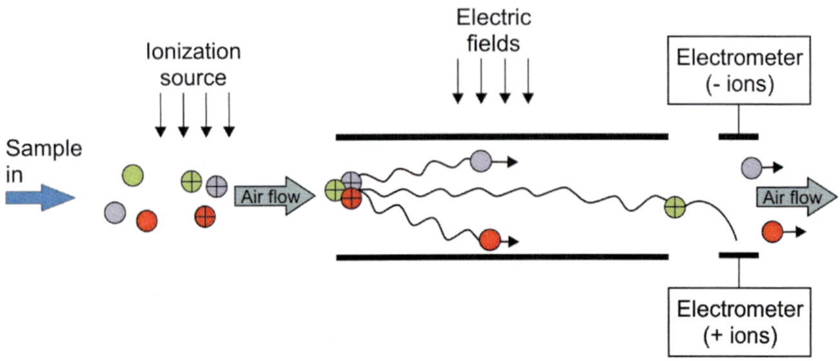

Fig. 8.19 Schematic of the differential mobility detector.

to pass through the filter will collide with the electrode plates and lose their charge. The compensation voltage is used to tune the correct ion species to the detector.

The DMD is equipped with two detectors, one for positive ions and one for negative ions. The system is 'tuned in time' to detect certain species by setting compensation voltages to accommodate transmission of the compound of interest in that particular part of the chromatogram. This 'time-segmented' detection thus leads to maximum selectivity for the application of interest as the matrix is eliminated.

8.15 Fourier transform-infrared (FT-IR)

Fourier transform infrared spectroscopy is a well-known and firmly established technique in analytical chemistry and more specifically in gas analysis. Many different types of instruments with varying degrees of complexity are available for applications ranging from the sophisticated laboratory to on-line measurements in production facilities. Fundamentally, infrared light generated by an IR source is passed through an optical system to divide the light into a spectrum. Molecules in the infrared beam absorb specific wavelengths of light by rotational, vibrational and oscillation movements inside the molecule and the amount of radiation absorbed is directly proportional to the amount of substance according to the Lambert-Beer law. The detector determines which wavelengths were absorbed as well as how much of each wavelength was absorbed. The pattern of absorbed wavelengths and relative intensities at the different wavelengths provide an absorption spectrum which is characteristic of each compound and can therefore be used as a 'fingerprint' of the

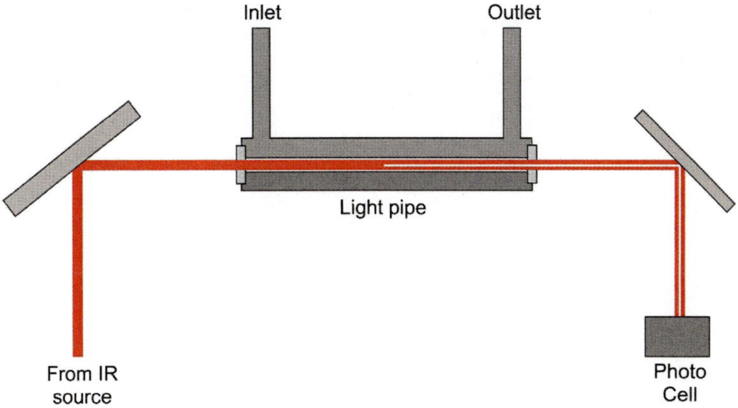

Fig. 8.20 Schematic of an FTIR beam path illustrating the absorption of a wavelength characteristic of a compound.

compound class. The amount of radiation absorbed at specific wavelengths can be used to quantitate the amount of the compound present in the sample. Not all compounds are IR active and gases like N_2 and O_2 have no IR activity while CO_2 and water vapour strongly absorbs IR radiation.

It is possible to use an FTIR as a GC detector with a configuration as in Fig. 8.20 but there are two extremely limiting factors that makes it impractical for many applications. These factors are the path length and the scan time required to detect the amounts typically eluting from GC columns. To get enough sample for FTIR detection, wide-bore, micro-packed and packed columns are most often used with an IR cell path length of typically 10–50 cm. Such a cell has a large volume and small, narrow GC peaks will therefore mix in the detector volume even though they were separated in the GC column. GC-FTIR is therefore limited to fairly high concentration samples and useful for packed or micro-packed column analysis of gaseous samples. Advances in data processing have however made analysis of fairly complex mixtures possible by FTIR on its own, without the need for prior separation by GC. Infra-red spectroscopy is a very powerful technique in the identification of isomers.

8.16 Detector summary

A summary of the detectors and some of their specifications are given in Table 8.3.

Table 8.3 Detector summary.

Type	Detector	Selectivity	Typical detectivity	Linear dynamic range
Mass dependent detectors (destructive detectors)	FID	Selective for compounds that can burn	pg carbon/s	10^7
	FPD	Selective for S or P	20 pg S/s	10^4
	PFPD	Selective for S, P and 25 other elements	<1 pg S/s	10^4
	MS	Universal in full scan mode	ng/s full scan	10^5
	MS	Selective in selected ion mode	pg/s SIM	10^5
	SCD	Selective for S	1 pg S/s	10^5
	NCD	Selective for N	1 pg N/s	10^5
Concentration dependent detectors (non-destructive detectors)	TCD	Universal, detects everything except carrier gas	ppm gases	10^6
	ECD	Selective for gas phase electrophores	fg Cl/s	10^4
	PID	Selective for UV ionisable compounds	pg C/s	10^6
	PDHID	Universal, detects everything except carrier gas	ppb gases	10^5
	FTIR	Selective for IR absorbing compounds	1 ng strong absorber	10^3
	AED	Selective, tunable for any element	0.1–20 pg/s	10^4
	DMD	Selective, tunable for some elements		$<10^3$

References

[1] Restek Corporation, Restek Searchable Chromatogram Library, Restek Corporation, Bellefonte, PA, USA., 2017 [Online].

[2] N. Ray, Gas chromatography. I. The separation and estimation of volatile organic compounds by gas-liquid partition chromatography, Journal of Applied Chemistry 4 (1) (1954) 21–25.

[3] Engineering ToolBox, Thermal Conductivity of Common Materials and Gases, Engineering ToolBox, 2003 [Online]. Available: https://www.engineeringtoolbox.com/thermal-conductivity-d_429.html.

[4] A. Toonen, R. van Loon, Agilent Application Note, Hydrogen Detection with a TCD Using Mixed Carrier Gas on the Agilent Micro GC, vols. 5991-3199EN, Agilent Technologies, Inc, USA, 2013, pp. 1–4.

[5] W. Wentworth, S. Vasnin, S. Stearns, C. Meyer, Pulsed discharge helium ionisation detector, Chromatographia 34 (5–8) (1992) 219–225.

[6] Advanced Industrial Chemistry Corporation, Ionization Potentials of Selected Compounds, Advanced Industrial Chemistry Corporation [Online]. Available: http://gcsrus.com/pdf/ionization_pot.pdf.

[7] R. Scott, New detector for vapour-phase partition chromatography, Nature 176 (4486) (1955) 793.

[8] J. Harley, W. Nel, V. Pretorius, Flame ionisation detector for gas chromatography, Nature 181 (4603) (1958) 177–178.

[9] I. McWilliam, R. Dewar, Flame ionization detector for gas chromatography, Nature 181 (1958) 760.

[10] K. Schofield, The enigmatic mechanism of the flame ionization detector: its overlooked implications for fossil fuel combustion modeling, Progress in Energy and Combustion Science 34 (2008) 330–350.

[11] W. Dietz, Response factors for gas chromatographic analyses, Journal of Gas Chromatography 5 (1967) 68–71.

[12] Activated Research Company, The Polyarc System, Activated Research Company, 2018 [Online]. Available: https://www.activatedresearch.com/technology/.

[13] J. Luong, Y. Hua, R. Gras, M. Hawryluk, In situ methanation with flame ionization detection for the determination of carbon dioxide in various matrices, Analytical Methods (10) (2018) 1275–1279.

[14] J. Luong, X. Yang, Y. Hua, P. Yang, R. Gras, Gas chromatography with in-situ catalytic hydrogenolysis and flame ionization detection for the direct measurement of formaldehyde and acetaldehyde in challenging matrices, Analytical Chemistry (2018), https://doi.org/10.1021/acs.analchem.8b04563. Downloaded from, http://pubs.acs.org.

[15] Activated Research Company, The Jetanizer, Activated Research Company, 2018 [Online]. Available: https://www.activatedresearch.com/jetanizerproduct/.

[16] S. Brody, J. Chaney, Flame photometric detector: the application of a specific detector to phosphorus and sulfur compounds sensitive to subnanogram quantities, Journal of Gas Chromatography 4 (1966) 42–46.

[17] S. Cheskis, E.A.A. Atar, Pulsed flame photometer - a novel gas chromatography detector, Analytical Chemistry 65 (1993) 539–555.

[18] T. Chasteen, Light-based detection systems in modern gas chromatography, in: The Spectrum, Center for Photochemical Sciences, Bowling Green State University, OH, USA, Spring 2002, pp. 7–11.

[19] A. Garcia-Campana, W. Baeyens, Sulfur chemiluminescence detector, in: Chemiluminescence in Analytical Chemistry, Marcel Dekker, Inc., New York, USA, 2001, p. 379.

[20] R. Hutte, R. Sievers, J. Birks, Gas chromatography detectors based on chemiluminescence, Journal of Chromatographic Science 24 (11) (1986) 499–505.
[21] J. Driscoll, J. Clarici, Ein neuer Photoionisationsdetektor für die Gas-Chromatographie, Chromatographia 9 (11) (November 1976) 567–570.
[22] Syft Technologies Ltd, SIFT-MS is the Leading Real-time Analytical Technique for Comprehensive Gas Analysis to Ultra-trace Levels, Syft Technologies Ltd [Online]. Available: https://www.syft.com/sift-ms/.
[23] JEOL, Gas Analysis: JMS-MT3010HRGA INFITOF Multi-Turn Time-Of-Flight Mass Spectrometer, JEOL, 1996–2018 [Online]. Available: https://www.jeol.co.jp/en/products/detail/JMS-MT3010HRGA_INFITOF.html.
[24] J. Hoker, F. Obersteiner, H. Bönisch, A. Engel, Comparison of GC/time-of-flight MS with GC/quadrupole MS for halocarbon trace gas analysis, Atmospheric Measurement Techinques 8 (2015) 2195–2206.
[25] C. Arkin, T. Griffin, A. Ottens, J. Diaz, D. Follistein, F. Adams, W. Helms, Evaluation of small mass spectrometer systems for permanent gas analysis, Journal of the American Society for Mass Spectrometry 13 (8) (2002) 1004–1012.
[26] G. Eiceman, Z. Karpas, H. Hill, Ion Mobility Spectrometry, third ed., CRC Press, Taylor and Francis Group., Boca Raton, FL, USA, 2013.

CHAPTER 9

Data system and data handling

Contents

The acquisition and processing of chromatographic data has come a long way since the days where peaks were recorded on a strip chart recorder. To determine the relative peak areas, a planimeter was often used and many chromatographers cut the peaks out and weighed them to get a numerical estimate of the peak area. This fell into disuse rapidly with the advent of the electronic integrator and with the emergence of personal computers (PC's) these were also rapidly replaced. In any chromatography laboratory nowadays one would find a computer next to every GC or several GC's networked to a single data system that is in many cases connected to a laboratory information management system (LIMS) that collects the results and reports them in a customised format.

Apart from controlling the GC operating conditions the function of the data system is to collect and process the signals emanating from the GC into

A practical guide to gas analysis by gas chromatography
ISBN 978-0-12-818888-0
https://doi.org/10.1016/B978-0-12-818888-0.00009-7

a set of results. Most data systems will display the chromatogram in real time and some even allow the changing of some parameters during run time. This is particularly useful when developing a method, for example, when the exact time at which a divert valve should switch needs to be determined. All data systems will save the data collected to enable post run reprocessing using different processing parameters to obtain a results report that satisfies the requirement of the analysis.

The rapid developments in data systems and data processing technology means that whatever one writes now on this topic will probably be outdated before too long. Yet, it is useful to understand some of the fundamental principles common to many data systems and chromatography software packages. It is strongly recommended that anyone who uses chromatography software acquaint themselves with the way in which their data system works. It is important to learn how to manipulate the various parameters, such as peak width, baseline correction, threshold values and many more in order to obtain the maximum benefit from the inherent power of the software to provide useful, meaningful and repeatable results. The system default parameters are based on the 'average' GC laboratory and as we have said before the gas analysis laboratory is a special case.

9.1 Data acquisition and processing parameters

There are many parameters that should be set correctly to ensure that meaningful data is recorded by the data system. If data was acquired using the incorrect parameters there will be no way to adjust it later and it is therefore important to get it right from the outset. Understanding how the data acquisition is done can help in setting the optimum acquisition parameters. The most vital piece of information that must be stored in the data system is a peak width value. Only the operator knows what type columns are being used, what carrier gas is used, what the sample is and therefore what chromatograms will be produced. The data system needs to differentiate between a peak and a spike or baseline drift. Even the noise measured by the data system is based on the peak width setting. Since gas analysis often uses isothermal column oven temperatures peaks will get broader with increasing retention times. Fortunately most data systems are programmed to expect this increase so the correct peak width setting is actually the initial peak width setting unless the first peak is broad like a matrix peak.

9.1.1 Analogue to digital conversion

The detector signal is analogue and is essentially continuous in both time and amplitude although the amplitude changes with time. The data system is digital i.e. it works with an 'on' or 'off' state. Therefore the signal requires a translation from analogue to digital using some form of ADC (Analogue to Digital Converter). The digitised record is discrete in both time and amplitude. It is discrete in time because there are a finite number of data points, each data point representing the average of the signal in the preceding interval. It is discrete in amplitude because the detector signal has been converted into an integer; each data point is the result of rounding off of the average signal in each time segment as shown schematically in Fig. 9.1. Therefore, the digitised record or raw data file does not contain all of the information in the original detector signal. The amount of information lost during the process of digitisation must be minimised, so this conversion is done as close to the detector as possible to avoid electrical interferences. Depending on the sampling rate and peak width settings a certain amount of digital bunching will be done to effectively smooth the data. Furthermore, the typical ADC has a range of −5 to +1000 mV. If the zero is set too low one will only see the tops of the peaks which will look like a perfectly smooth baseline with really sharp peaks. In other words when the chromatography looks too good there is a serious problem.

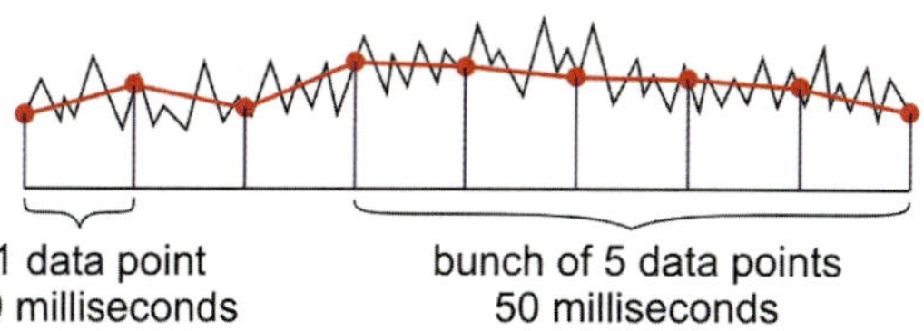

Fig. 9.1 Schematic showing the analogue signal (*black*) above the baseline and the digitised points at the set frequency (*red dots*) to give a digital signal (*red line*).

9.1.2 Baseline, zero and noise

Every detector has a certain background signal when it is operating. Depending on the detector, this background current, for example for a FID, is usually of the order of 10^{-14} A if good quality fuel gases and low bleed columns are used. However the PDHID has an acceptable background signal of 1 nA (1000 pA). To allow the data system to operate within its range, this background signal must be zeroed. Many modern instruments automatically zero the detector but older systems and standalone detectors must be manually set close to true zero or slightly above.

Operating any detector with a high background signal will compromise and reduce the linear range of that detector and requires immediate attention, see Chapter 8 and 12. The actual data system zero setting can be confirmed by turning the detector electronics off.

As soon as one zooms in to the signal, chromatograms contain noise and drift as well as chromatographic peaks. The signal noise is sampled every few milliseconds using the 50 or 60 Hz mains frequency as a time reference. Some forms of chromatography require sampling rates as high as 400 Hz but this will not normally be needed in gas analysis. The term 'noise' refers to the variations in the chromatogram signal that do not represent the elution of analytes through the detector and are variations that occur on a time scale smaller than the peak width. Contributions to noise include flicker noise within the detector itself, noise caused by fluctuations in the flow through the detector and saw tooth noise caused by line-frequency interferences. The likely source of the noise may be recognized by the appearance of the noise on the chromatogram trace, as shown in Fig. 9.2.

Flicker noise appears random. Flow noise can appear in a wave form, all alike, and regularly spaced. Saw tooth noise is also distinguished by repeated fluctuations, but they tend to be more ragged, less regular and slower. Combinations of these noise patterns can also be present.

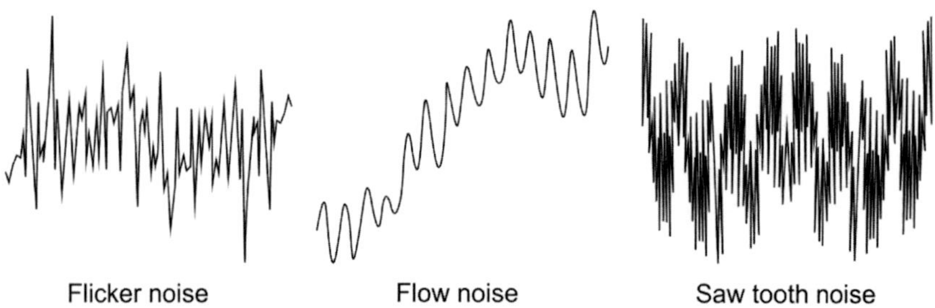

Fig. 9.2 Examples of the types of noise that may be present in a chromatogram.

Saw tooth noise and flow noise may be eliminated, but flicker noise is normal. Most detectors produce flicker noise on the order of a few microvolts and this noise usually determines the height of the smallest peak that can be detected. In the presence of noise, the baseline is never really flat, so tiny peaks may become indistinguishable from the noise and become hard to quantify.

A special case of noise is a 'spike' which usually has only a single data point above or below the baseline and hence will have zero area as shown in Fig. 9.3. Spikes can be caused by particulates in a gas line, electrical

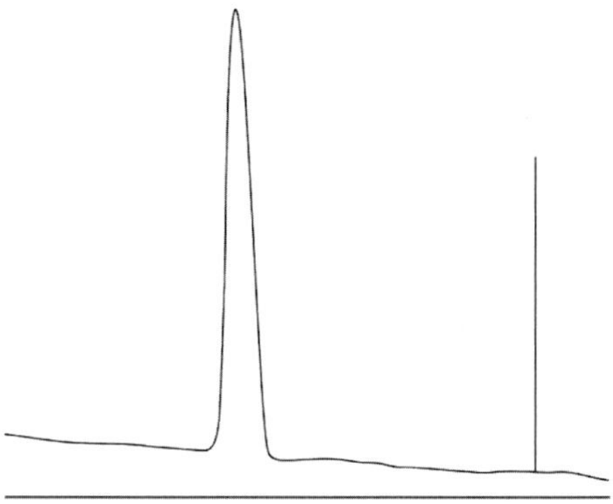

Fig. 9.3 A peak has width versus a spike that has no width on a baseline that shows drift.

induction, column particulates or mechanical disturbances. Some detectors are so sensitive to room pressure that opening and closing of the laboratory door can cause spikes on the detector signal.

After the peak width setting, the next important piece of information that is required by the data system is how much noise is present at a specific time and what threshold setting will distinguish a peak from this noise. Most modern data systems are able to automatically measure noise. This is based on the peak width selected previously and a common 'rule of thumb' is that the signal-to-noise ratio should be at least 10:1 for accurate quantification. With proper adjustment, data systems may detect peaks with a signal-to-noise ratio as low as 2:1 but the accuracy may be poor. These values are true signal to noise ratios and not some calculated value such as root mean square (rms) noise.

Some systems require the setting of the threshold while others will require a setting for signal to noise (S/N) ratio. The threshold can usually be calculated by the data system from a noise measurement and from the S/N ratio setting. If this value is too low, hundreds of retention times will indicate that just about every noise peak is being reported as a peak. If the value is too high, small peaks will not be integrated and the peak start and end times will be incorrectly determined which means that the peak areas will not be correct. It is important that this be optimised to ensure good quantitation. If small peak areas are to be rejected this can easily be done by setting a minimum area to be reported as attempting to achieve this by juggling the peak threshold may lead to quantitation errors.

Drift refers to any slow variation in the chromatogram that occurs on a time scale larger than the peak width. There is no clear-cut boundary between noise and drift. The contributors to noise listed above can also cause drift. Other sources of drift include changes in ambient temperature and loss of a portion of the stationary phase at high column temperatures due to column bleed. Drift seldom causes analytical problems and, except when caused by column bleed, is easily reduced by allowing sufficient time for the chromatographic system to stabilise. Column bleed, on the other hand, is something that happens during every programmed run and can only be made to look better by subtracting a blank chromatogram.

9.1.3 Range and attenuation

Two parameters that together determine the size of the peaks that can be detected is the detector range and attenuation. The range refers to the amount of amplification and will determine the minimum and maximum signals that will be recorded. The minimum can usually be set at a specific value, such as 10^{-11} or 10^{-12} Amp full scale. The lower the value, the smaller the amounts that will elicit a response from the detector but that also means that the noise will be amplified as well. Rendering the detector more sensitive by using low values for the minimum range therefore will not necessarily provide lower detection limits and the signal to noise ratio may remain the same or even deteriorate at lower range settings. Attenuation is the process of reducing the signal to fit on the screen or report so that the peaks can be observed. Changing of the range will effect a step change in the peak area integrated whereas attenuation only affects the visual output.

While the maximum of the range is usually determined by the point at which saturation of the detector signal will occur, some data systems can detect this point and will then automatically change ranges, to reduce the detector signal by factors of 2 in order to prevent saturation of the signal. Saturated signals are seen as flat topped peaks and are usually only observed for the solvent peaks in liquid chromatography and the matrix gas in gas analysis. The data system can keep track of the attenuation used and can calculate what the signal would have been without applying the attenuation. In this way both small and very large peaks can be determined in a single chromatogram. On some systems the attenuation can be set by the user and when the sensitivity in a method is given as 8×10^{-11} FS (full scale) it means the range was 10^{-11} Amp and the attenuation set at $2^3 = 8$.

9.1.4 Peak identification

As discussed, the first thing that the data system needs to know is what the expected width of peak is. This will depend on the type of chromatography being done. Packed columns with nitrogen carrier gas may give peaks of 10–15 s wide at half height while narrow bore capillary columns with hydrogen as carrier will give peaks less than 1 s wide. This value gives the integration system the possibility of differentiating between a real peak, a noise spike or baseline drift. It is always better to err on the side of taking too many data points over a peak than having a distorted picture of the peak due to a limited amount of data points as can be seen in Fig. 9.4. As a guideline one should have a minimum of 10 data points across the narrowest peak in the chromatogram. Too many data points do no harm

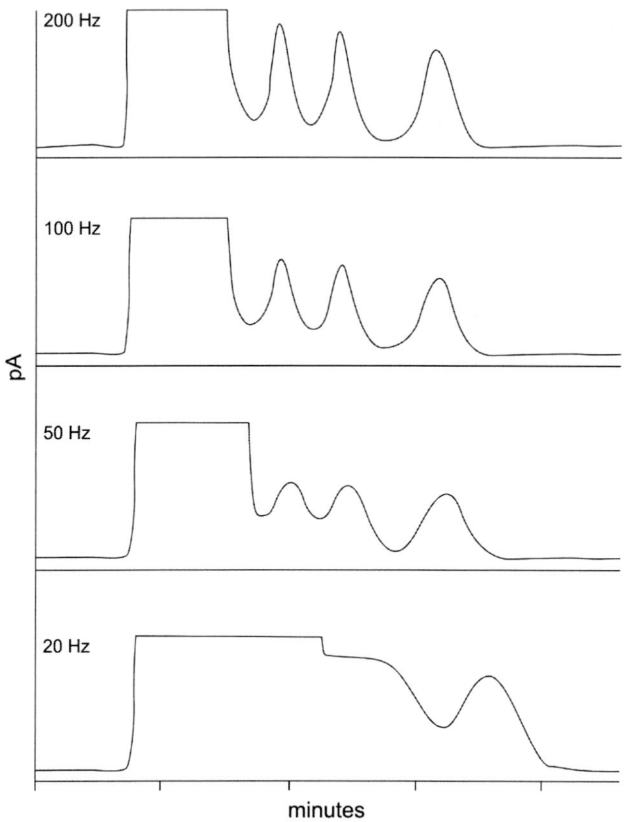

Fig. 9.4 Comparison of results at different data collection rates showing the peak distortion due a sampling frequency that is too low. The quality of data diminishes as the data collection rate decreases.

except that the data files become rather large and require more time for processing or reprocessing. Even the automatic noise evaluation done by the data system is calculated from the peak width set.

The next important role of the data system is to accurately determine the retention time (t_R) of each peak. Although chromatography is a separation technique, components are 'identified' because of the repeatability of the retention time under certain reproducible operating conditions. The stability of the retention time measurement of any analyte may vary considerably due to changing environmental conditions or sample concentration.

To integrate, or determine the area under the peak both the start and the end of the peak must be determined accurately. The onset of a peak is sensed by successive data points increasing beyond the pre-set threshold value as shown in Fig. 9.5 and similarly the end of the peak can be determined when successive data points fall below the threshold. Several data samples can be added together as a bunch or setting the slope sensitivity to create digital filtering. During acquisition, a number of values are stored in memory before being written to a file, so that it is easy for the data system to extrapolate backwards and determine the exact onset of the peak.

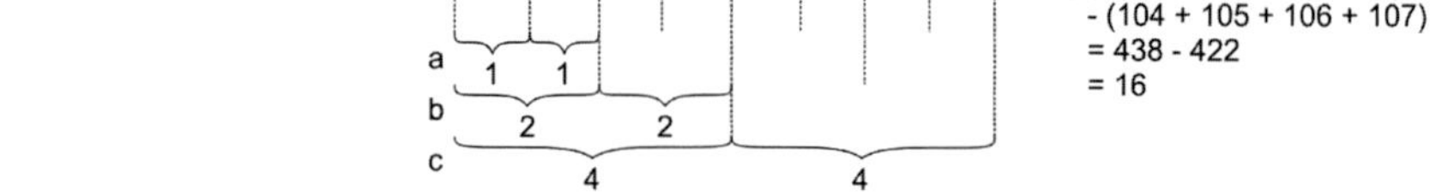

Fig. 9.5 In this figure the successive data points are shown. If the difference between one data point and the next exceed the threshold value of 1 then a peak start is recorded (A). With a bunching factor of 2, the difference between the first two and next two data points (B) is used to determine the effective slope to register a peak start and similarly a bunching factor of 4 would use the difference between the first four and next four data points (C) to determine the effective slope.

As the peak elutes the signal will increase to a maximum value and then decrease back to the baseline on a well separated component. The point at which the signal reaches its maximum value is the retention time. To determine this point, the difference between one data point and the next is monitored and will be positive at the beginning of the peak. At the apex this difference will become negative and this signals the peak maximum, as

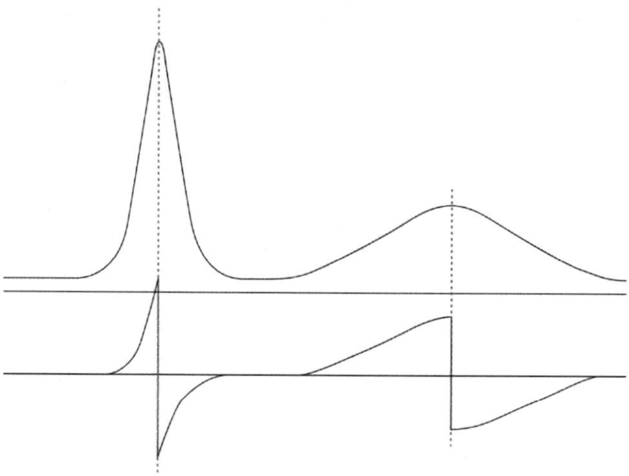

Fig. 9.6 A partial chromatogram with two peaks (top) and their first derivatives (bottom) showing how the retention times are determined as the times when the smoothed first derivative signal passes through zero.

in Fig. 9.6. However, should there be any noise close to this retention time, it could be incorrectly reported. The data system is generally set up to monitor the slope of the signal and to smooth the data, using the weighted average of 5 or more points, while calculating the first derivative of the signal to accurately determine the time that this passes through zero.

9.1.5 Peak integration

The amount of response that is observed for a certain concentration of analyte will vary widely based on sample size, detector used and carrier gas used. It is important to measure this response with high precision and reliability. Commonly the area under a peak is integrated and measured in units such as pA.s although peak height in μV may be used in certain circumstances. To achieve this, the heights of the data points, that is the differences between the values of the data points and the baseline, are added to provide a numerical value for the total peak height which is then multiplied by the time difference between the start and end of the peak as in Fig. 9.7. The peak height is determined by the difference between the value of the data point closest to the apex of the peak and the baseline.

Often the start and end points of the peaks are difficult to determine accurately, mostly because of a noisy baseline. However, since the peak height can usually be determined accurately, a better approach to integrate or calculate the area under the peak is to approximate the peak to a triangle.

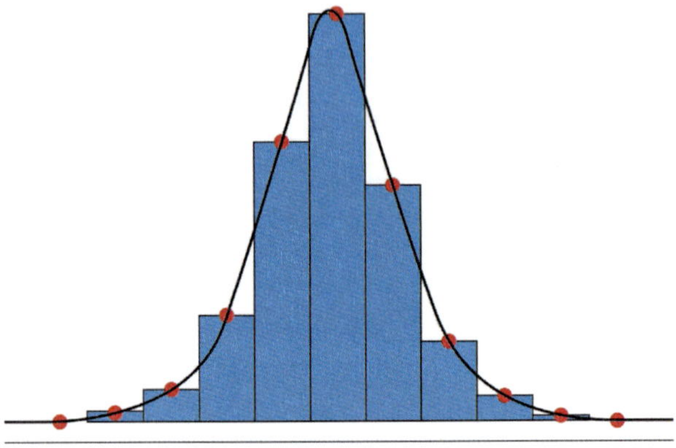

Fig. 9.7 Schematic showing how the peak area is determined by adding the heights of the individual data points and multiplying the heights by the time difference between successive data points between the start and end of the peak.

Although chromatography peaks are known to be Gaussian, sharp peaks closely resemble triangles. The area of a triangle is half of the base width times the height ($^1/_2$ B × H), which can be rearranged to give the area of peak by the height multiplied by the width at half height (W × $^1/_2$ H).

Most chromatography data systems include peak area as well as peak height in the results report. If peak width at half height is not directly reported a very close estimate is obtained by dividing peak area by peak height. The units must be in seconds. For example if the peak area is 500,000 μV.s and the peak height is 250,000 μV, peak width at half height is 2 s. This is a diagnostic value which must agree with the peak width set in the method.

9.1.6 Compound identification

The identification of analytes in a chromatogram is based on their gross retention time. The retention time is determined by column length, type of stationary phase, density of column packing, type of carrier gas, flow of carrier gas and column temperature.

All data systems allow control of the peak identification window within which the analyte is expected to elute. This could be in absolute time or as a percentage of total retention time or a sum of the two. For example the peak window is set to 0.5 min (±30 s) as well as 10% of the retention time. For a peak at 5 min (300 s) this will give an identification window of 30 s plus 10% of 300 s = 30 s or 60 s in total. Therefore any peak eluting between 4.0 and 6.0 min could be identified as this compound. Should there be more than one peak falling within this window, the analyst is effectively asking the data system to make its own identification. Many data systems will allow the user to choose whether to identify the peak nearest the centre of the retention time window or the peak of maximum concentration within the window.

Clearly the identification window must be set narrow enough to try and avoid this situation. At the same time the window should allow for the run to run or day to day fluctuations in retention time. Typically in gas analysis we expect retention time reproducibility to be around ±0.1% or better, provided that similar concentrations are always measured. As a starting point, set the data system to 1% so that in the above example the peak identification window will be from 4.95 to 5.05 min.

9.1.7 Unresolved peaks

There is no substitute for good separation and any peaks not fully resolved will cause some error in the area despite all the manufacturer's claims that their software has patented algorithms to handle fused peaks. The amount of integration error will vary enormously with relative peak size as well as the baseline allocation. The most important parameter is that whatever correction is applied on the standard must be the same on the sample. For example, if a tangential baseline correction is done on the standard chromatogram and a valley to valley correction is done on the sample chromatogram, the results will be meaningless. Some examples of such computer aided 'separations' are show in Fig. 9.8.

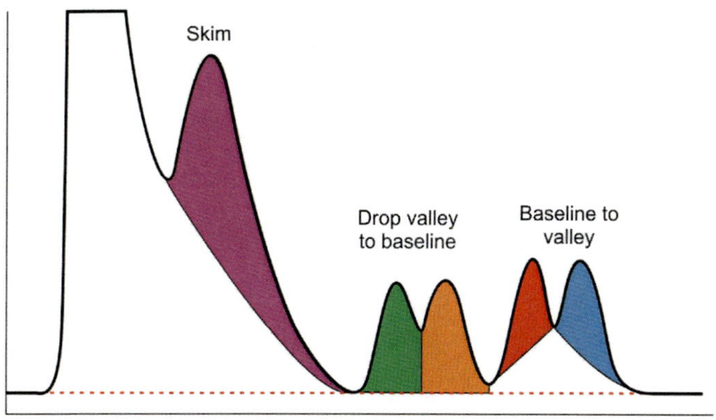

Fig. 9.8 A simulated chromatogram showing the effect of using software to obtain values for the areas of co-eluting peaks. Clearly each baseline assignment adds some error to the result, in some cases this may be almost 100%.

Computers do exactly what we tell them to do and it is therefore of vital importance to set the acquisition and processing parameters to values that will give consistent results. It has been said that if the computer software is so clever; why not use a connecting tube in place of the column so that zero separation takes place and all the calculation is done by the software!

Any small peak on the tail of a major peak causes quantitation problems. According to software used there may be ways of performing corrections but it is imperative that any such corrections be the same on standard as well as sample in order to obtain reasonable results. The best solution is to find a column or a system that elutes the trace component before the major peak and with baseline separation.

9.1.8 Timed events

In chromatography, for integration purposes, it is generally inadvisable to used timed integration events since there will always be some shift in peak retention time. Irrespective of the perfect flow controllers that compensates for fluctuations in ambient temperature and pressure, we know that columns change with time and that retention time also varies with sample size. Unlike spectroscopy where a certain element emits light at a known and fixed wavelength, chromatographic variations are best handled by the data system calculating points of inflection. The correct setting for noise, threshold and peak width will normally allow good integration without the use of timed events. Manual integration must be avoided as it makes the results subjective and difficult to reproduce or to validate the method.

However, in gas analysis there are valve switches that are controlled by the data system and therefore occur at fixed times. It is advisable to stop integration just before each valve switch and to restart with a new baseline sometime after the valve switch. No matter how well the system is balanced, some small baseline shift at each valve switch may occur. It is also common to inhibit integration at injection time because of the valve switch and to start it after some time, but well in advance of the first eluting peak.

Depending on the system configuration, peak widths may change after the valve switch. This will necessitate a timed event to optimise the new peak width. Also it may be necessary to change noise or threshold after switching a signal from one detector to another. If the noise level of the same detector changes after the valve switch, then there is almost certainly a leak in one part of the system. Most modern data systems allow the user to measure the noise and threshold at any point in the chromatogram. It is acceptable to zero a detector as it is switched into operation.

Although this should be avoided by optimal system design, in some cases it may be necessary to allow negative peaks to be integrated and these polarity changes can also be set in the timed events table.

Any event such as forcing of valleys that is dependent on chromatographic separation should not be set at a fixed time since the software is quite capable of calculating a valley based on the derivative of the signal passing through zero. As the retention times gradually change this point of inflection will vary accordingly and the valley points will be correctly allocated. Some of the more commonly used timed events are described in Table 9.1.

Table 9.1 Utilising timed events.

Event	Application
Baseline now	Used to set the baseline to whatever the signal is at that point. This is useful to reset the baseline after a valve switch but before peaks elute. Also used prior to switching detector signals on.
Baseline hold	The baseline is kept at the current value for the rest of the chromatogram. Useful for using a constant baseline when baseline separation of peaks is not possible.
Baseline back	This the similar to the baseline hold but works in reverse. This forces the same baseline to be used for all the peaks eluting before the event and is also used when baseline separations are not attainable.
Baseline next valley	Similar to baseline now, this event will set the baseline at the valley between the next two peaks. This can be used when the baseline changes after elution of a peak such as the matrix gas.
Baseline all valleys	This will set the baseline at the valleys between all the peaks and is sometimes used if proper separation cannot be obtained. This will result in lower peak areas and cannot be used if quantitative values are required.
Negative peaks	Turn detection of negative peaks on or off. Sometimes there are negative peaks, for example when using a TCD that should either be integrated or not. On most systems negative peaks can be inverted and detected as positive peaks or the detector polarity can be changed to show negative peaks as positive.
Manual integration	Used to manually set the start or end points of a peak and thereby also the baseline, overriding the automated integration parameters. Unless the run to run variability is 0 which it will never be, this will result in incorrect integration and should not be used in automated analysis. This can only be done post run on most data systems in the reprocessing mode.

Continued

Table 9.1 Utilising timed events.—cont'd

Event	Application
Area reject	Prevents the inclusion of peak areas less than this value in the results and can be used to omit peaks that are not of interest, such as peaks resulting from septum bleed, from being reported.
Solvent peak	Ignores a peak, typically arising from a solvent used or in gas analysis, for the gas matrix from being used in results calculation and the results report.
Tangent skim	Will estimate the baseline for smaller peaks eluting on the tail of a larger peak, such as solvent or matrix gas in order to get peak areas for the smaller peaks. This is inherently flawed and should not be used for quantitative results.
Baseline drop	As an alternative to tangent skimming the baseline will be held constant as in baseline hold and peaks integrated by dropping a vertical line from the valley between two peaks to the baseline. This will result in larger peaks areas but will not necessarily be more accurate.
Area sum	Used to combine two or more peaks and report them as one. For example, if the butane isomers are separated on a column but a total for butane and its isomers must be reported.
Peak width	The peak width is typically set at the beginning of a chromatogram when peaks are relatively narrow. Towards the end of the chromatogram peaks may however be significantly broader and require an adjustment of the peak width. Also after a valve switch, peaks may have different peak widths.
Noise	Also set at the beginning of a run but may change when switching to a different detector or carrier gas during a run. When using dual detectors make sure that the correct values for each detector are used.

9.1.9 Using timed events

In Fig. 9.9 an imaginary chromatogram is shown to illustrate the use of some of the timed integration events. There are often more ways than one to get the desired result but since the events can be added and changed during post run processing it means one can try different options until it works without having to re-run the sample. The following discussion refers to the labels in Fig. 9.9. The events described may be called by different names in different software packages but similar events will be present in most.

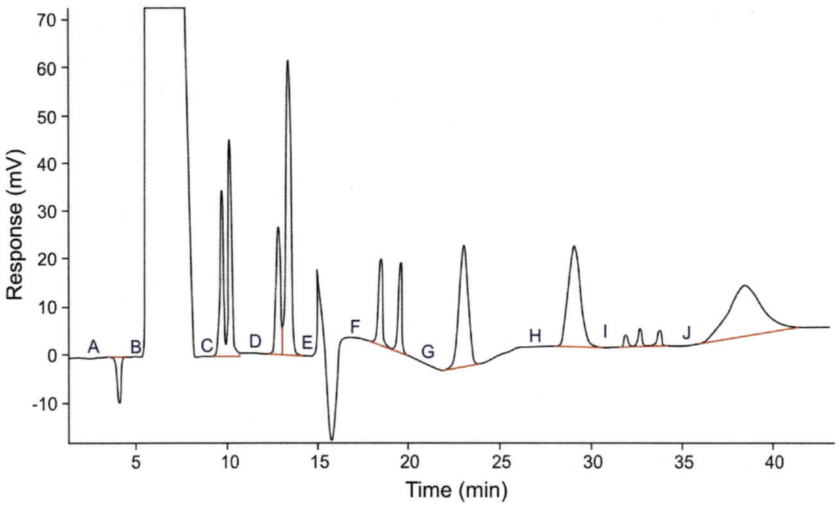

Fig. 9.9 Simulated chromatogram illustrating the use of timed events to properly integrate the peaks to be determined.

At A the baseline is flat and the noise more or less the same as in the rest of the chromatogram, so it might be the place to set the noise threshold. The next peak is negative, so one would enable negative peak detection at A and disable it at B. The next peak is the matrix and must not be integrated so one would turn integration off at B or use the inhibit integration or solvent peak event. At C the integration is turned on and the sum peaks event is used to integrate the peak pair as one since they are isomers of the same compound and must be reported as one. This event is turned off at D as the next peak pair must be reported separately although there is some co-elution. The baseline should be kept constant possibly by using the baseline forward event and the peaks split by dropping the integration line to the baseline. At E the integration is turned off again so that the disturbance caused by a valve switch is not integrated. At F the integration is turned on again and it might be necessary to use the set baseline now event as the baseline is now much higher than at the start. If this switch included a detector switch it may be necessary to do a noise update as well as a new peak width setting. The next two peaks elute on a downward sloping baseline and setting the baseline at all valleys can be used to ensure they are integrated correctly. This event will also ensure the peak after G will be integrated despite the change in direction of the baseline but the peak after G is broader than the earlier eluting peaks so that it might be necessary to change the peak width setting. If set wide enough it will also apply to the peak after H. The three small peaks after I are of no interest and the area reject value can be used to eliminate them from the results. The broad peak after J might be difficult to integrate as it is very broad and on a sloping baseline. Many software packages will split this peak and it might be necessary to sum the parts of the peak or to set the start and end of the peak using events. Setting a wider peak width at J may help to integrate this more reliably. Even so, it might be necessary to manually integrate this one. Apart from the time based area sum function described above most software packages allow summing of any identified peaks to give a value of say total C2's irrespective of their elution times.

9.2 Sequence and method files

The method file usually contains all the information on chromatographic settings such as temperatures and flows. It also contains the list of analytes and the units in which it should be reported as well as all the processing parameters described in the previous section. Some software includes the calibration data in the method file while in others a link to a separate calibration file will be used by the method. Method files may also have an auto-incrementing raw data file prefix but generally make use of default settings about where this data is stored and other actions. On systems with more than one channel, the method would usually contain the processing parameters for all the channels.

All data systems offer some form of an analysis list or sequence file as well as method files. A sequence file is essentially a work list which allows the setting of acquisition and processing parameters for each run from each sample on one or more channels. This allows that several injections of the same or different samples can be run under the same or different conditions.

The sequence files of most data systems are set up for the analysis of liquid samples using an auto sampler and typically, each entry of each channel requires the information for a vial number, a data storage folder, a raw data file base name and cycle number, a sample name, a method file name, and a report format. Optional parameters include an export file format, an alarm action, a calibration level if relevant, processing parameters including a calibration file, sample weight, injection volume, dilution factor and internal standard amount, any number of user fields and the option for running external data processing programs prior to or after acquisition.

Clearly some of these options are of very little use in gas analysis but many of them can be used to automate and greatly facilitate the turnover of samples. As described earlier, there are automated gas dilution systems commercially available that allow the successive dynamic blending and dilution of gas standards to deliver calibration standards of different concentrations. By setting this up in a calibration sequence table, the gas blending and dilution system can prepare the first standard, issue a start signal to the GC and after injection continue with preparing the next calibration level. By combining the dilution system and GC software sequence file, as many calibration levels and replicates at each level as required can be analysed and processed with little or no intervention from the analyst. Some software even allows for this type of processing to start at a specific time, say midnight on the day before, to have the system ready at start of work in the morning. Another very useful modification is to connect a stream selector valve to the GC in such a way that the selected stream coincides with the vial number in the sequence file. Several samples can then be connected to the stream selector valve and the sequence file can be set up to analyse each sample in turn, either once or as many replicates as needed. In applications where a number of process streams must be routinely monitored a similar set-up can be used to analyse the different streams according to a set schedule.

Calibration for gas analysis is typically done at a specific pressure. However, gas samples may come in containers at different pressures and if a correction for the sample needs to be made this can be included in the sequence file as the sample amount. Upon processing of the data a correction will then be made for the difference between the sample and calibrated pressure.

Sequence files are also very useful for automated batch reprocessing of raw data files. For example, when it has been established that a new calibration needs to be applied on a batch of analysis files that have already been acquired, to get revised results determined using this new calibration.

9.3 Calibration, data and results processing

Most chromatographers will be able to identify with the situation where a customer asks for a qualitative result rather than a quantitative result. Especially when told how long an analysis will take or how much it will cost, considering the need for standards and calibration. Following analysis, the analyst reports that the compound or compounds that the customer is interested in are indeed present in the sample. The immediate next question is always, and without exception; 'how much?' Following an explanation that the result is qualitative, the next customer question is a variant of 'surely you can estimate the amount'. Such analyst guesstimates, in order to satisfy the by now irate customer, usually changes over time into an absolute truth that eventually comes back and bites the analyst.

Bad quantitative analyses do not equate to semi-quantitative analyses.

9.3.1 Results processing

Once the peaks have been integrated and identified, the next step is to determine how much response a unit of mass or concentration gives at the detector for each different component. If we assume that the response of the detector is directly proportional to the amount of material emerging from the column at that time, we will require a constant of proportionality, or response factor. This number will differ from one component to another and from one day to another. If this assumed proportionality were true, only a single data point is required for calibration. This is often used in on-line GC analysers where the range of concentrations that are measured varies very little from the calibrated value. Unfortunately this assumption is not a valid one and for quantitative analyses over a wider range it is imperative to use multiple calibration points that cover the entire range of concentrations to be analysed. It is also good analytical practice to do more than one replicate at each calibration level. The overlaid calibration curves for two components are shown in Fig. 9.10.

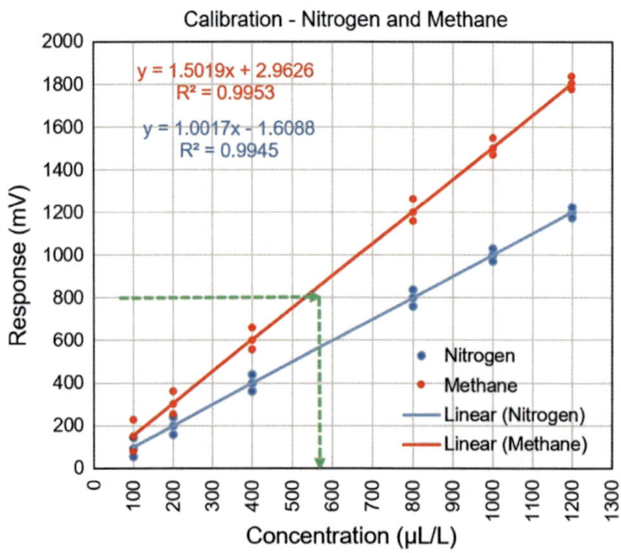

Fig. 9.10 Examples of calibration curves showing the direct proportionality between the detector response and compound concentration.

All detectors used in chromatography have a limited linear dynamic range and it is necessary to use multiple concentration levels for calibration to ensure that the system is operating within this range. Outside the linear range quantitation is simply not possible as the proportionality between detector response and concentration that applies in the linear range simply does not apply when the calibration line deviates from a straight line as illustrated in Fig. 9.11. The range between the lowest and highest calibration points that still falls within the linear range defines the lower and upper quantitation limits. Furthermore, any statistical or mathematical calculation to provide values outside this range will be, at best, an estimate and not a quantitative value.

The limits to the measurement range mean that it is virtually impossible to determine very high and very low concentrations using a single injection and detector. For example, to certify the purity of a pure gas, it is customary to determine the impurities as accurately as possible and subtract the total impurities from 100%.

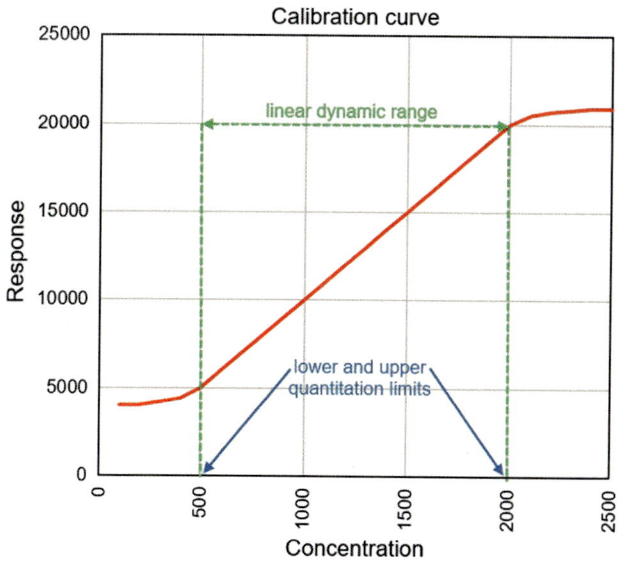

Fig. 9.11 A simulated calibration curve showing the linear dynamic range within which valid quantitation can be done. Clearly detector responses that results in concentration values below 500 and above 2000 will not be valid.

Certain detectors respond exponentially such as the Flame Photometric detector in sulphur mode. Most data systems can electronically convert the exponential signal to a linear signal by using the square root of the signal and plot that against concentration for calibration. The response curve should therefore be in the format $y = mx + c$ and the result can be obtained by reading the concentration corresponding to the square root of the sample response from the calibration curve.

Generally, unless there is a physical or chemical reason for deviation from a straight line, all calibration curves should be straight lines. It is rather rare for the calibration curve to pass through the origin. This is the reason that a single concentration level calibration cannot realistically be extrapolated through the origin. It is also obvious that two calibration points will always give a straight line and therefore the absolute minimum number of calibration levels cannot be less than three. As a general guideline, the lower the linear dynamic range of the detector, the more important it is to check that the detector is linear over the full analytical range. Also, if the calibration range spans several orders of magnitude the higher values will determine the slope and deviations from linearity at the lower end may be completely obscured.

9.3.2 Quantitation methods

9.3.2.1 Area percent

Area percent quantitation is arguably the most popular and universally used type of quantitation probably because it is by far the easiest to do, requires no standards and is the default on many data systems. Following peak detection and integration, the peaks and their areas are listed, all the peak areas added together and the area of each peak converted into a relative amount by dividing with the total area of all the peaks. After normalising, that is multiplying with one hundred, this gives the relative percentage of each peak area but not, as is often accepted, the relative amount of each compound. An example of a chromatogram and the area (and height) percent report generated for it is given in Fig. 9.12 and Table 9.2.

$$Relative\ amount_{(x)} = \frac{Area_{(x)}}{\sum Area_{(i)}}$$

The same can also be done using peak heights instead of peak areas, especially if the peak areas cannot be determined due to co-elution or a

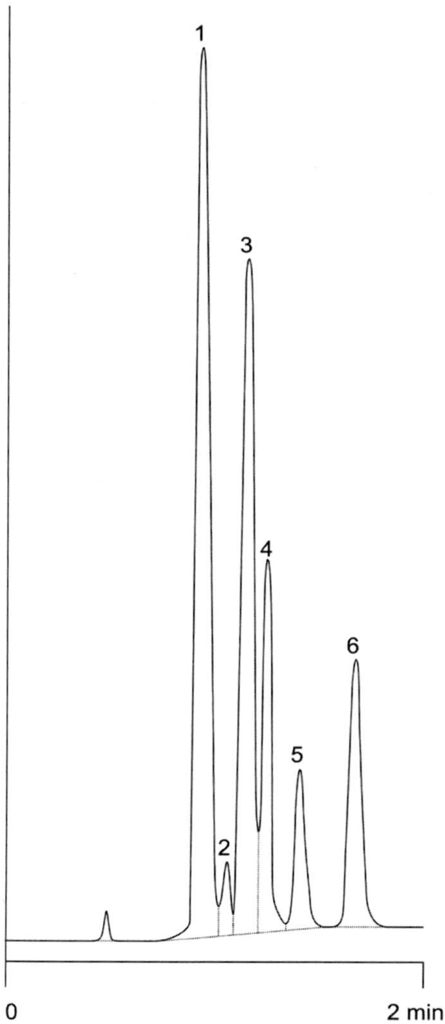

Fig. 9.12 The chromatogram for which the integration data is given in Table 9.2.

Table 9.2 This is a typical area% report for the chromatogram in Fig. 9.12 with a column added for the height% showing that the values obtained for area% and height% do not differ significantly.

Peak	Rt	Area	Height	Area%	Height%
1	0.27	98,725	4931	38.23	38.22
2	0.56	835	42	0.32	0.33
3	1.06	67,896	3395	26.29	26.32
4	1.36	38,399	1915	14.87	14.84
5	1.54	19,146	957	7.41	7.42
6	1.83	33,226	1660	12.87	12.87
Totals		**258,227**	**12,900**	**100**	**100**

noisy baseline [1]. When using peak heights the numbers are smaller and therefore will appear to be less sensitive but it is still an extremely useful and often preferred alternative to using peak areas. Since this is a purely arithmetical calculation, the total is always 100%. However, there is no detector that gives identical responses for the same concentration of different compounds. For this reason alone, quantitation requires that the response per unit quantity for each compound be determined.

In order to get closer to actual concentration values, the peak areas are often multiplied with a relative response value [2]. For mass dependent detectors, such as the FID, the response of a compound such as methanol can be related to the response of methane. An amount of methanol could conceivably give a peak area that is exactly half of the response obtained for an equimolar amount of methane. Therefore the peak area of the methane peak can be multiplied by two to correct for the lower response that the detector has for methanol compared to the response for methane. Similarly, the relative responses of compounds with other functional groups can be determined and used to compensate for the differences in the detector response towards these compounds. After applying the relative response values, the recalculated peaks areas are normalised and expressed as mass percentages or mass% and is only valid if normalised.

$$Relative\ amount_{(x)} = \frac{Area_{(x)} \times Relative\ response_{(x)}}{\sum_{i=1}^{n}\left(Area_{(i)} \times Relative\ response_{(i)}\right)}$$

There are published lists of such relative responses and it is quite commonly used in industry [3]. However the relative response can be affected by many factors including, differences in sample composition, FID's from different manufacturers, using different quality hydrogen, make up and air gases or flows for the FID, detector age and contamination; all of which makes this technique pseudo-quantitative at best. Apart from that, it is only really valid for mass dependent detectors and cannot, and should not, be applied to concentration dependent detectors such as the TCD. Clearly, incorrect peak identification will cause incorrect relative response values to be applied for quantification of that analyte. For this reason it is good practice to report results in amounts before and after normalisation.

The major problem with the area percent type of quantitation can be seen in a very simple example. If there are only four peaks of equal area, each will be 25%. If for some reason the next sample contains only three such peaks, the areas will be 33% and if there are five such peaks in another sample, the areas would all be 20%. Clearly none of these has any relation to the absolute amounts present. Yet area percent quantitation has value for comparative applications such as process monitoring provided the entire sample is analysed and the integration and peak identification is done in exactly the same way every time. The more peaks there are, the smaller the errors would be, making this type of quantitation the only recourse in the analysis of very complex samples where it becomes impossible to calibrate for all the individual compounds. This is seldom the case in gas analysis where samples are seldom so complex that the use of area percent is warranted. One exception would be the analysis of gas samples containing relatively large amounts of volatile organic compounds as encountered in environmental analyses as well as samples originating in chemical and petrochemical processing plants.

9.3.2.2 External standard

In gas analysis this is the most commonly used method of quantitation since it is usually not practical to use an internal standard. It requires a series of standards containing different concentrations, covering the required ranges, of all the component gases that must be determined [4]. The standards may be in separate cylinders or can be obtained through dynamic dilution or other means as described in Chapter 4. The standards are analysed, preferably three or more times each to provide the different calibration levels for each analyte. A calibration curve, similar to the one in Fig. 9.10, is constructed and the concentrations of the samples are determined by rearranging the linear regression equation, $y = mx + c$ into the form $x = (y - c)/m$ where y is the detector response, c the regression line intercept, m the regression line slope and x the concentration to be determined. Response factors are not required as these are determined each time calibration is done. The response factor is defined as the detector response divided by the concentration which is exactly what the slope of the linear regression equation provides.

A less rigorous and also less accurate form of external standard quantitation is accomplished by comparison of the peak area of a standard with those of the same components in the sample. Response factors need not necessarily be known as it can be safely assumed that the response for the compound in the standard will be the same as the response for the same compound in the sample, barring matrix interferences. Calculation of the concentration is then accomplished by a simple equation of the ratio of the peaks area of the standard versus the area for the same compound in the sample [2].

$$Concentration_{(x)} = \frac{Area_{(x)} \times Concentration_{(standard)}}{Area_{(standard)}}$$

Reasonable results are sometimes obtained when relative response factors are known and not all standard compounds are available. For example, the response of oxygen and argon are very similar on a TCD and elute practically at the same time on a Molecular sieve column at above ambient temperatures. If oxygen must be determined as well as flammable components, nobody would prepare a gas standard that contains oxygen as well as the flammable components. Argon is then often substituted for oxygen in the standard mixture and the argon calibration used to quantify the sum of the oxygen and any argon that may be in the sample. Unidentified compounds in environmental air samples are sometimes quantified using the same calibration curve as methane and these compounds are then typically reported as a concentration 'as methane'.

Identical analysis conditions for both the standards and the samples are imperative. For gas analysis, reproducibility of sample volume, temperature and pressure is particularly important. An often (conveniently) forgotten effect, is the contribution of water vapour that may be present in samples. The amount of water vapour in a gaseous sample may increase with increased temperature and reduced pressure or decrease at lower temperatures and higher pressures. Most calibration gas standards are provided dry and do not contain any moisture. It is clear that after calibrating with a dry gas and then analysing a water saturated sample will result in incorrect results as it affects the partial pressures of the components in the sample loop, apart from the deleterious effect it will have on moisture sensitive columns. If required, the gas standard can be humidified by passing it through a water trap prior to injecting it into the GC. Obviously the water trap must be held at a constant, known, temperature and the gas standard should not contain components that are water soluble. Given that most gases will dissolve to some extent in water, this should be applied with care.

The results are reported in the same units as the standard, usually as volume per volume which with ideality assumed equates to mole per mole, and will not add up to exactly 100%. Provided all components are being quantified the result should still be between 99 and 101%. Many analysts prefer to normalise these results so that they add up to 100%, but as before, incorrect peak identification will cause incorrect quantification of that analyte and it is good practice to report results in amounts before and after normalisation.

A quick check on the validity of the calibration can be done by reprocessing the files acquired for calibration as if they were samples and then to compare the known and measured values. Quality control samples should also be analysed to ensure that the new calibration provides values consistent with those obtained from previous calibrations.

9.3.2.3 Internal standards

The method of preference for chromatographic quantitation uses an internal standard (IS). However, meeting the criteria for an internal standard in gas analysis is difficult in practice. An internal standard that exhibits the same physical and chemical properties as the components of interest should be chosen. Ideally the internal standard peak should be positioned towards the middle of the chromatogram in an area where co-elution with sample components is unlikely [5]. Clearly these criteria cannot be readily met in the majority of gas analyses. Even if a costly noble gas such as xenon was used this would not have the same chemical properties as the gases analysed.

As with external standard calibration, a set of calibration standards is also needed but each standard as well as all samples should contain the same amount of internal standard. Such standards are very difficult to obtain but can be made using dynamic or static blending equipment. Another way is through adding a fixed amount of the internal standard with a gas tight syringe to the calibration gas and gas samples that are contained in glass bulbs through the septum of the side ports of the glass bulbs. These methods are tedious and will not necessarily be any more accurate than external standard calibration. However, should an internal standard be available in a separate cylinder a 10 port valve can easily be plumbed to automatically add a fixed volume of internal gas standard to each gas sample injected using a second sample loop as described in Chapter 6. This is probably the only way in which a standard can be added accurately without the use of specialised blending equipment.

As with external standard calibration, the calibration standards containing the components to be determined at different concentration levels as well as the internal standard at a constant, known concentration are analysed and preferably with three or more replicates at each level. When setting up the calibration curve, the ratio between the peaks areas of the calibrated compound and the internal standard are plotted against the concentration or amount ratio of the calibrated compound and the internal standard. In this way the response of each component can be determined relative to the internal standard response. Most chromatography data systems will calculate relative responses for varying amounts of internal standard and therefore determine the ratio between the response factors of the compound and internal standard. This is valid provided the ratio between the response factors remain constant over the concentration range of both the calibrated compound and internal standard. The concentration of the compound to be determined in a sample is calculated using the same calculations in reverse but luckily most data systems will plot relative response factors versus relative concentration in internal standard quantitation mode and will do the calculations automatically.

$$Concentration_{(x)} = \frac{Area_{(x)} \times Concentration_{(IS)}}{Area_{(IS)} \times Slope_{calibration\ curve}}$$

Alternatively, when the response factors of the analyte and internal standard are known, the sample amount can be calculated using these factors.

$$Concentration_{(x)} = \frac{Area_{(x)} \times Response\ factor_{(x)} \times Concentration_{(IS)}}{Area_{(IS)} \times Response\ factor_{(IS)}}$$

A calibration check similar to what was described for external standard calibration as well as the quality controls must be done following internal standard calibration.

9.3.2.4 Standard addition

The principle of standard addition is most often used in recovery studies where losses during sample preparation are established. It can also be used to estimate the concentration of target compounds, typically in situations where the result does not warrant calibration due to time or resource constraints. Using a sample and a single component standard, a known amount of standard is added to a known amount of the sample and rerun. The original concentration can be calculated from the increment in the peak area as illustrated in Fig. 9.13.

$$Fraction_{(x)} = \frac{Volume\ fraction\ added}{\frac{Area_{(after\ addition)}}{Area_{(before\ addition)}} \times (1 + Volume\ fraction\ added) - 1}$$

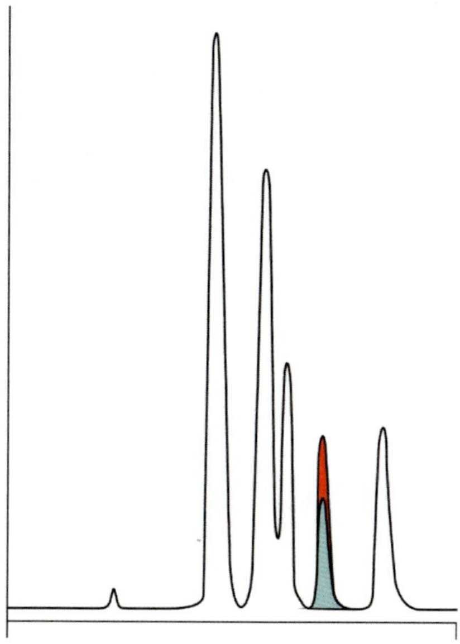

Fig. 9.13 Overlaid chromatograms of the same sample before and after standard addition showing the increment in the target compound amount.

The main problem with this is that the result depends on the accuracy of the addition and the slightest variation will give an erroneous result. Another approach that may be more accurate involves adding different amounts of the standard to the sample and reanalysing it after each addition. The peak areas are then plotted against the added concentration and a regression line drawn through the points. The line is then extrapolated through the y-axis until it intercepts the x-axis or the x-intercept can be calculated from the regression equation for $y = 0$. This will be at a negative value and its absolute value is taken as the original concentration. This approach assumes that the area versus concentration curve will pass through zero and as stated earlier, this is seldom the case.

In gas analysis this is not really practical unless dynamic blending equipment is available. The standard can be added to the sample through the septum in the side port of a glass sample bulb and is useful to confirm the identity of peaks in qualitative analysis. However, standard addition is inherently flawed and should not be used for critical quantitative analyses or to replace proper multi-point calibration using certified and traceable standards.

9.4 Validation of results and measurement uncertainty

Regardless of the quantitation method used, results must make sense from first principles. In the preceding discussions on inlets, columns, detectors and sampling we mentioned many things to look out for that may indicate that the performance of the entire system is not optimal. More 'common sense' evaluations are described in Chapter 12, especially Section 12.4.4, that can ensure that results are sensible and that the numerical values obtained from quantitation are realistic.

The final step before reporting results is to ensure that the method and the results obtained with it are fit for purpose and that implies that the method should be validated to provide evidence to this effect. There are two, sometimes conflicting approaches. On the one hand the analytical scientist strives for perfection and would spend a lot of time, money and effort to ensure that a method is as good as it can be. They are then often frustrated since there never seems enough time to do it right, but somehow there is always time to do it over. On the other hand the customer may be perfectly happy with an uncertain result on time rather than an absolute value too late. The additional time and cost to attain perfection are often not warranted in terms of the method purpose and at some point the method must be declared as good enough. To determine when the method and results obtained with it is good enough; all methods should be validated to provide documented evidence that the method is fit for the purpose. This is especially important in methods used for product certification and to show compliance with legislation.

Regardless of the type of calibration used, confirmation that the calibration will provide valid values must be done. The regression coefficient alone is not enough as it does not provide sufficient 'goodness of fit' evidence and additional statistical evidence is often required. An example of this is shown in Fig. 9.14. The range and detection limit can be obtained from the calibration although the latter is more realistically determined in the same matrix as the sample. Once again this is very difficult to do in gas analysis.

ISO 5725 uses two terms 'trueness' and 'precision' to describe the accuracy of a measurement method. 'Trueness' refers to the closeness of agreement between the arithmetic mean of a large number of test results and the true or accepted reference value. 'Precision' refers to the closeness of agreement between test results' [6]. The precision (repeatability and reproducibility), is not too difficult to determine as a representative sample

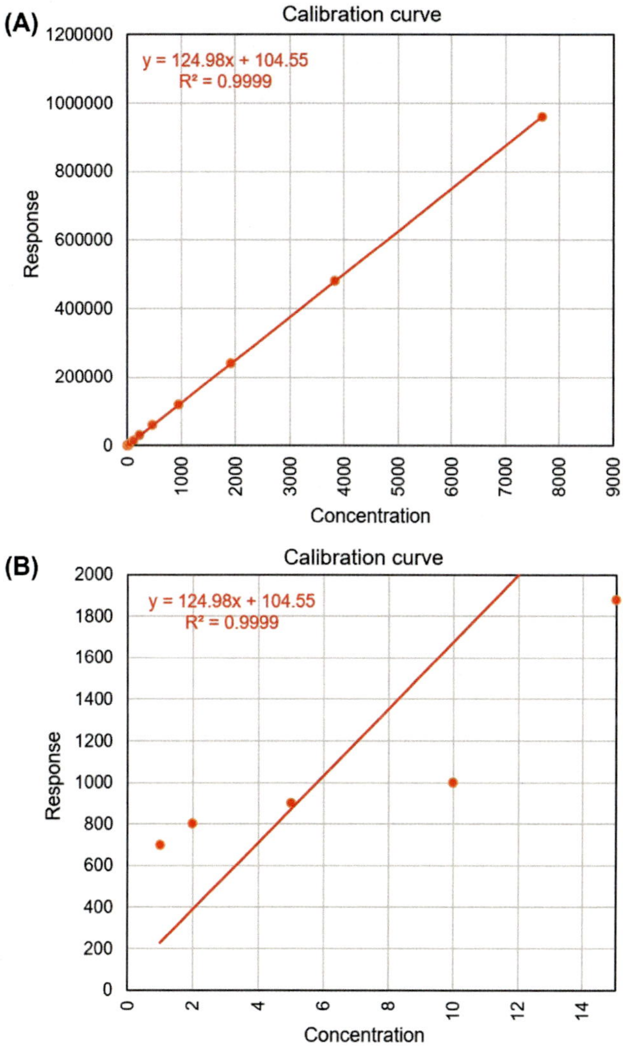

Fig. 9.14 A seemingly near prefect calibration curve (A) which deviates considerably at the lower end when that part is enlarged (B).

or samples can be analysed a number of times over a short or extended time, by the same or different analysts, without too much trouble. It is much more difficult to confirm the trueness of gas analyses mainly because of the difficulty in obtaining suitable certified reference standards.

Finally the measurement uncertainty of all results should be determined since ISO defines a result as consisting of a number, units and uncertainty.

In gas analysis the major sources of uncertainty should be carefully evaluated as it highlights the areas in which method improvements will have the greatest benefit towards lowering the expanded uncertainty.

The major source of uncertainty is usually the gas standards used for calibration as given on the certificates of analysis or calculated for in-house prepared standards or blends. Some commercial gas blending systems will calculate the uncertainty of the blends which makes life somewhat easier. It is also important to determine the uncertainty of the calibration line that is used to predict the value of the analyte in the sample from the calibration. For this a rather scary looking statistical equation is used but once incorporated into laboratory management software or a spreadsheet program, it becomes a very useful tool [7]. For example, the two terms, $1/n$ and $1/m$ refers to the number of calibration levels and number of replicates respectively and it is obvious that adding more levels and replicates will greatly reduce the calibration uncertainty for the predicted analyte concentration.

$$S_{x_0} = \frac{S_{y/x}}{b}\sqrt{\frac{1}{m} + \frac{1}{n} + \frac{\left(y_0 - \overline{y}\right)^2}{b^2\sum_1\left(x_i - \overline{x}\right)^2}}$$

where $S_{y/x}$ is the residual standard deviation, n and m the number of calibration levels and replicates respectively, b the slope of the calibration curve, y_0 the mean of the sample replicate measurements, $\overline{y}$ the mean of the y values of the calibration standards, x_i the value on the x-axis and $\overline{x}$ the mean of the x_i values.

There is an often neglected or ignored uncertainty associated with the amount of sample injected, especially when a fixed size sample loop is used that ensures that the same volume of gas is injected every time. The same volume does not equate to the same amount as the volume of gas in the sample loop is significantly affected by the temperature and pressure as explained in Chapter 1. Even if the sample loop pressure is controlled using pressure regulators and the temperature is held relatively constant in a valve oven, it is still well worth the trouble to calculate the effect of the pressure and temperature fluctuations on the amount injected. This becomes even more important if the sample loop is held at ambient pressure and temperature as ambient conditions can vary considerably during the course of a day.

Another easily missed uncertainty is the effect of gases that are not determined on the amount injected. The most common of these is probably water vapour, but other gases can contribute to this uncertainty. For example, when using helium as carrier and a TCD, low concentrations of hydrogen may be missed and helium at any concentration will not be seen at all. Similarly, air and many other gases present in a hydrocarbon gas will be invisible when analysed by FID only.

Since the trueness or bias may be difficult to determine, it is often ignored but a reasonably good estimate can be obtained by simply reprocessing the calibration points as if they were samples and use the difference between the known and measured values as a trueness value. Precision is usually fairly easy, albeit a bit time consuming, to determine and preference should be given to using reproducibility rather than repeatability precision as it will provide the more conservative uncertainty value.

There are many excellent books, guidelines and articles on method validation and it is not within the scope of this book to discuss it in detail. However, it is important to establish and specify the requirements of any method as what may work for one sample matrix may not apply to a different matrix. Method validation will establish if a method is sufficiently selective to determine the analytes in the specific matrix, accurate enough and sufficiently robust to be applied routinely.

For more on validation and uncertainty estimation we found the Eurachem guides [8,9] very useful but there are many, many more publications on this topic and more are added regularly. A quick Internet search will uncover these.

References

[1] G. Guiochon, C. Guillemin, Chapter 15. Quantitative analysis by gas chromatography measurement of peak area and derivation of sample composition, Journal of Chromatography Library (1988) 629–659. Elsevier.

[2] G. Guiochon, C. Guillemin, Chapter 14. Quantitative analysis by gas chromatography response factors. Determination, accuracy and precision, Journal of Chromatography Library (1988) 587–627. Elsevier.

[3] W. Dietz, Response factors for gas chromatographic analyses, Journal of Gas Chromatography 5 (1967) 68–71.

[4] G. Rhoderick, Analysis of natural gas: the necessity of multiple standards for calibration, Journal of Chromatography A 1017 (2003) 131–139.

[5] J. Dolan, When should an internal standard be used? LCGC 30 (6) (2012) 474–480.

[6] International Organization for Standardization, Accuracy (Trueness and Precision) of Measurement Methods and Results — Part 1: General Principles and Definitions, ISO, Geneva, Switzerland, ISO 5725-1:1994(en).

[7] T. Taylor, How to Estimate Error in Calibrated Instrument Methods - and Why We Have Stopped Doing It!, LCGC chromatographyonline.com, August 2015, 2015

[Online]. Available: http://www.chromatographyonline.com/how-estimate-error-calibrated-instrument-methods-and-why-we-have-stopped-doing-it.
[8] EURACHEM workgroup, The Fitness for Purpose of Analytical Methods. A Laboratory Guide to Method Validation and Related Topics, LGC (Teddington) Ltd, Teddington, UK, 1998.
[9] EURACHEM/CITAC Working Group, Quantifying Uncertainty in Analytical Measurement, second ed., CITAC, 2000. Guide number 4.

CHAPTER 10

Multidimensional gas analysis

Contents

10.1 Introduction

For gas analysis, gas chromatography systems must be capable of meeting several goals many of which are in conflict. Achieving baseline separation on a single column could, for instance, result in excessive analysis times that are unacceptable. Incompatibility between column material or some detectors and certain analytes may require diversion of such analytes away from the column or detectors. To overcome these limitations, it is necessary to use multi-column switching techniques, otherwise known as multidimensional chromatography. Multidimensional GC using two-position valves allows the combination of columns with different selectivities to achieve difficult separations or to prevent some analytes from reaching the detectors.

A practical guide to gas analysis by gas chromatography
ISBN 978-0-12-818888-0
https://doi.org/10.1016/B978-0-12-818888-0.00010-3

There are many different column switching configurations that can be employed depending on the complexity of the analysis undertaken. Analytical systems often have several different configurations within one GC. Here we will limit our discussion to the basic configurations which include, backflush-to-vent, backflush-to-detector, heart cutting, series-bypass and column sequence reversal.

The most common method of switching GC columns is by the use of multi-port slide or rotary valves. These two-position valves are driven by pneumatic, electric or micro-electric actuators, controlled by timed events from the instrument software or another control system. Valves with 4, 6, 8, 10, 12 or even 14 ports are available, suitable for automatic gas or liquid sample injection and all the types of column switching described above [1]. Only the five configurations listed above will be discussed in detail and shown with 4 or 6 or 10 port valves where applicable. We have also adopted a convention that shows the sample loop on certain ports, but clearly all connections could be moved to different port numbers and the functionality of the configuration maintained. In most valve diagrams so far we have changed the flow direction in the sample loop when switching from the load to inject positions. This is purely a convention and does not necessarily affect the performance of the system.

In order to achieve stable and repeatable chromatographic results, the multi-port valves must be temperature controlled. They are therefore normally mounted in a separate valve oven or inlet system, since mounting them in the column oven would restrict the use of high temperature and risk exceeding their maximum operating temperature. However for isothermal analyses at a modest temperature, the valves may be mounted inside the oven. Rotary valves can be selected with very small dead-volumes, making them suitable for packed, wide-bore (0.53 mm i.d.) or narrow bore capillary columns. The valve must be selected with ports matching the internal diameter of the column and all interconnecting tubing must be chosen to prevent any diameter changes. In multi-column applications it is best to keep the column internal diameters the same or very similar and the flow as close to the optimum as possible. Although it is not impossible to use columns with different internal diameters in series, the optimum efficiency will have to be sacrificed on one or both of the columns. In multi-column systems one sometimes has to use the optimum flow for one column and accept less efficiency when another column is brought in line after a valve switch.

An alternative method of column switching uses column pressure-balancing, controlled by solenoid valves located outside the heated zone. This is known as a Deans switch named after its inventor, David R. Deans and falls within the discipline of microfluidics [2]. This is explained in detail in Chapter 5, but is briefly explained here again. It can be used for column back-flushing and heart-cutting, although it is not convenient for the other configurations. In this technique, which is particularly well suited to capillary columns, the pressure at the column mid-point between the first and second columns is maintained by a supplementary pressure controller at just above the pressure supplied by the column inlet carrier gas controller. When the inlet controller is shut off by a solenoid valve and a vent solenoid is simultaneously opened, the mid-point controller maintains the flow through the second column to the detector and also supplies the flow to backflush or heart-cut the first column. If a split injector is installed on the system, increasing the auxiliary pressure at the mid-point can reverse the flow in the first column and flush the remainder of the first column contents to the injector split vent. A typical flow configuration for an analytical system for column backflush, using a Deans pressure balanced system, is illustrated in Fig. 10.1.

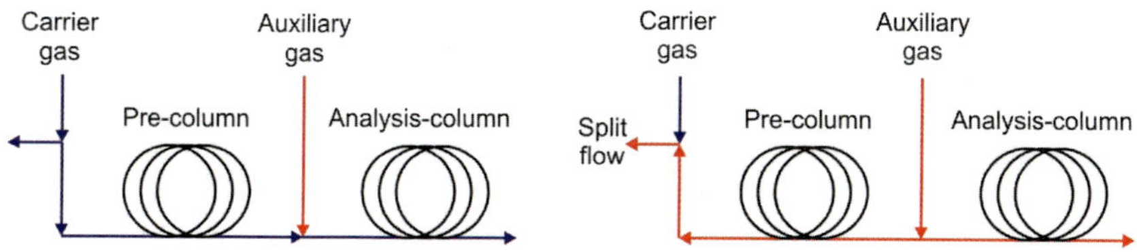

Fig. 10.1 Schematic showing how increased pressure at the mid-point between two columns can be used to reverse the flow in the first column.

While it is emphasised in the chapter on columns that the elution order of gases on a particular column is invariant, the timing and valve configurations described in the next sections allow the chromatographer to select the final elution order of analytes to some extent. This has significant advantages, for example, to allow sharper peaks on earlier eluting components for better detection of trace impurities.

10.2 Backflush-to-vent or detector

This technique is used to selectivity remove late eluting components from the sample being analysed by reversing the carrier gas flow in a pre-column. Examples of such components include compounds which would deactivate the analysis column such as water vapour on a Molecular sieve column. Compounds like carbon dioxide having unacceptably long retention times on a Molecular sieve column can also be eliminated in this way. Contaminants which would gradually accumulate on the column, changing column polarity or creating a noisy baseline or interfering with measured components such as high molecular weight, high boiling matrix compounds which are of no interest for the gas analysis application can be expelled.

10.2.1 Principle of operation

The standby position has the pre-column in the backflush position to ensure that the column is kept clean when not in use. As the sample is injected, the pre-column ('backflush' or 'stripper' column) is connected in series with the analytical column. They remain in this configuration until all the components that need to be measured have eluted onto the analytical column as illustrated in Fig. 10.2. When the valve is actuated, the carrier gas flow in the pre-column is reversed, backflushing the unwanted components to vent. The separation of the components that have eluted onto the analytical column and their elution and detection continue in the normal way. It is clear that a second flow controller is required to implement this configuration. While the suggested configuration requires only one 10-port valve it could be done with two valves. These steps are illustrated in the series of plumbing diagrams in Fig. 10.3.

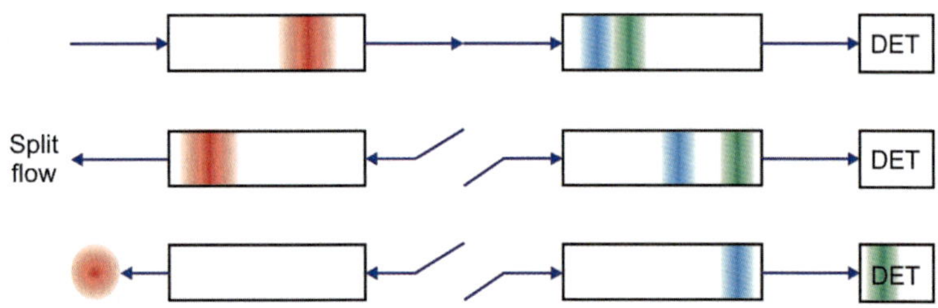

Fig. 10.2 Schematic of the backflush principle. As soon as the analytes to be determined have entered the analytical column but while the matrix or unwanted analytes are still in the pre-column, the flow in the pre-column is reversed to flush the unwanted components out while the analytes to be determined continue through the analytical column to the detector.

Setting up of the system is somewhat empirical. Due to the compressibility of the carrier gas, the elution time of an analyte will be different if the pre-column is in series with a second column or open to vent. Correct backflush time is adjusted by running known compounds and adjusting this time to get all the unwanted components flushed back.

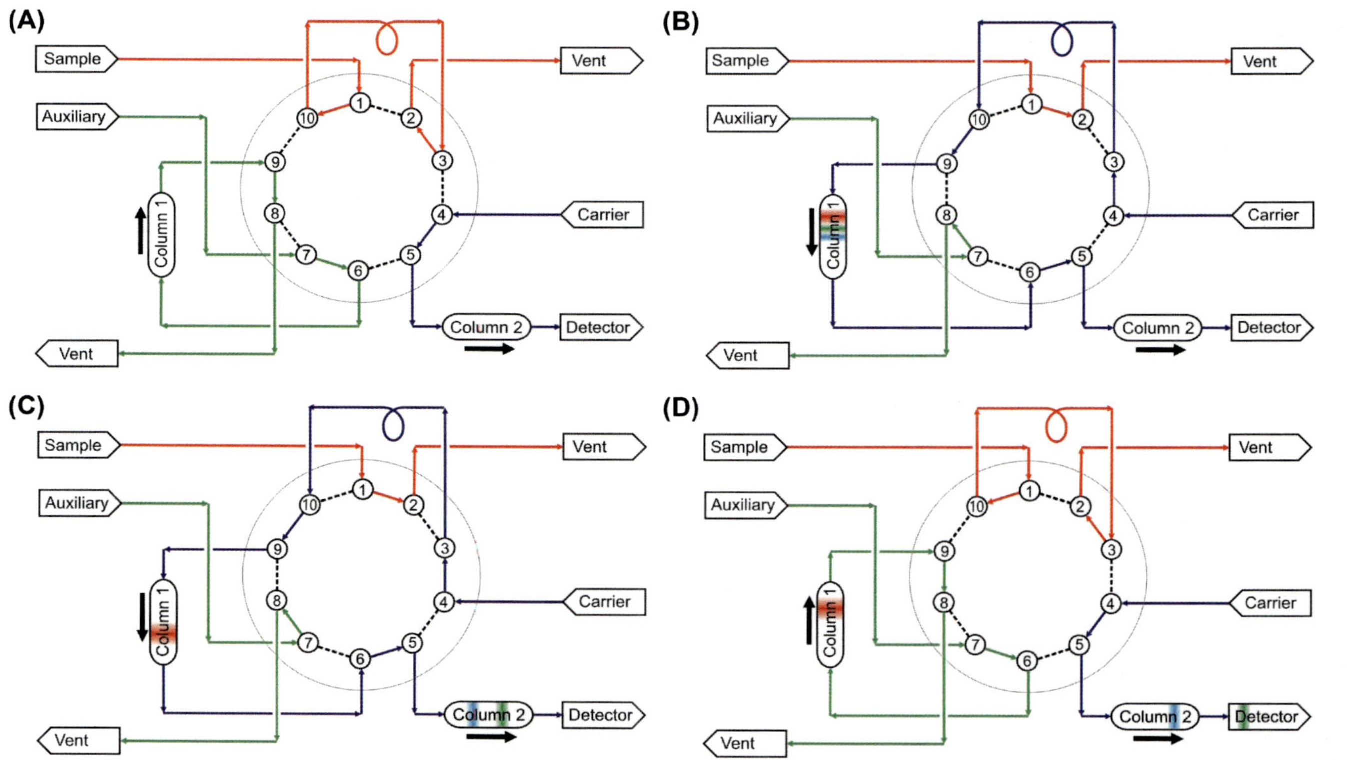

Fig. 10.3 Four steps in the backflush to vent configuration. (A) With the valve in the load position carrier gas flows through the analytical column to the detector and auxiliary gas through the pre-column to vent. (B) Upon injection the carrier gas transfers the sample to the pre-column. (C) When the analytes to be measured have entered the analytical column and while the unwanted compounds are still in the pre-column the valve is switched back to the load position. (D) The analytes to be measured are transferred to the detector by the carrier gas while the auxiliary gas flushes the unwanted components back to the vent.

It is important to look carefully at the plumbing diagram in Fig. 10.3. When the sample is injected the complete volume of auxiliary gas in the pre-column is also injected on to the analysis column. This volume is many times larger than the volume of the sample loop so that any tiny impurities from leaks or gas impurities can cause ghost peaks in the chromatogram. They will often elute just before any true analyte as they have a somewhat shorter retention time and will probably be rather broad. If there was an air leak at port 6 a sharp peak could be expected provided the detector is able to respond to air components. The quality and cleanliness of the auxiliary gas is of prime importance to avoid any ghost peaks.

The same functionality can be obtained using a Deans switching system shown in Fig. 10.1 using differential pressures. This has the advantage of very low dead volumes for narrow-bore capillary column operation.

10.2.2 Typical applications

A Beverage Grade Carbon Dioxide Analyser is configured for backflush-to-vent operation in order to eliminate the carbon dioxide matrix from the chromatogram allowing the trace impurities to be separated on a Molecular sieve column. The measurement of carbon dioxide itself (the bulk gas) is not relevant but several other possible contaminants must be determined [3]. The column switching operation is executed through a two position 10-port valve configured as shown in Fig. 10.3. Another important application is to backflush water vapour to vent to protect columns such as Alumina and Molecular sieve from changing polarity and retention times with usage. The chromatogram obtained from the analysis of high purity carbon dioxide is shown in Fig. 10.4.

10.2.3 Backflush-to-detector

This is identical to Backflush-to-vent except that port 8 is connected to a detector so that the backflushed components elute in reverse direction and are detected as a single peak. Gas chromatography systems configured in this way are used in the analysis of refinery gases, where the hydrocarbons C6 and above are backflushed to the detector and presented as a single peak. An example of a typical analysis is shown in Fig. 10.5.

10.2.4 Backflush time and flow

In a well-designed system the normal flow and the backflush flow would be set the same which implies that in an isothermal run, backflush time

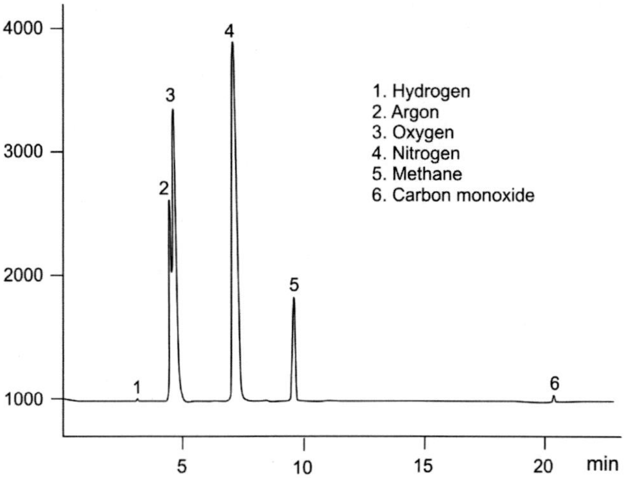

Column: Molecular Sieve 5A, 50 m x 0.53 mm; 50 μm film at 25°C

Fig. 10.4 Chromatogram of permanent gas impurities in beverage grade carbon dioxide obtained on a Molecular sieve column with backflushing of the carbon dioxide matrix.

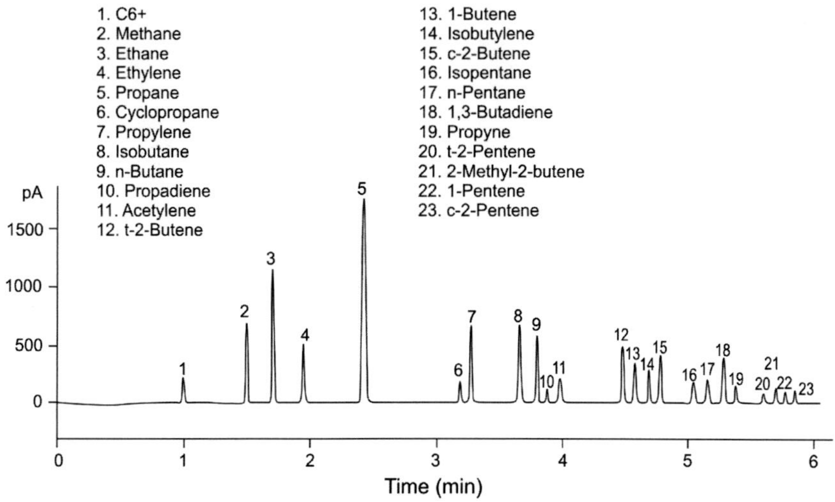

Fig. 10.5 Representative chromatogram from a RGA analyser showing the elution of all hydrocarbons with carbon numbers 6 and above as a single peak at the beginning of the chromatogram.

must exceed forward flow time. Some users try to reduce cycle time by using higher flows for backflushing. While this will help in backflush-to-vent it can cause some separation on the backflush-to-detector application. One would expect that during backflushing all separated compounds will recombine and elute as a composite from the column. In practice though, a certain amount of separation will occur and the lighter components will elute from the backflush slower than heavier components.

The reason for this is that carrier gas is compressible and therefore the velocity through the column is slowest at the inlet end and fastest at the detector. For example, when the average linear velocity is 15 cm/s and the column head pressure is 100 kPa, the carrier gas is compressed to probably half its volume at the inlet and the velocity will be about half of the velocity at the detector end. When the valve is switched to backflush, the inlet and exhaust ends are reversed. Heavy components that have moved only a short way down the column have only seen the region of slow velocity but are backflushed in the high velocity region. An analyte that was at the point of elution has to travel back the full length of the column and starts in the region of low velocity and will take exactly the same time to elute in this direction as it would have eluted in the forward position.

When temperature programming is used, it is not necessary to have a backflush time as long as the forward flow time as the elevated temperature will elute the backflushed components more rapidly. However it is always prudent to ensure that the backflush time is longer than the forward flow time.

When backflushing to the detector there will always be some disturbance due to the flow changes caused by the valve switch. Flow sensitive detectors like TCD and PDHID will need a few seconds to stabilise while the FID would need some baffle to prevent extinguishing the flame. Timed events are used to inhibit integration at the valve switching time.

10.3 Heart cutting

Unlike backflush-to-vent, heart cutting is used when the matrix peak is not the last eluting peak from the pre-column. In this configuration, which is predominantly used for trace analysis, a pre-column is used to selectively remove or cut a matrix peak from the sample being analysed as illustrated in Fig. 10.6. This means that the trace components can be separated by the analytical column without interference from the main component, even when relatively large sample volumes are necessary for reasons of sensitivity [4].

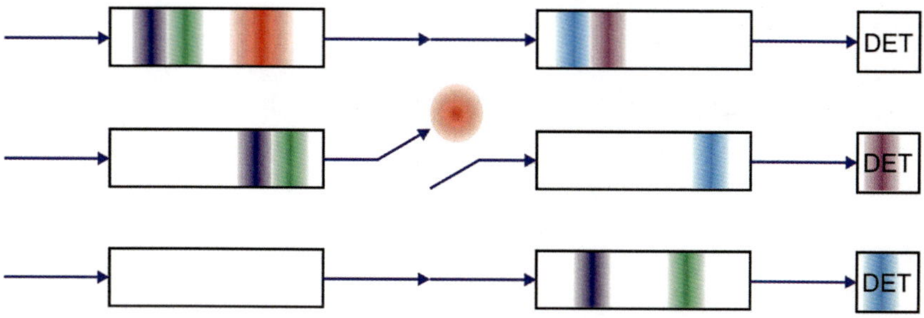

Fig. 10.6 Schematic of the heart cut principle. As soon as the analytes to be determined have entered the analytical column but while the matrix is still in the pre-column, the flow from the pre-column is directed to vent while the analytes to be determined continue through the analytical column to the detector. After venting of the matrix, the columns are switched back in series to allow the remainder of the analytes to enter and separate on the analytical column and pass to the detector.

10.3.1 Principle of operation

The sample is injected with the columns connected in series and, when the analyte to be eliminated has almost gone through the pre-column, also known as the pre-cut or heart-cut column, it is connected to vent. The flow to the analysis column is maintained by an additional flow controller. When the unwanted analyte has been vented or cut, the columns are switched back to the series position. In this way analytes from either side of the matrix peak can be analysed.

This configuration needs two independent valves, one sample injection valve and a vent valve as in Fig. 10.7. Two flow controllers are also required. Since the pre-column is not maintained in series, the primary flow controller has to adjust to the rapid change of column resistance when the valve is switched. Although it would be possible to repeat the heartcut several times in a single run it is not advised as this technique is quite difficult to set up reliably. There are two critical switching times that have to be optimised which makes this a more difficult analysis. Because of the compressibility of the carrier gas, the elution time of any analyte will not be the same through the pre-column when connected to vent as opposed to being in series with the analysis column. This would be easiest to set up if a non-destructive detector was used between the two columns, but this would also allow the matrix on to the analysis column which may destroy that column. Alternatively a needle valve could be set to have the same resistance as the analysis column and then plumbed in place of it. Baseline disturbances are to be expected with this configuration and it is good

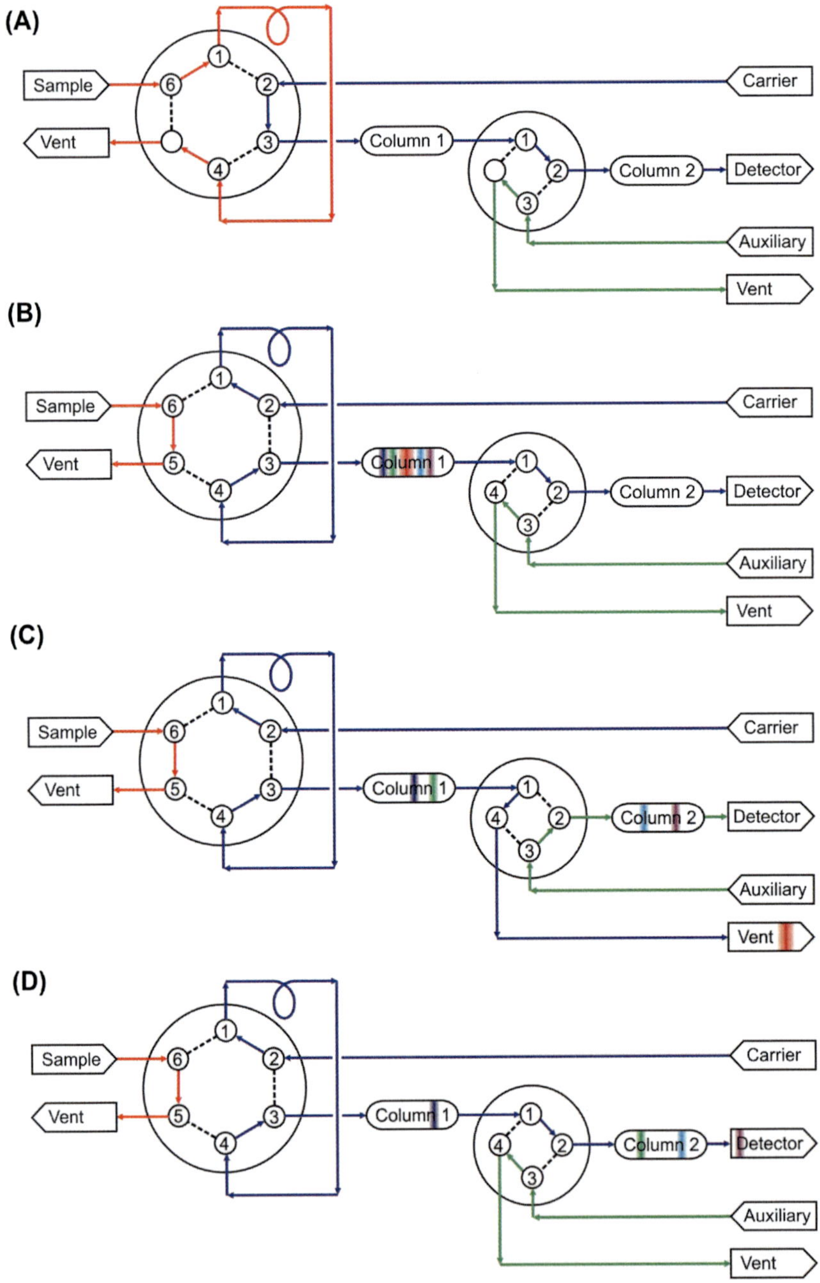

Fig. 10.7 The steps in the heart cut configuration. (A) The load position with carrier gas going through both columns in series and the auxiliary gas to vent. (B) Upon injection the sample is separated from the matrix and later eluting compounds on the pre-column. (C) The four-port divert valve is used to cut the matrix out to vent. (D) After venting the matrix the divert valve is returned to its original position and the remainder of the analytes pass on to the analytical column and detector.

practice to use integration inhibit functions in the data handling software at these switching times.

10.3.2 Typical application

Commercial gas analysers configured in this way are used for the analysis of trace impurities in gases where the bulk gas will elute after some components to be determined but before some other components that also need to be determined, for example the analysis of trace impurities in NF_3. Fig. 10.7 shows a schematic representation of this technique while Fig. 10.8 shows the chromatogram obtained.

10.3.3 Heart cut to column or detector

Heart cutting can also be used to further separate compounds that cannot be separated on one set of columns by passing a group of peaks that are not separated on the pre-column and also co-elute on the analytical column to a third column of different polarity on which the separation can be achieved. This requires replacing the four port valve in Fig. 10.7 with a six port valve as well as the addition of another column and a second detector as shown in Fig. 10.9. The second detector can be different from the first detector so that compounds that are not well detected on the first detector

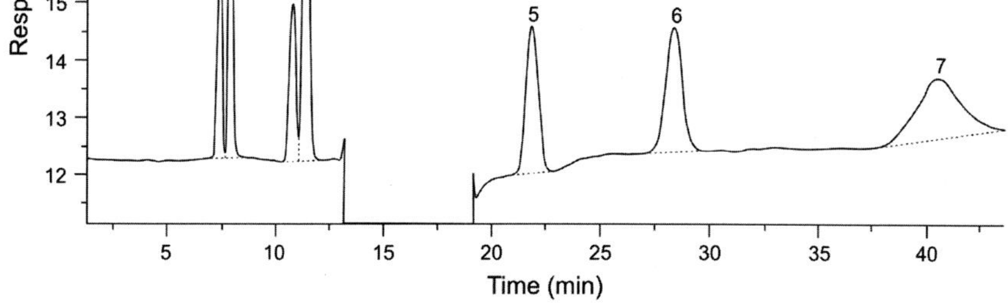

Fig. 10.8 Chromatogram of impurities in NF_3. The change in baseline when the matrix elutes between CF_4 and CO_2 is clearly visible. Integration events are used to ignore the baseline disturbance for integration purposes [5]. *(Reprinted from J.P. de Coning, J.M. Swinley, Optimisation of a gas chromatographic method for trace gaseous impurities in nitrogen trifluoride by column sequence reversal, Journal of Chromatography A 1180 (2008) 151–158.)*

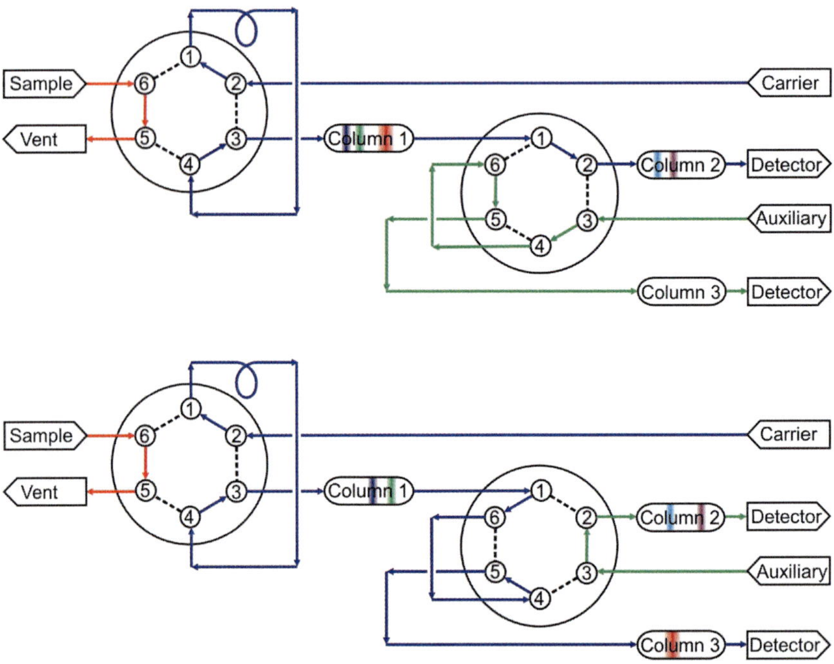

Fig. 10.9 Heart cut to a column or detector using two six port valves, one for sample introduction and the other for column and detector selection. The second valve directs the analytes either to column 2 (top) or to column 3 (bottom).

can be passed to a detector with different selectivity and sensitivity. While this offers many possible solutions for difficult separations, it is also rather complex to select the correct columns and hardware components to construct a robust system.

In Fig. 10.8 the separation between CH_4 and CF_4 is adequate at low concentrations but clearly if one needs to determine small amounts of CF_4 in CH_4 the small CF_4 peak would be hidden in the tail of the CH_4. Using heart cut to a different column where CF_4 elutes before CH_4 could be one solution or alternatively heart cutting to an ECD would detect the CF_4 well while giving almost no response for the CH_4.

This can be done using differential pressure switching instead of a six port valve [6] and is often used for really difficult separations such as volatile organic compounds in heavily contaminated air. The second analytical column can also be mounted in a secondary oven which allows the use of a temperature program on the second analytical column that is independent of the primary column oven.

10.3.4 Column switching with a Deans switch

Agilent described an interesting application using a Deans switch to send parts of a natural gas sample to different columns [13]. The sample is injected into a PLOT-Q pre-column connected in series via a Deans switch to a Molecular sieve column. As soon as the air and methane have entered the Molecular sieve column the rest of the sample is diverted to another PLOT-Q column for the separation of the remainder of the compounds. Conditions were selected so that there the retention times did not overlap on the two columns which allowed combining the effluent from the two analytical columns using a T-piece and sending it to a single TCD detector. This application would normally require two detectors, one for each column. The chromatogram obtained is shown in Fig. 10.10. The observed elution order appears to be very different from those shown in the chapter on columns because the parallel columns effluents are combined at the detector end. In practice this would be difficult to maintain as the elution times may vary with changing concentrations and with extended usage.

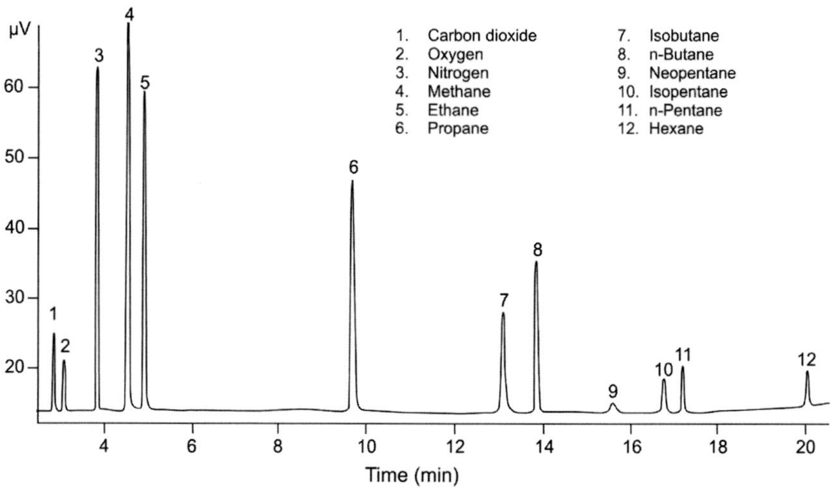

Fig. 10.10 Combined effluent from two different columns with carbon dioxide eluting from the PLOT-Q column, oxygen, nitrogen and methane from the Molecular sieve column and the later eluting compounds all from the PLOT-Q column. ©Agilent Technologies, Inc. 2013. Reproduced with Permission, Courtesy of Agilent Technologies, Inc.

10.4 Series bypass

This configuration, Fig. 10.11, allows a column to be switched in (series) and out of line (bypass), to selectively prevent analytes from entering a column that is not compatible with that component. This technique cannot be used with differential pressure switching.

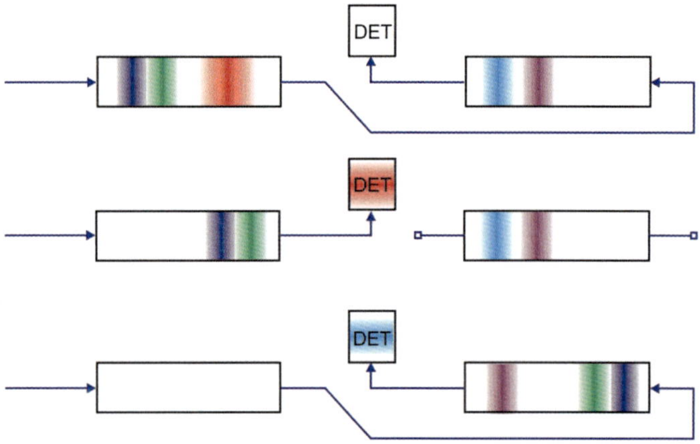

Fig. 10.11 Schematic of the series bypass principle showing how some compounds are stored in the second column while incompatible compounds are passed directly to the detector.

One valve is used for sample injection and a second valve is used for bypass operation as shown in Fig. 10.12. While this can be done with a 4-port valve it is recommended to use a 6-port valve with a balancing needle valve restrictor. This needle valve should be mounted in a heated environment so that there is no chance of sample condensation on a cold surface. The needle valve is adjusted to have the same resistance as the analytical column so that the flow controller sees exactly the same effective resistance in the series and bypass positions. Only one flow controller is required as the analytes are 'parked' in the second column without flow for a while during the bypass stage.

Typical applications that require this technique include the analysis of permanent gases which are only separated by a Molecular sieve column while some of the components which would deactivate the column (such as water or carbon dioxide) are bypassed. Another application would be the analysis of low-boiling compounds in a mixture containing higher-boiling

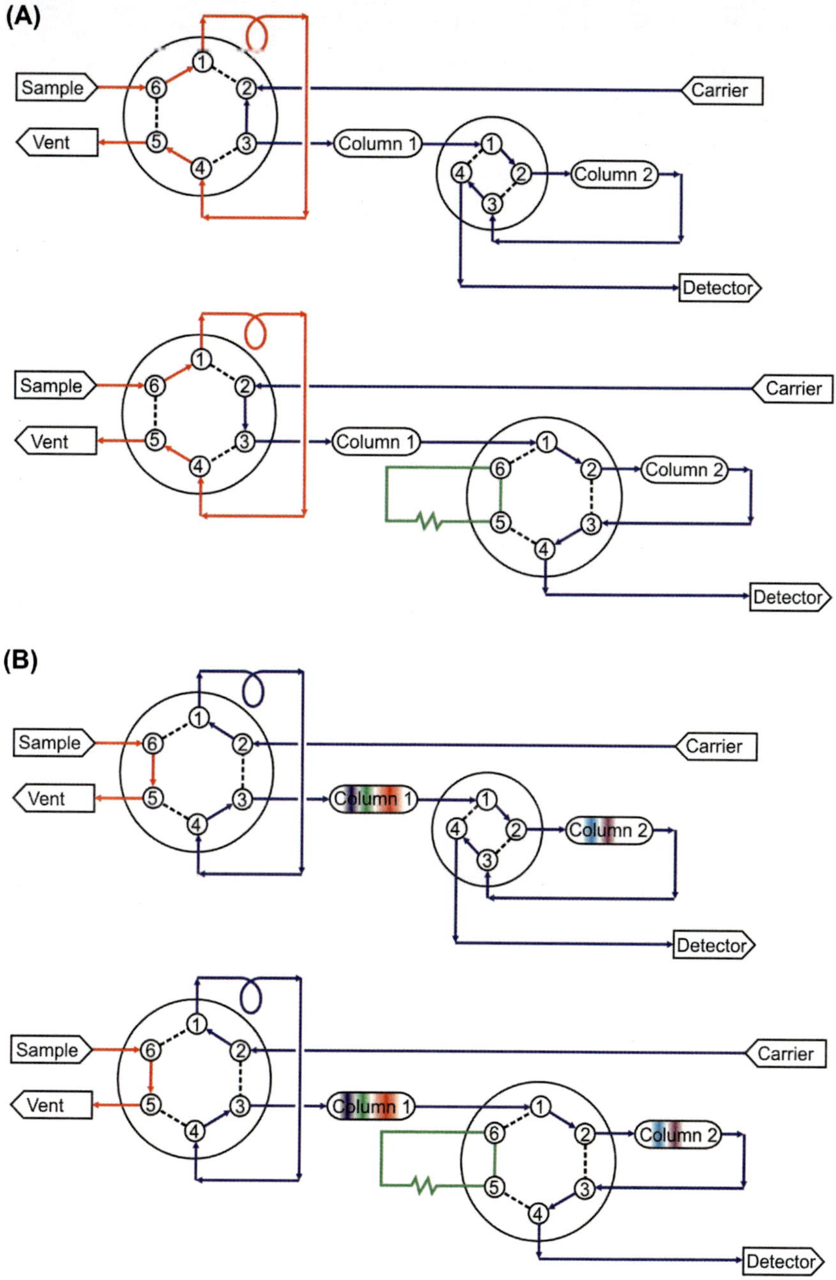

Fig. 10.12 The steps in the series bypass configuration. In each segment the configuration using a four port valve is shown at the top with the configuration using a six port valve below. In the load position (A) carrier gas goes through both columns. After injection and just before the compound that must bypass the second column elutes from the first column (B), the valve is switched passing it to the detector while the compounds that have already passed into the second column are stored (C) with no flow through the analytical column. After the compound that bypassed the second column eluted, the valve is returned to the original position and the remainder of the compounds already separated on the second column (D) are swept to the detector.

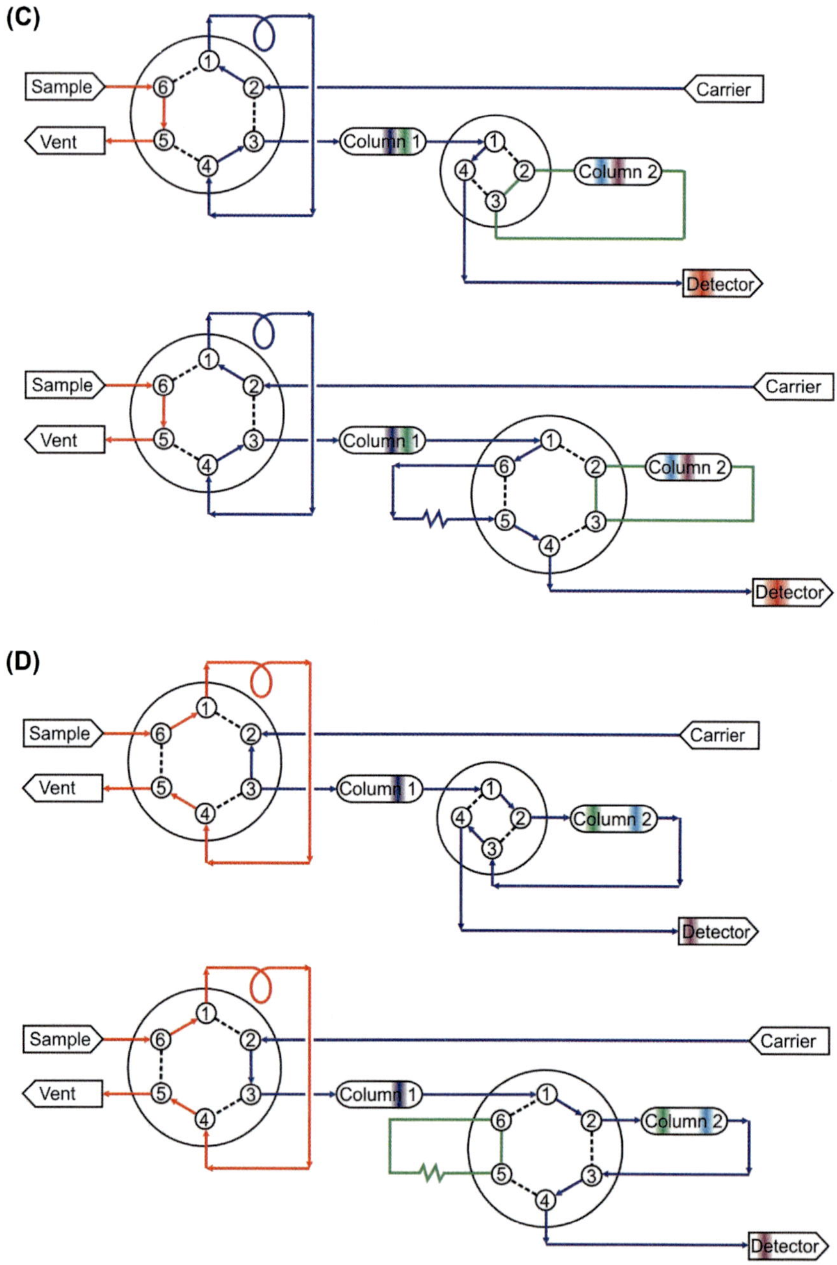

Fig. 10.12 Cont'd

matrix components. Only the first group pass through the bypass column for separation, while the higher-boiling components are bypassed.

10.4.1 Principle of operation

The sample is injected with the two columns in series, allowing some of the analytes to elute into the second column. Just before the analyte that is not compatible with the second column elutes from the first column, the bypass valve is actuated to store or park certain components in the second column under stopped flow conditions. All analytes eluting from the first column go directly to the detector through the balancing needle valve. When the valve is actuated again the parked analytes will continue to separate and to elute from the second column. The whole bypassing sequence can be repeated if necessary to allow further groups of components to bypass the second column.

This system is one of the easiest configurations to optimise. The first setting is the adjustment of the needle valve. Monitor the column head pressure in the series position and set the system to bypass. Adjust the needle valve until the column head pressure is the same as in the series position. Inject the sample with the system set in the bypass mode. The elution times will give direct values for valve switching as the delay volume through the needle valve should be minimal.

10.4.2 Typical application

The most common use of this technique is for a separation that requires baseline separation of oxygen and nitrogen on a Molecular sieve column but with carbon dioxide also present in the sample. Transformer oil gas has been analysed by this technique for many years. A porous polymer column is used ahead of a Molecular sieve column. The peaks of H_2, O_2 and N_2 elute through the series columns but not CO and CH_4 before the CO_2 peak is due to elute from the porous polymer column. The bypass valve is switched trapping CO and CH_4 in the Molecular sieve until the CO_2 and the C_2 hydrocarbon gases are eluted directly to the detector. Thereafter the columns are again switched in series so that the CO and CH_4 are the last peaks eluted. The bypass position is again used to elute the C3's. This is how the ASTM transformer gas analyser is plumbed and Fig. 10.13 shows the chromatogram from this analysis. An alternative approach to this analysis is shown in Chapter 11.

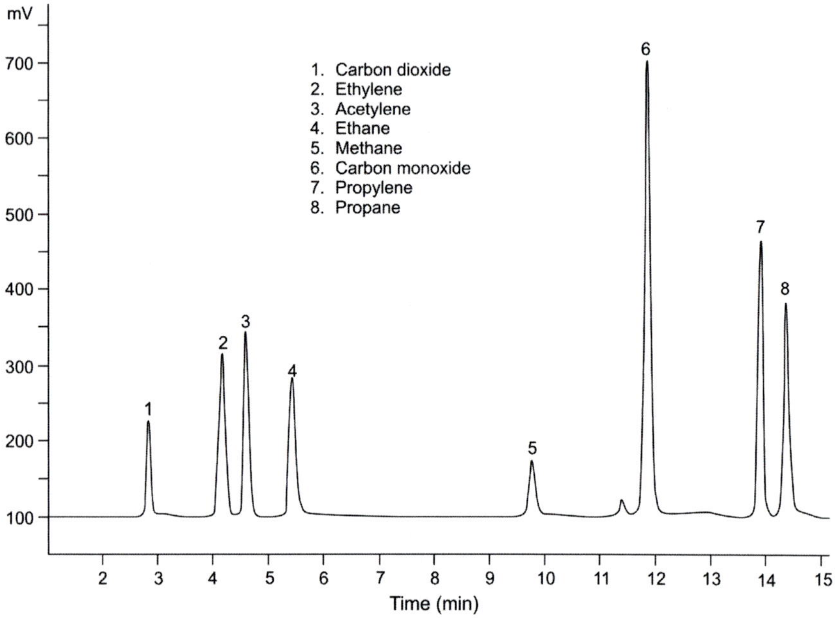

Fig. 10.13 Example of a representative chromatogram of the methaniser FID side of a transformer gas analyser showing the elution of carbon dioxide and the C_2 hydrocarbons before methane and carbon monoxide that would normally elute first.

10.5 Column sequence reversal

As an alternative to bypassing, or to increase the effective length of column for a group of selected components, the sequence of two different columns can be reversed. In this technique, certain peaks pass only through the first column while others also pass through the second column and some may go through the first column twice as shown in the schematic in Fig. 10.14.

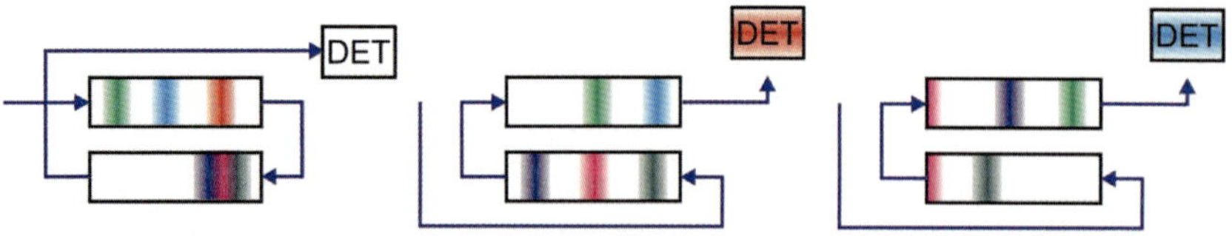

Fig. 10.14 Schematic of the column sequence reversal principle showing how the sequence in which compounds encounter the first or second columns can be reversed with a switching valve. This allows peaks that normally elute later to be eluted earlier than the compounds it would normally follow. There is no reversal of flow direction in either column.

Since the two columns are switched around, only one flow controller is required and the column head pressure will be the same in both positions of the valve in a leak-tight system. This application cannot be done with differential pressure switching. One important use of this technique is to artificially change the elution order of some compounds to improve peak shape and thereby detection limits. Another use of this configuration is to convert a heart cutting application into a backflush application which is much easier to set up because it has only one switching time to optimise. This can also be used to change elution order where a trace compound would get lost in the tail of a major concentration.

10.5.1 Principle of operation

The sample is injected with both columns in series, allowing the desired group of peaks to elute through the first column for example a porous polymer column and into the second column like a Molecular sieve. The valve is then switched so that the second column is now ahead of the first column, allowing late eluting analytes from the first column to elute directly to the detector. This effectively bypasses the second column for these components. The analytes already moving through the second column continue in the same direction and often pass through the first column for a second time before reaching the detector. This configuration is illustrated in Fig. 10.15 and a feature of it is that there is no change in flow direction through either column.

The setting of valve switching times is similar to backflush-to-vent systems in that there is no direct measurement of retention times. However, close approximations can be achieved by injecting a sample and a few seconds later switching the valve. This will give retention data for the first column which will not be very different for the two columns in series.

In the example shown in Fig. 10.16, conditions have been carefully chosen to change a heart-cut analysis into a potential backflush-to-vent analysis as the matrix peak is now the last component to elute. In this way a much more robust and reliable system, that has only one valve switching time to optimise, was obtained.

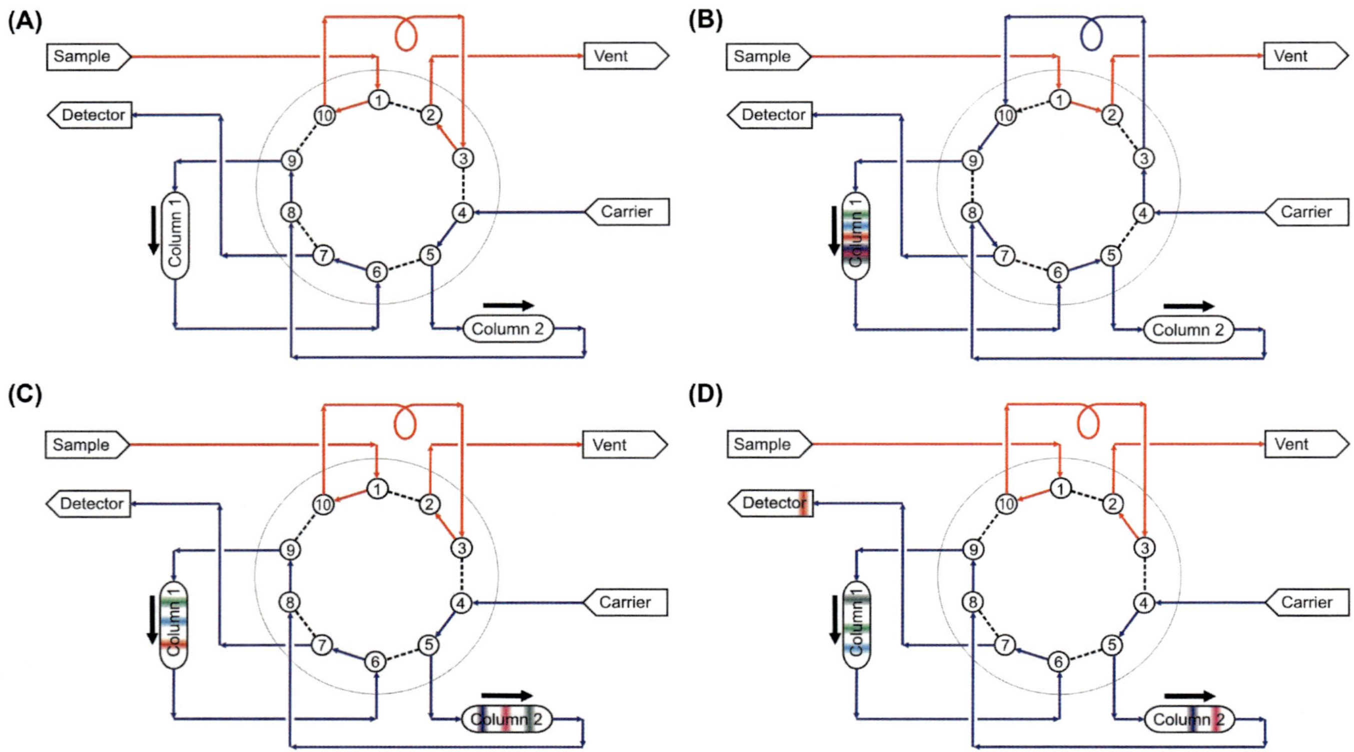

Fig. 10.15 On injection all compound are transferred to the first column (A) and after the fast moving compounds have entered the second column (B) the valve is switched to allow the slower moving compounds to elute first (C) after which the fast moving compounds elute having to traverse the first column for a second time (D).

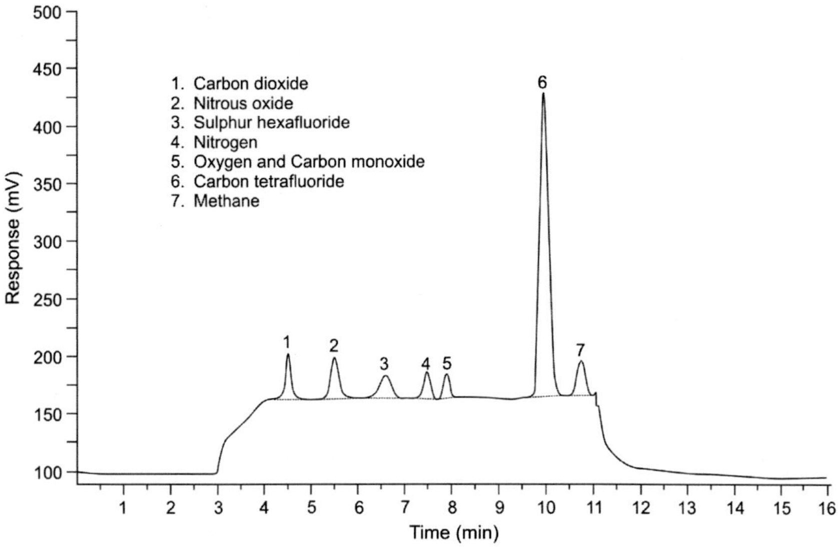

Fig. 10.16 In an improved configuration using column sequence reversal the analysis shown in Fig. 10.8 has much better peak shapes for CO_2, N_2O and SF_6 as these compounds pass through the first column only once while not compromising compounds that pass through the first column twice. The total run time has also been reduced by a factor of 3 [5]. *(Reprinted from J.P. de Coning, J.M. Swinley, Optimisation of a gas chromatographic method for trace gaseous impurities in nitrogen trifluoride by column sequence reversal, Journal of Chromatography A 1180 (2008) 151–158.)*

10.5.2 Typical application

Apart from the example in Fig. 10.16 sequence reversal is also utilised in natural gas analysis. The natural gas sample is injected onto the Shincarbon column in series with a Molecular sieve column. The elution order of CO and CH_4 is opposite on the two columns but, since the Shincarbon separation dominates, CO elutes before CH_4 and the sequence is reversed so that CO_2 elutes directly to the detector from the Shincarbon column. Note that this application elutes the sample through each column only once. Such an analysis requires that the hydrogen channel will use a 10-port valve with backflush-to-vent and argon or nitrogen as carrier gas on a dedicated TCD while a 10-port valve will be used on the other channel having helium carrier gas with a second TCD. An example is given in Fig. 10.17 in which a signal switch is done after the H_2 peak.

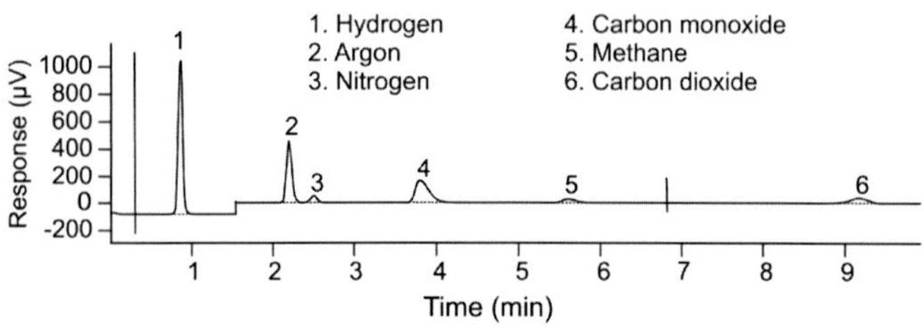

Fig. 10.17 Chromatogram showing the column sequence reversal using ShinCarbon and Molecular sieve columns. The signal for hydrogen is switched at 1.5 min from the second TCD to be displayed on the same axis as the compounds detected on the first TCD.

10.6 Practical guidelines

When one looks at very complex chromatography systems, sometimes using as many as nine valves, it is intimidating to see the plumbing diagrams. However, on closer inspection it is possible to isolate portions of these and note that many of the above configurations are combined to make up these systems.

The retention time of a component is influenced by the column length, the column packing or stationary phase, the mesh size and density of packing in packed columns, the film thickness in capillary columns, the carrier gas, the operating temperature and the density of the sample matrix gas. Clearly when using more than one column several of the above will be interactive. For example, on a Hayesep Q column nitrogen elutes slightly before oxygen as well as CO before CH_4 but on Molecular sieve the order is reversed. When using column sequence reversal all these go once through the Molecular sieve once but often twice through the Hayesep Q. Clearly the dominance of one column over the other will affect the elution order and, if not selected correctly, co-elution will result. Specifically for column sequence reversal, it is a matter of experience to select all of the above parameters correctly to ensure a robust working system. It is a lot easier to balance these systems if each column is in a separately heated oven.

When using multidimensional systems it is imperative to allow the analysis cycle to run to completion otherwise valves will not be reset to initial positions. Certain compounds could well still be trapped in some part of the system and will appear as broad ghost peaks in the next analysis. Even after a power outage it is advisable to start a run with a blank sample and let the system run a full cycle.

Swept volumes can be used to delay sample injection for up to 5 min for packed columns without noticeable peak broadening. This is done by using up to 10 m of $^1/_8''$ tubing as a delay column in series with and ahead of the analytical column. For PLOT columns delay columns up to 60 m, 0.53 mm deactivated tubing can be used.

When a multidimensional system is designed it is important to make sure that a power interruption will not be catastrophic and introduce samples that may destroy columns. For this reason we have shown all our sampling valves the 'load' position when not actuated, that is when the power is shut down. Likewise a backflush-to-vent system should be plumbed to be in the backflush position when the power or air supply fails. A heart-cut system should vent all sample eluting from the pre-column in the event of power loss. Column sequence reversal systems would typically switch back to 'load' while series bypass systems would default to the bypass position. With all configurations it is important to run a full cycle following a power failure to ensure that all valves are in their correct positions. A calibration check sample should also be run to ensure that everything has returned to normal.

Multidimensional gas chromatography has to be used in a wide variety of applications to improve the chromatographic performance by reducing analysis times coupled with better resolution and therefore improving quantitation. For reliable results it is essential that full automation of the chromatographic system be applied so that separation of complex 'real-world' samples can be analysed in a single timed run. A single run is better than trying to combine results from injections on two or more different analytical systems that will require much more sample and increase the uncertainty of the analysis.

Multi-column systems must be designed for robust operation which means that critical switching times must be avoided. The longer the baseline signal is between peaks the easier it is to set a reliable switching time. For example, if there is only a 10 s baseline and the system is set to switch in the middle of this you have a window of ±5 s before there will be doubtful results. Differences in sample to sample concentration as well as slight environmental changes could easily make this retention time change enough so that different switching times may be necessary for day and night operation. Also calibration will need to be checked very regularly and might require recalibration twice a day for the two methods.

With a 60 s baseline it is clear that a 5 s time change will not affect the quality of the results and calibration will be stable over a much longer period. We have seen systems that, although daily calibration checks are done, the actual calibration is stable for at least a year. The goal of all systems is to be reproducible. It does not make sense to have an analytical system that can only be used by one individual and then only on days when the sun rises at a specific time. Repeatability is defined as the precision obtained by one operator on one analytical system over a short time span whereas reproducibility is defined as multiple operators using one or more systems over an extended time. Obviously both are essential for obtaining reliable results.

The advantages of retaining sample integrity by using at-line and on-line analysis means that multidimensional techniques are gaining ever-increasing importance in gas analysis by gas chromatography. Finally the only practical solution is to design a tailor made solution for a specific set of analytical needs. An example of such a system is given in Fig. 11.6 and the resulting chromatogram in Fig. 11.7.

Do not forget that the sample determines the correct analytical setup.

10.7 The Micro-GC

A major shortcoming of gas chromatography has always been the length of time necessary for separation and much development in the fields of fast and ultra-fast GC has taken place. The first and simplest change was to switch to hydrogen carrier if the detector system allowed. Hydrogen carrier gas is approximately three times quicker than nitrogen. The next logical step would be to trim column length to the point where just sufficient separation takes place. However this is now heading in the direction of dedicated analysers such as Process GC's and not a versatile gas analysis system.

Many diverse developments have been made in the quest to speed up analyses. We know from the theory that the optimum carrier gas velocity should not be exceeded too much otherwise separating efficiency will be lost. In other words a short column operated just above optimum carrier gas velocity will give much better results than a long column run at double the optimum carrier gas velocity. Although the solution seems to be to cut the column length, the next analysis might require the extra length for a different separation. Using smaller and smaller diameter columns and fast temperature programming rates has reduced run times considerably but these columns have vastly reduced sample capacity. Many companies have developed various techniques to try to reduce thermal mass and increase sample throughput including direct resistive heating of columns to give very high temperature ramping rates. These developments have had considerable success in reducing run times by a factor of 10–50 but these successes are predominantly in the analysis of higher molecular weight and therefore higher boiling point samples. As we have said before, the analysis required is dictated by the sample and in gas analysis column capacity plays an important role hence the use of packed and micropacked columns.

Furthermore, from the discussion in Section 7.6 it is also clear that small diameter columns with thin films will not give a phase ratio that would be applicable for low boiling samples such as gases.

For gas analysis reduced cycle times using micro-machining technology has been much more successful. Like many Process GC's the initial Micro-GC's were only available for isothermal operation. Modern systems can be temperature programmed. This technology has seen miniaturisation of all components and this integrated design is optimised for rapid results. The components are etched into wafers so that very fine control of operating parameters can be achieved. Where the sample size is typically in the microliter or even milliliter range on a laboratory system, inlet volumes in the nanoliter range are used with Micro-GC systems. The detector volumes are therefore also in the nanoliter range. Since the Thermal Conductivity Detector is concentration dependent, the sensitivity is not compromised and low or sometimes sub ppm levels can be quantified. Since a large number of gas analyses can be done using a TCD the Micro-GC soon established itself as a rapid system for monitoring gas mixtures such as natural gas. The introduction of the Differential Mobility Detector (DMD) further increased the market for Micro-GC systems. For really trace analysis and for a full choice of selective detectors a laboratory GC has many advantages over the Micro-GC. Furthermore, when using micro-fluidics (Deans switches) as discussed in Section 5.14 it is not possible to use Micro-GC's for all multidimensional configurations but backflushing and stream switching can be done very well.

Generally Micro-GC's are purchased for a specific application as they are much less flexible for general purpose gas analysis. To change a column in a micro-GC requires replacement of the whole analytical module which includes injector and detector. Also micro-GC's are typically used to sample a stream having a low pressure continuous sample flow to waste. For example a four channel micro-GC can be configured as a Refinery Gas Analyser provided the concentrations of the components do not change significantly and that the stream pressure can be adapted [7]. They can also be used for various other applications such as the analysis of biogas [8], natural gas [9] and the composition of fuel cell gases [10]. We have not discussed process GC's as the subject is covered in existing books [11,12].

10.8 Comprehensive two dimensional GC × GC

In comprehensive two dimensional GC × GC two columns of different polarities, lengths, internal diameters and film thickness are typically used. A relatively long column which can be either polar or non-polar is connected to a short narrow bore column using a modulator. The eluate from the first column is stored for a brief time in the modulator by cooling (cryogenic modulation) or trapping in a loop (flow modulation) before being released into the second column that has a different polarity. The second column is often housed in a separate oven that may be situated inside the GC oven. Cooling of the cryogenic modulator is done by liquid nitrogen or carbon dioxide and requires rapid cooling and heating cycles to modulate the first column eluate. The advantage of cryogenic modulation is that it focusses the eluate from the first column before injecting it into the second column as a narrow band. With flow modulation the trapping is done in a length of tubing that is filled and emptied using microfluidics as described in the sections of Dean's switches but refocussing of the first column eluate is not possible. Repeated trapping of the first column eluate and injection onto the second column results by using specialised software in a two dimensional chromatogram with the retention time of the first column typically on the x-axis and the retention time in the second column on the y-axis. In terms of the explanation of heart-cutting earlier in this chapter, GC × GC can be seen as continuous heart-cutting of peaks eluting from the first column.

By connecting two columns of different internal diameters together, the optimum column linear velocity will be a compromise that will be a bit too slow for the first column and too fast for the second column. Despite this, good separation is still obtained which is a testament to the quality of modern capillary columns. With the multidimensional gas analyses configurations described earlier in this chapter optimum linear velocity is also often sacrificed to obtain results in a reasonable time.

For gas analysis the cryogenic modulation is not really useful because their very low boiling points make it very difficult to obtain satisfactory modulation of the first column eluate. Even with liquid nitrogen it is only possible to get efficient focusing of C4 and higher compounds which fall outside our definition of gases. Flow modulation could overcome this problem but for most of the gases discussed in this book GC × GC hardly seems worth the additional cost and increased complexity of the instrument and data processing given that there are easier, more robust and less costly alternatives as described in this chapter. Most permanent gas samples are not complex enough to justify the use of GC × GC although future developments may render this statement false.

However, for the analysis of volatile organics in gas, headspace and adsorbent samples that are much more complex than the permanent gases, GC × GC becomes the method of choice because of its ability to provide separation of the multitude of compounds that may be contained in these samples. GC × GC can be used with any of the fast detectors such as the FID, SCD and TOF-MS and is rapidly becoming an indispensable tool in environmental, petrochemical, biomedical and many other applications.

References

[1] Valco Instruments Company Inc, Valco VICI Cheminert Catalog 65 INT, VICI AG International, Switzerland, 2012.

[2] D. Deans, A new technique for heart cutting in gas chromatography, Chromatographia 1 (1) (1968) 18–22.

[3] Wasson-ECE Instrumentation, Beverage Grade Carbon Dioxide, Broduct Brochure (not dated).

[4] Y. Weijun, Trace analysis of impurities in bulk gases by gas chromatography-pulsed discharge helium ionization detection with "heart-cutting" technique, Journal of Chromatography A 1167 (2) (2007) 225–230.

[5] J. de Coning, J. Swinley, Optimisation of a gas chromatographic method for trace gaseous impurities in nitrogen trifluoride by column sequence reversal, Journal of Chromatography A 1180 (2008) 151–158.

[6] J. Luong, R. Gras, H. Cortes, R. Shellie, Multidimensional gas chromatography for the characterization of permanent gases and light hydrocarbons in catalytic cracking process, Journal of Chromatography A 1271 (2013) 185–191.

[7] C. Duvekot, Agilent Application Note: Fast Refinery Gas Analysis Using the Agilent 490 Micro GC Quad, Agilent Technologies, USA, 2012.

[8] R. van Loon, Agilent Application Note: Analysis of Biogas Using the Agilent 490 Micro GC Biogas Analyzer, Agilent Technologies, USA, 2011.

[9] R. van Loon, Agilent Application Note: Fast Analysis of Natural Gas Using the Agilent 490 Micro GC Natural Gas Analyzer, Agilent Technologies, USA, 2012.

[10] S. Amarasinghe, K. Ahmed, Agilent Application Note: Fast Gas Composition Analysis for Fuel Cell Development and Testing Using an Agilent Micro GC, Agilent Technologies, USA, 2014.

[11] J. Love, Process Automation Handbook: A Guide to Theory and Practice, Springer Science & Business Media, 2007.

[12] B. Liptak, Analytical Instrumentation, Routledge, 2018.

[13] K. Beard, K.S. Coleman, Agilent application note: S. Single Detector Deans Switch Natural Gas Analysis with Resolution of Oxygen and Nitrogen, Agilent Technologies, USA, 2013. https://www.agilent.com/cs/library/applications/5991-2063EN.pdf.

CHAPTER 11

Practical guidelines for selected gas analysis applications

Contents

A practical guide to gas analysis by gas chromatography
ISBN 978-0-12-818888-0
https://doi.org/10.1016/B978-0-12-818888-0.00011-5

It is simply not possible to provide a detailed discussion of all the possible applications of gas analysis by gas chromatography. There are many applications and creative chromatographers are adding novel applications all the time, helped by developments in improved column technology and detector sensitivity and selectivity. Given the rapid development and constant improvements in the field of gas chromatography, some of the information provided here may be somewhat dated although the general principles will always apply. This section therefore tries to summarise some of the applications in different analytical fields and hopefully will offer guidance in setting up specific analytical methods or assist in laterally applying the principles in the development of related work. A number of examples where these principles are applied to solve specific problems are included.

11.1 Practical examples

When looking for a solution for any given analytical setup, the old adage of KISS (keep it suitably simple) must be applied. The simpler the solution, the more robust and reliable the system will be. We strongly advocate that any system be optimised for the analysis required and that a so-called universal system which, by definition, would be very complicated and therefore be used under compromised conditions is not ideal. As a general observation, many analyses could be adequately done in less time if we were willing to shorten the columns.

11.1.1 Example 1

You are required to separate H_2, O_2, N_2, CO and CH_4 at percentage concentrations but O_2 and N_2 are not in a 1:4 ratio of air.

Referring to Chapter 7 you would choose a Molecular sieve column because O_2 and N_2 must be fully separated and all the components will elute within a reasonable time. You would probably start at a temperature of 50–70 °C depending on the type of column used. As previously discussed this could be a packed, micropacked or PLOT column using any

Molecular sieve but 13X would shorten the analysis time. Referring to Chapter 8 you would choose a TCD with argon or nitrogen carrier at a velocity of 12–15 cm/s for H_2 quantitation and although some signal will be obtained when the other gases elute a second TCD with helium or hydrogen carrier gas at approximately 30 cm/s will be required for quantitation of the other gases. A PDHID would be too sensitive unless an impractically small sample size is used. These concentrations are also too high for a methaniser. From Chapter 6 you could choose a 10-port valve for injection and backflushing on the H_2 channel and a 6-port gas sample valve for the other channel. Alternatively one 10-port sample valve with two sample loops could be used to simultaneously inject onto the two different columns in parallel as shown in Fig. 6.19.

In a case such as this, calibration will most probably be done using a high pressure cylinder of blended components of these five gases. A limited range of multilevel calibration points could be done using different sample pressures or by some dynamic dilution. Having multiple cylinders of different concentrations of these gases is rather expensive. Also bear in mind that the gas suppliers will only blend concentrations that are calculated to be safe. Refer to Section 4.3.2.2 for calibration using different pressures with a suitable gas inlet as shown in Fig. 6.14.

Since this application will be done using external standard calibration, as described in Section 9.3.2.2, the results will be in percent (v/v) and while this will add up to very close to 100% for the standards, in practice, it will always be less than 100% for samples. This is because not all components in the sample will be quantified as they will not elute from the system or will not be detected. For example from the above analysis, there will not be peaks for water vapour, C2 and higher hydrocarbons or CO_2 if present. The amount that the total calculated concentrations differ from 100% is diagnostic of how complete the analysis of the gas really is. For this reason, results are usually normalised to 100% and then reported. The discussion of calibration and data handling will not be repeated for each of the following examples.

11.1.2 Example 2

You are required to separate H_2, O_2, N_2, CO, CH_4 and CO_2 at percentage levels which again implies that the O_2 and N_2 are not air.

This example is the same as Example 1 except that CO_2 has been added. The CO_2 will not elute from Molecular sieve within a reasonable time.

Any carbon Molecular sieve will not adequately separate these components except at sub-ambient temperatures. We must conclude that no single column will work successfully and therefore a multi-column system must be used as described in Chapter 10 on multidimensional applications. Once you have decided on a configuration, for example, series bypass or column sequence reversal you will need to refer to Chapter 5 for the appropriate valves to use. Here we would want to use a TCD with argon or nitrogen carrier at a velocity of 12–15 cm/s for H_2 quantitation and a TCD with helium or hydrogen carrier gas at 20–25 cm/s for quantitation of the other gases. This application is discussed at length as an example of column sequence reversal in Section 10.5.

11.1.3 Example 3

You are required to check for low percentage oxygen impurities in the nitrogen atmosphere in food products such as potato crisps.

The first consideration here is how to get a representative sample without leaking in oxygen from the air; refer to Chapter 6. We cannot stick a gas tight syringe through the packet and suck in a sample without allowing oxygen leakage. Ideally the packet should only be opened in a nitrogen environment such as a larger sized sampling bag so we could purge this bag with nitrogen and then pierce the food packaging with a syringe or a needle attached to our sample valve. By gently compressing the food packet a positive flush of the sample loop could be achieved so that a representative sample will be taken.

The actual analysis is really trivial. Since oxygen elutes before nitrogen on a Molecular sieve 5A column there is no tail to interfere with the oxygen. A short packed column less than 0.5 m or a Plot column less than 2 m would be recommended at a temperature of 50–70 °C depending on the type of column. One would use helium or hydrogen carrier gas at 20–25 cm/s with a TCD.

Since the nitrogen elutes after oxygen you could suggest that the nitrogen be backflushed to vent. While this would work, it would complicate this simple analysis which will only take a few minutes on this short column under the conditions recommended.

11.1.4 Example 4

You are required to quantify trace (ppb levels) of H_2, CH_4 and CO_2 in air samples.

Referring to Chapter 7 this is a relatively easy separation for a carbon molecular sieve such as Carboxen or porous polymer such as Hayesep or PLOT Q. We know that trace analysis requires a large sample which gives a practical problem – a massive air peak between the analytes of interest. Since this is trace analysis you would choose a PDHID which will overload with the air peak. There is no easy way to change the elution order, so you will need to use a bypass valve to vent the large air peak to protect the detector. Referring to Chapter 10 an auxiliary flow to the detector must be supplied during the bypass switch. From Chapter 6 you will choose a 6-port sample valve with a sample loop that is small enough not to vastly overload the system but large enough to give the required sensitivity and a 4-port bypass valve. You will also run your column for good separation and will use helium carrier gas at 20–22 cm/s and set the temperature to 30–35 °C. The run time may be shortened by increasing the column temperature if too much separation is available.

11.1.5 Example 5

You are required to quantify CO_2 over the range of 1 ppm–60% in a mixture containing O_2, N_2, CO and CH_4.

Firstly let us look at what detector will handle this range. TCD will handle the percentages, but will not detect concentrations below 10 ppm even with a large sample size. Although a methaniser FID would be sensitive enough, the methaniser would saturate with the higher concentrations. Likewise a PDHID could handle low concentrations, but not percentages. The solution is to run a TCD with hydrogen or helium carrier gas in series with a methaniser FID using the TCD for high concentrations and the methaniser FID for the low concentrations. The sample loop does not need to be too large due to the good sensitivity of the methaniser FID and must also allow the TCD to be linear at 60%. A sample loop size of 250 μL would be a good starting point but a smaller loop will probably still give sufficient sensitivity. Clearly there will be separate calibration data on each detector. The TCD would be calibrated for the 1–60% range and the methaniser FID from 1 ppm to say 20,000 ppm (2%) which will give adequate overlap. Agreement between the results on any values in the low percentage range will also confirm good calibration on both channels.

Now let us decide on a column system for this mixture. Since only CO_2 needs to be quantified it is only necessary to separate CO_2 from the matrix. Any porous polymer column could be used and a rather short column should be used. Because of the large dynamic range required for CO_2, column capacity could be important so a packed or micro-packed column would be preferred, again keeping it simple and therefore reliable and robust. In practice you would look in the GC oven to see if there may already be a suitable column installed then look in the store cupboard as there are many different phases and dimensions that could adequately satisfy this separation requirement.

11.1.6 Example 6

You need to determine impurities of O_2 and N_2 in the range of 10–100 ppm in a hydrogen stream.

Since we expect this to be purity analysis, there is no need for separating oxygen and argon. From Chapter 7 we would want to choose a Molecular sieve column but we also know that it is difficult to quantify peaks on a sloping baseline and H_2 elutes ahead of the impurities. However, if we use a TCD with pure hydrogen carrier gas we can inject as large a sample as we like and the H_2 peak will not be seen in the analysis and therefore a short Molecular sieve column will be adequate. The carrier gas should be purified and flow control using crimps and all-metal connections must be used where possible. This will help to lower impurities in the carrier gas, which will in turn, effectively decrease background noise on the TCD and improve the signal to noise on the low concentration peaks. We would also need to look carefully at sampling techniques to avoid ingress of air into the sampling system as described in Chapter 6. Remember that in gas analysis it is common practise to report gas purity by subtracting the concentrations of the contaminants from 100%. For example, if this is to be sold as hydrogen 4.5 then the total impurities must be less than 50 ppm. Calibration could be done using static standards in a cylinder but a dynamic dilution system would be preferable as static standards of oxygen and nitrogen at 10 ppm levels would quickly degrade with use.

11.1.7 Example 7

You are required to quantify O_2 and CO from 0.1 to 50 ppm in CO_2.

In gas analysis, we could define trace and ultra-trace analysis as quantifying impurities at levels lower than those present in gases of 5.0 purity, in other words, low ppm and ppb concentrations. This presents a whole new set of challenges especially when O_2 or N_2 are involved. We will consider these as hardware, sample handling, column selection, calibration and data handling.

Firstly all the sampling and switching valves must be fitted with purged housings as in Section 5.8. It is also imperative that the carrier and ionisation gas be purer than the levels to be quantified otherwise peak inversion will occur and a non-linear calibration curve will be produced. Needless to say, the gas cylinder regulator must be electronic grade and all fittings must be metal-to-metal as motivated in Chapter 2. This also implies that getters must be used and all flow controls should be done with crimped tubing as described in Chapter 3. The sample valve must have a long thin tube on the outlet port to minimise back diffusion of air while purging the sample loop.

This analysis requires a large sample as the sample system needs to purge extensively. Ideally a liquid sample of CO_2 should be taken which means that it must be sampled at a pressure of at least 6 MPa and using a liquid sample valve. It is also necessary to run standards and samples many times to ensure that the concentrations measured are stable.

In the previous examples we have assumed that any stationary phase that can do the separation may be used. However, when we work at these very low concentrations the true picture appears. Depending on the column material, stationary phase and carrier gas purity, low concentrations of certain gases are 'eaten' by some stationary phases. Even our old faithful Molecular sieve needs to be conditioned with multiple injections of low O_2 samples after baking. Certain porous polymers are known to irreversibly absorb O_2 below 10 ppm. Before starting up a new column system it is common to allow the system to run overnight while injecting a low concentration mixture to neutralise any active sites. In this example, it is also important that the CO gives a sharp peak and therefore deactivated stainless steel tubing such as Silcosteel must be used for packed or micro-packed columns.

It is clear that calibration will also present a new approach. Gas suppliers have difficulty in certifying high pressure cylinders at such low concentrations with any traceability. One must therefore have a dilution system to generate on-site standards. Multiple dilutions may be required to cover the whole range required as described in Chapter 4.

Since the specification calls for O_2 quantitation, O_2 and Ar must be separated and we know this should be possible at these low concentrations on a long thick film Molecular sieve PLOT column at low temperature. However, the concentration range of Ar is unknown. In the event that the O_2 is not fully separated from the Ar the use of peak height rather that peak area may give more reliable results. This is often true for any small signal that approaches the noise level.

Now back to the challenge of quantifying O_2 and CO in CO_2 at these low levels. We could consider a porous polymer column plumbed for backflush of the CO_2 to vent in series with a Molecular sieve to separate O_2, Ar and CO. When this long column is operated at the low temperature needed for O_2 and Ar separation, CO will elute very late and be a wide and flat peak which will be difficult to integrate correctly. Since we need to use a PDHID temperature programming is not a viable option. The most practical solution is to use a short Molecular sieve for CO analysis and a long Molecular sieve for the O_2 and Ar separation in parallel using a valve to switch between these two columns. Obviously the column not in use still needs carrier gas flow. Alternatively, if a second detector (PDHID or methaniser) is available, dual injection could be done to cut down the overall analysis time.

11.2 Oxygen, argon and nitrogen

As we have seen in previous chapters and in the above examples it is relatively easy to separate oxygen and nitrogen using a short Molecular sieve 5A or 13X column at above ambient temperature. What happens when we need to determine traces of nitrogen in oxygen? What about separating argon and oxygen? What about traces of these gases? Each of these pose problems of their own but can be resolved using different column configurations.

When the matrix is oxygen the first major concern is the detector. We cannot use the solution proposed in Section 11.1.6 above because oxygen may not be used as carrier gas as it will oxidise the stationary phase. Also there is a strong probability that the filaments of a TCD may be incompatible with large amounts of oxygen as it will react with the hot filaments shortening the lifetime of the TCD. It is also not practical to time program the filament current as a stable baseline cannot be reached very quickly. Another concern is that argon will co-elute with oxygen and traces of nitrogen will be lost in the tail of the oxygen peak. One solution is to use an oxygen removal system which typically consists of a heated gettering alloy that selectively absorbs oxygen but allows the impurities to pass through for further separation on a suitable analytical column. The oxygen trap is mounted in a separate temperature controlled oven between the sample loop and the column. For routine applications, it is useful to use two traps in separate ovens and connected on a ten port valve so that one can be used to trap oxygen while the second is being regenerated by baking at a higher temperature. As with all catalytic conversions care must be taken to ensure that the impurities to be analysed are not significantly altered or removed by the catalyst.

Argon is the third most abundant gas in the atmosphere and is always difficult to separate from oxygen. Typically if air is injected into a GC the result will show 79% nitrogen and 21% oxygen since the 0.9% argon co-elutes with oxygen. There is a small difference in molecular size and this is the reason that these two gases can be separated on a Molecular sieve column at ambient temperature. It is important to have sufficient stationary phase to effect this separation and enough to avoid overload. The best column configuration here is a 0.53 mm i.d. PLOT column coated with Molecular sieve 5A with a film thickness of 50 μm run at optimum average velocity. The sample size must be kept to a minimum that is needed to give sufficient sensitivity. If an injector is used the smallest volume insert and a split must be used to keep the injection band as narrow as possible. Provided that the concentrations are low, baseline separation can be expected at a temperature of 25–30 °C.However if the concentrations of the components are relatively high, sub-ambient cooling of the oven will be required to obtain adequate separation as in Fig. 11.1.

Problems arise when there is a much larger concentration of argon than oxygen. Since we try to elute the low concentration component before the large or matrix component the elution order can be reversed by using a

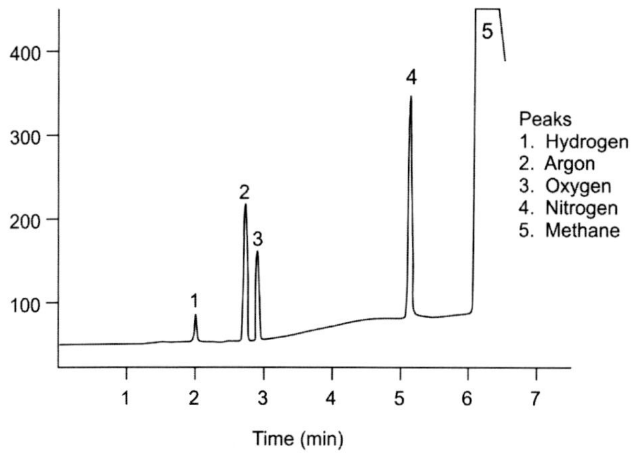

Column: Rt-Molecular sieve 5A, 30 m, 0.53 mm ID, 50 μm. Injection, VICI 6-port sample valve with purge housing. Sample loop: 250 μL. Oven Temperature: 0°C for 3 min then increased at 10°C/min to 50°C. Carrier Gas He, constant flow. Detector VICI PDHID.

Fig. 11.1 Chromatogram of hydrogen, argon, oxygen and nitrogen in natural gas.

porous polymer but this requires sub-ambient operation to achieve separation. The elution order on a Plot Q column will be nitrogen, oxygen and argon but even at −20 °C the peaks will elute very close together and baseline separation might not be possible especially if the relative concentrations are high. With this elution order it is clear that we can analyse traces of nitrogen in oxygen or argon as well as traces of oxygen in argon but in the absence of nitrogen. An alternative approach would be to use either Hayesep D or DB which both elute nitrogen before oxygen or argon at ambient temperature.

11.3 Hydrogen

The quantitation of hydrogen at various concentrations is a very important application in fields such as the petrochemical industry and alternative energy.

From Table 8.1, it is clear that helium and hydrogen have similar thermal conductivity approximately ten times higher than nitrogen or argon. Since the thermal conductivity detector responds to differences in thermal conductivity, it is clear that, for a fixed amount of hydrogen, the response will be about ten times larger when the carrier gas is nitrogen than when helium is used. Argon carrier gas will give a further improvement in signal intensity. It is well known that the sample size and sample concentration of hydrogen can cause unpredictable inverted peaks when using helium as carrier. Although the detector response will vary with the amount of hydrogen using helium carrier gas, this is not linear and is therefore not quantitative. This effect and a way of countering it was discussed in Section 8.2.6.

For the above reasons most Natural Gas (NGA) and Refinery Gas Analysers (RGA) have a dedicated hydrogen channel that uses a thermal conductivity detector with nitrogen or argon carrier gas. Usually this channel uses a Molecular sieve pre-column with backflushing to prevent high boiling compounds entering the analytical column.

The linear dynamic range of the TCD is somewhat limited as compared to other detectors so the size of sample loop will dictate this range. As a guideline, a sample loop of 1000 μL will allow low ppm (v/v) to low percentage (v/v) determinations while a 20 μL loop will allow quantitation from low percentages to 99%. It is possible to use a 4 port valve as a switchable sample loop that can be automated by the GC's software to accommodate both ranges on the same instrument as shown in Fig. 11.2.

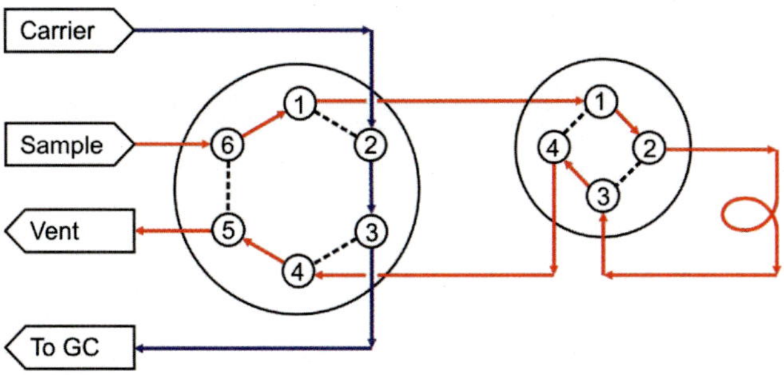

Fig. 11.2 Valve configuration allowing either large or small volume injections. The tubing connecting the two valves constitutes the smaller loop and by switching the four-port valve the larger sample loop is brought in line for large volume injections. All tubing must have the same internal diameter.

In some applications it is necessary to separate helium from hydrogen. Although helium has a higher mass it has a slightly smaller molecular diameter which means that helium will elute first on a Molecular sieve 5A column as shown in Fig. 11.3.

The determination of trace levels of impurities in hydrogen is required for fuel-cell applications and a good summary of the methods that may be used was compiled by A.S. Brown and others [2]. Some of the requirements for fuel quality hydrogen are given in Table 11.1, according to ISO/DIS 14687-2. From the table two potential problems are evident; the number of different components to be analysed and the very low amounts to be determined. Some of these will be addressed in the following sections dealing with other matrices but the principles will be the same.

When faced with an analysis requirement such as this, it is clear that no single analytical method will be capable of measuring all the components and sample sizes should therefore be large enough to perform the required measurements. A possible approach to this would be to use a GC-FID with a methaniser to determine the CO, CH_4, CO_2 and the higher carbon numbered hydrocarbons and with ultra-pure hydrogen as carrier gas so that relatively large sample volumes can be injected. Formaldehyde and formic acid can be accommodated on the same system although a PDHID would be superior to the methaniser FID for these two compounds but the hydrogen would overload the detector. The permanent gases can be done using a Molecular sieve column with back-flushing of the heavier components and a PDHID detector, but may require sub ambient temperature to separate the argon and oxygen. The sulphur compounds would require an inert flow path and a sulphur-specific detector such as the PFPD, FPD,

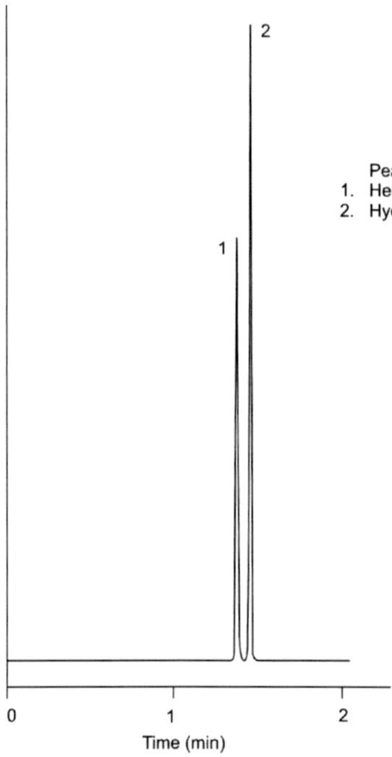

Column: Rt®-Msieve 5A, 30 m, 0.53 mm ID, 50 μm (cat.# 19723).
Sample: helium/hydrogen gas mixture. Injection volume: 25 μL split 10:1.
Liner: 2 mm split Precision® liner w/wool (cat.# 20823).
Injector temperature: 200°C. Split vent flow rate: 40 mL/min.
Oven Temperature: 30°C.
Carrier gas: nitrogen, constant pressure (7.0 psi, 48.3 kPa).
Linear velocity: 38 cm/s at 30°C. Detector: TCD at 200°C.
Make-up gas: nitrogen. Data rate: 20 Hz. Sensitivity mode: N_2/Ar.
Instrument: HP5890 GC.

Fig. 11.3 Chromatogram showing the separation between helium and hydrogen on a Molecular sieve column using argon as carrier gas [1]. ©Restek Corporation. *(Reproduced with permission, courtesy of Restek Corporation.)*

Table 11.1 Hydrogen fuel specification for Type I grade D [2].

Component	Max. impurities (μmol/mol)	Component	Max. impurities (μmol/mol)
Water	5	Total hydrocarbons	2
Oxygen	5	Helium	300
Nitrogen	100	Argon	100
Carbon dioxide	2	Carbon monoxide	0.2
Total sulphur compounds	0.004	Formaldehyde	0.01
Formic acid	0.2	Ammonia	0.1
Halogenated compounds	0.05	Particulates	1 mg/kg

SCD or AED. The trace halogenated compounds would be best determined using an ECD and helium determined with a TCD using argon or nitrogen as carrier gas. Ammonia and water would be best determined using a spectroscopic method such as FTIR with a long path gas cell or cavity ring down spectroscopy and the particulates can obviously not be done by GC at all.

11.4 Water vapour

There are many methods for determining the water content of gases, but since they are not always usable or available, it is prudent to look at the strengths and weaknesses of Gas Chromatography as a tool for water vapour analysis. Although water has a boiling point way above ambient temperature and is therefore outside of our definition of gas analysis, most gas samples contain some water vapour. Any attempt to remove this before injection is very likely to compromise the integrity of the sample, for example CO_2 is highly soluble in water and trapping out water will lower the concentration of CO_2 reaching the injector. The colder the water trap the more water it will remove and the more CO_2 will dissolve in the collected water. The solubility of some gases in water is given in Table 11.2.

Many robust gas analysers use backflush-to-vent to ensure that water vapour does not get on to columns such as Molecular sieves which are hydrophilic and will change their separation characteristics. When such a system is perfectly calibrated it is common that real samples will report less than 100% due to the presence of water vapour. If there is liquid water in the sample container and the GC system backflushes water vapour, then the amounts of compounds determined will be affected by the temperature of the sample container and connecting lines.

Table 11.2 Grams of gas dissolved in 100 g of water when the total pressure above the solution is 101.3 kPa [3].

Compound	Solubility	Compound	Solubility
Acetylene	0.117	Ethylene	0.0149
Ammonia	52.9	Hydrogen	0.00016
Bromine	14.9	Hydrogen sulphide	0.385
Carbon dioxide	0.169	Methane	0.0023
Carbon monoxide	0.0028	Nitrogen	0.0019
Chlorine	0.729	Oxygen	0.0043
Ethane	0.0062	Sulphur dioxide	11.28

Since water vapour is highly polar the only way to obtain a reasonably sharp peak is to use a polar column such as a PLOT U that gives an acceptable peak shape for water as shown in Fig. 11.4. Other columns may give reasonable peak shapes but water vapour will show moderate to severe tailing on most phases. Carboxen 1006 separates water from all the C2's and on a PoraBond Q column; water vapour will elute before the H_2S peak.

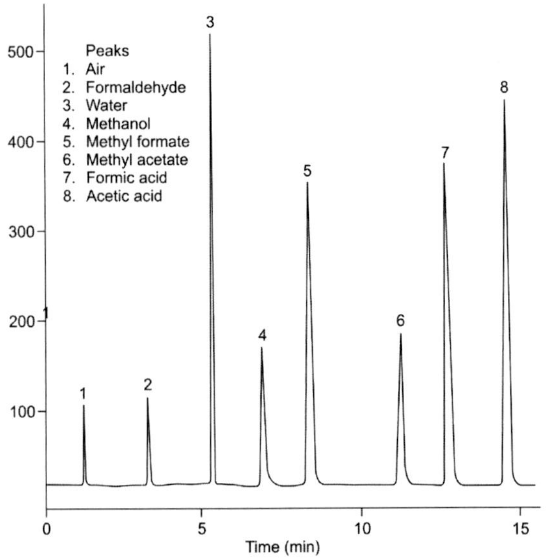

Fig. 11.4 Good peak shape of water on the polar HP-PLOT U column.

Perhaps the most difficult part of water vapour analysis is calibration and the most practical way of doing this is using permeation. All interconnecting tubing must be deactivated metal tubing because absorption of water on the tubing walls will take place. Even so, obtaining equilibrium can take several hours depending on the length of connecting tubing and the concentration of water vapour in the calibration gas and samples.

Given the difficulties in the accurate determination of water in gaseous streams it is usually easier to use Fourier transform infrared (FTIR) spectroscopy or one of the many other techniques that are more suited for this analysis. Trubyanov et al. [4] described the quantitative conversion of water to acetylene using calcium carbide prior to GC-FID analysis of trace moisture in electronic grade ammonia.

11.5 Natural gas, refinery gas and liquid petroleum gas

One of the major applications of gas analysis is found in the petrochemical industry in which the analysis is considerably complicated by the complex nature of the samples to be analysed. As there are a multitude of possible isomers, it is practical to refer to groups of hydrocarbons by the number of carbon atoms they contain, the so-called carbon number, for example C1 is methane, the C2's are ethane, ethene (ethylene) and ethyne (acetylene), etc. Production streams that are mainly gaseous at the temperatures used in the production plants may liquefy when sampled into containers that are cooled down after sampling. Even more problematic are mixed phase samples that may contain hydrocarbon gases and liquids as well as water in the same sample container, often at high pressure. Many volatile compounds become entrained in the gas streams and although they are not gases at ambient conditions, are present in the gas samples and provision must be made to separate them. Some samples like liquefied petroleum gas (LPG) will liquefy under their own vapour pressure and will often arrive at the laboratory in a high pressure sample vessel containing both liquid and gas phases. Sampling from such a container is problematic in that the composition changes depending on the relative amounts of liquid and gas in the sample vessel as was described in Chapter 6.

Refinery and natural gas samples may arrive at the laboratory at pressures as low as 100 kPa, or even at sub-ambient pressure and as high as 20 MPa. Simply connecting such samples to the GC with stainless steel fittings without considering the compatibility of the hardware will almost certainly lead to loss of sample integrity or even hardware failure. Trying to sample from a cylinder containing 20 MPa of gas with a gas tight syringe will turn the plunger into a potentially lethal projectile and the syringe barrel into a fragmentation grenade. Most gas sample valves can handle fairly high pressures but the GC sample inlets are usually not able to withstand pressures above 500 kPa or even lower as in the micro-GC that has a maximum sample pressure of 100 kPa.

In addition to the discussion of this in Chapter 6 consider that by simply lowering the pressure of a gas, it will cool down considerably due to the Joule-Thompson effect [5]. This may cause condensation of some of the components which could accumulate in the transfer tubing as liquids of even freeze and block the line completely. Heated pressure reduction systems containing heated pressure regulators are commercially available. Another means of pressure reduction is to use a length of heated capillary tubing to reduce the pressure and then expand the stream into larger bore tubing that is also heated to counter the cooling effect. The Gas Technology Institute in the USA also patented a system by which the gas is diluted with heated inert gas prior to cooling and pressure reduction [6]. Since compounds will only condense when their partial pressures exceed their vapour pressures, the dilution lowers the concentrations to a point where no condensation takes place at the lower temperature and pressure.

Samples such as LPG that typically contain hydrocarbon gases from C1–C4 which are all gases at room temperature are difficult to sample representatively at ambient pressure without discrimination due to their varying boiling points. These samples should be introduced while maintaining the sample as a liquid and if the sample pressure is too low to completely liquefy the sample it can be pressurised with an inert gas. The sample is usually taken from the bottom of the sample vessel and injected at full pressure using a liquid sampling valve situated close to the GC injector. The sample evaporates upon entering the GC and may expand considerably causing broadening of the sample application to the column. The sample size must therefore be small and as with all mixed phase sampling systems, the sample lines must be purged with an inert gas between samples to prevent sample carryover.

Liquid injection valves do not have an external sample loop but use a sample groove engraved into the rotor between the sample and waste ports that has a specific volume. Since this volume is extremely small it is wise to add a particulate filter into the sample line to prevent blocking the sample groove. The liquid injection valve can be used to inject directly into the GC injector using an auxiliary gas or by cutting into the carrier gas line leading to the injector as shown in Fig. 11.5. The connection between the

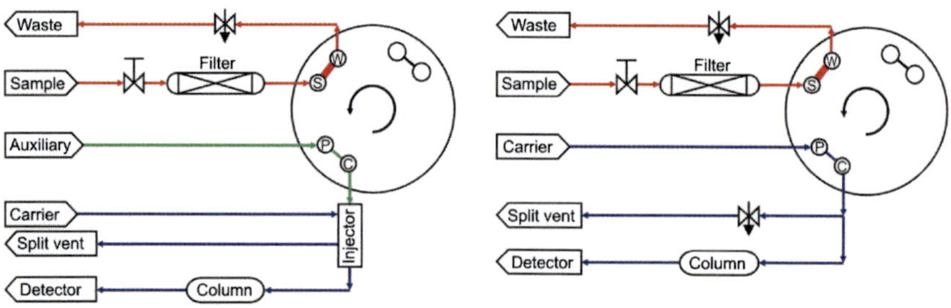

Fig. 11.5 Liquid injection valve inlet system. The liquid from the valve in injected into the GC injector using a very short piece of tubing fitted with a needle that passes through the septum into the injector (left). A similar configuration that does not use the injector but a manually operated split valve (right).

valve and injector should be a short as possible with the valve mounted as close to the injector as possible. The connection would typically end with a needle that pierces the injector septum so that the sample is injected when the valve is switched. While the injector is useful for splitting the sample when necessary it is also possible to bypass the injector altogether and add a manually operated split line between the valve and column.

There are many commercially available hardware configurations of dedicated analysers for natural gas which contains mainly methane and refinery gas that contains considerable amounts of nitrogen and in which all the C1–C5 isomers have to be separated. A possible RGA configuration that can also be used for the analysis of natural gas is given in Fig. 11.6 and the resulting chromatograms in Fig. 11.7.

There are many companies that specialise in providing customised, pre-configured systems that are supplied with guaranteed separations, full software control of valve switching, automated processing of the results and reporting in the format required by the customer in printed or electronic format compatible with laboratory information systems or plant control software. In most of these applications, a number of different columns are used with some columns often situated in separate temperature controlled ovens. An Internet search for the suppliers of Natural Gas Analysers (NGA), Refinery Gas Analysers (RGA) or Liquid Petroleum Gas analysers (LPG) will provide a list of companies active in this field.

11.5.1 Calculated values

Many applications require that the calorific value of a gas mixture be determined as this determines the price charged for that gas. End-use consumers of gas are interested in the heat energy that combusting the gas will generate as that determines the cost and the volumes to be used. Since the heat energy of the gas is related to the relative proportion of 'lighter' methane versus 'heavier' ethane, propane, butane, pentane, and other components, heat energy is not a constant value between different gas sources.

The heat energy of a particular gas stream is measured by units of calorific value, which is defined as the number of heat units released when a unit volume of the gas burns. Both SI and imperial units are still used in industry. Typical units of calorific value are Joule (J), British thermal units (BTU) and kilocalories (kcal) with 1 BTU = 1055 J = 0,252 kcal. The heat

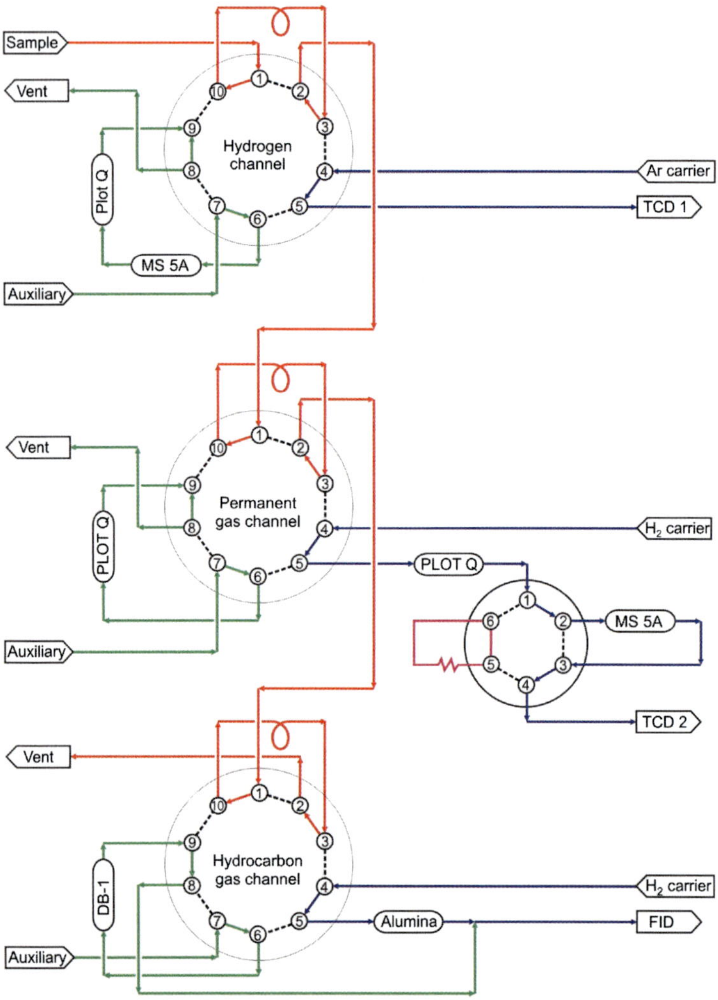

Fig. 11.6 Valve and column configuration of a system that can be used as an RGA. The hydrogen channel (top) uses argon carrier gas and two columns, a Plot Q column to retain everything that will elute later than CO and a Molecular sieve to retain the permanent gases. As soon as the hydrogen eluted into the first TCD, both columns are backflushed. The permanent gas channel (middle) also uses a Plot Q pre-column on the ten port valve to backflush everything that will elute after CO_2. The gases that elute before CO_2 are temporarily stored in the Molecular sieve column on the six port valve to allow the CO_2 to elute first after which the permanent gases are separated and eluted to the second TCD using H_2 carrier gas. On the hydrocarbon gas channel everything eluting after C5 is backflushed to the FID detector and recorded as a single peak named C6+. The hydrocarbon gases between C1 and C5 are separated on the alumina column.

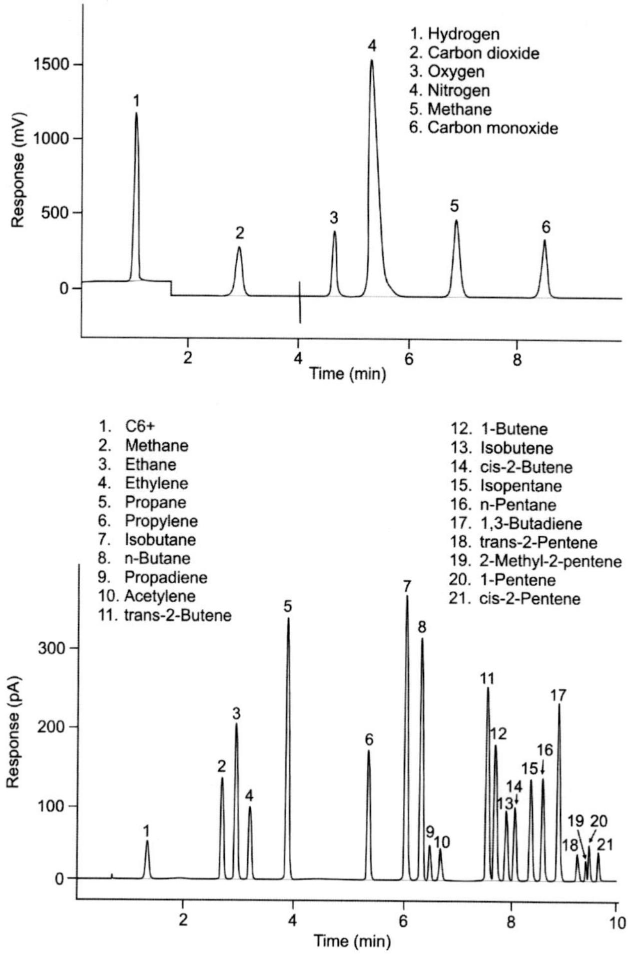

Fig. 11.7 Chromatograms from the TCD (top) and FID (bottom) obtained from a RGA with the configuration as in Fig. 11.6. The valve switch at 4 min between CO_2 and O_2 illustrates the robustness of having non-critical switching times.

energy is calculated according to International and National standards which should be consulted for a detailed explanation of the calculations [7,8]. The calculations are usually done automatically using dedicated add-on software that works with the integration software but in summary the calculations involve the following steps.

A typical analysis report will provide the results that the data system calculated from a previously established external standard calibration as a percentage of each. Data from more than one channel is usually combined and normalised to 100% to give the percentage of each component by volume.

$$C_{i_{Norm}} = \frac{[C_i]}{\sum_{i=1}^{n}(C_i)} \times 100$$

These amounts are multiplied by the molecular weight of each component, the figures obtained are added and the amount of each component expressed as a percentage of the total to give the molar percentage or mole fraction of each component [9].

$$C_{i_M} = \frac{[C_{i_{Norm}} \times MM_i]}{\sum_{i=1}^{n}(C_{i_{Norm}} \times MM_i)} \times 100$$

From here several physical characteristics of the gas can be calculated as described in detail in ISO 6976:2016 [7] and the book by Attari and Klass [10]. These include the dry molecular mass, the relative density or specific gravity, the heating value and Wobbe index. The heat of combustion of natural gas is in the range of 36–42 MJ/m^3.

11.6 Sulphur gas

Sulphur compounds are increasingly being targeted for monitoring to extremely low levels in the environment because they contribute to air pollution and are well known as catalyst poisons in many industrial processes. The damage caused to the environment by acid rain has been established and strict desulphurisation of effluent gases has been implemented in most countries. Catalytic reactors having a constant feed of gas containing only a trace impurity of a reactive sulphur impurity can quickly lose efficiency as the catalyst reacts continuously with the sulphur gas. With ever stricter limits for the sulphur content in fuels, engine exhaust and stack emissions being implemented, the analysis of trace quantities of sulphur compounds has therefore become more and more important. Most sulphur gases have bad odours and have the beneficial use as an additive to flammable gases such as natural gas and LPG to provide an olfactory indicator of leaks. On the other hand the objectionable odours of these compounds require their concentration in food and beverage products to be minimised and therefore monitored, for example, in beverage grade CO_2.

The two key aspects of successful sulphur analysis are maintaining the compound integrity throughout the system and then selectively detecting small quantities of these compounds usually in a complex matrix with much higher concentrations of other compounds. Arguably the most important factor in analysing these compounds is the fact that sulphur compounds are adsorbed onto, or react with, steel or stainless steel surfaces, especially when hot. The gases most commonly monitored include hydrogen sulphide, sulphur dioxide and carbonyl sulphide as well as other sulphides, disulphides, mercaptans and thiophenes. Being highly reactive the whole analytical system needs to be inert if meaningful results are to be reported. If the sample is collected in a glass container that has not been correctly passivated, the sulphur gases will adhere to the walls of the container and all sulphur gases may be reported as 'not detected' despite the fact that there is a distinctive sulphurous smell at the sample point. For this reason it is imperative that multi-level calibration be used and the linear range confirmed. Refer to Section 9.3 for more detail.

Several commercial companies have researched deactivation of metals and this has led to the introduction of patented materials such as Silcosteel™ Sulfinert™ and Ultimetal™. Materials deactivated in this way have highly inert surfaces and tubing, sample containers and column materials have been tested over long periods of time to show that there is no loss of sulphur compounds. These materials can, therefore be used to produce highly inert packed, micro-packed as well as PLOT and WCOT capillary columns with Sulfinert™ being the material of choice for sulphur gases. Clearly all connectors and fittings that are used in the system must also be deactivated. Experience has shown that a single brass union in a transfer line can effectively remove all low concentration sulphur compounds from the sample. When used at ambient temperatures thick walled polymer tubing such as PEEK can be suitable for sampling systems and results have also been reported using sulphur-doped carrier gas that has the effect of continuously deactivating the entire system.

The valves used for stream selection and sample injection also need to be corrosion resistant and as inert as possible. For applications that do not require the valves to be heated, a range of chemically inert valves, similar to those used in HPLC may be used. For applications that require using elevated temperatures, suitably deactivated valves with inert bodies made preferably of Hastelloy C or titanium should be used.

11.6.1 Columns

Historically glass or Teflon columns were packed with stationary phases including modified silica gels such as Chromasil 310 or 330 as well as acetone washed Poropak Q, called Poropak QS. The fragility of glass and the permeability and low temperatures tolerance of Teflon rendered these materials less useful and most analyses are done today using PLOT columns. Many applications use capillary columns with stationary phases consisting of derivatives of porous polymers such as PLOT Q and PLOT U [11]. Silica PLOT columns are employed when a different elution order of the sulphur compounds is required and WCOT phases such as methylsilicone are commonly used in many petrochemical production processes as it provides acceptable recoveries of H_2S, COS and SO_2. A summary of the elution order of sulphur and hydrocarbon gases is given in Table 11.3.

To determine very low sulphur levels, large volume injections are usually required and therefore wide bore columns with thick stationary phases are preferred, typically 0.53 mm columns with 7 μm or more film thickness.

11.6.1.1 Silica columns

This column gives good separation between H_2S and COS and separates COS and SO_2 from air and CO_2. The elution order is different from that of

Table 11.3 Elution order of sulphur and hydrocarbon gases on different columns.

Silica	Plot Q	Plot U	Methylsilicone
Methane	Methane	Methane	Methane
Ethane	Ethane	Ethane	Ethane
Carbonyl sulphide	Hydrogen sulphide	Hydrogen sulphide	Hydrogen sulphide
Propane	Carbonyl sulphide	Carbonyl sulphide	Carbonyl sulphide + Propane
Hydrogen sulphide	Propane	Propane	
Butane	Methyl mercaptan	Methyl mercaptan	Butane
Carbon disulphide	Butane	Butane	Methyl mercaptan
Methyl mercaptan	Ethyl mercaptan	Carbon disulphide	Ethyl mercaptan
Ethyl mercaptan	Dimethyl sulphide + Carbon disulphide	Ethyl mercaptan	Dimethyl sulphide
Dimethyl sulphide		Dimethyl sulphide	Carbon disulphide

the PLOT Q and non-polar WCOT columns. An example of a chromatogram that shows the separation of sulphur compounds can be seen in Fig. 7.17. The C2 and C3 hydrocarbon isomers show good separation on this column.

11.6.1.2 PLOT Q columns (divinylbenzene)

The PLOT Q is the least polar porous polymer column and provides good separation for the hydrocarbon gases up to C5. It provides separation of H_2S from ethane and propane but cannot separate carbon disulphide and dimethyl sulphide. It has a high maximum temperature and therefore often used for samples in which higher boiling compounds are present without having to backflush the latter. The C2 and C3 hydrocarbon isomers show some separation on this column.

11.6.1.3 PLOT U (ethylene glycol and dimethacrylate in divinylbenzene)

PLOT-U is the most polar column of this family. It separates polar compounds very well and also provides good separation of gaseous sulphur compounds, making it a very good choice for this analysis. Both H_2S and COS elute between ethane and propane such that it allows trace analysis of these compounds as impurities in ethylene, propylene and propane since propylene co-elutes with propane.

11.6.1.4 Methylsilicone WCOT columns

On these thick film columns H_2S and COS are well resolved from each other but COS co-elutes with propane. This is not serious if the propane concentration is low as in natural gas but in a propane or propylene matrix quenching of the detector due to the high concentration of hydrocarbons will compromise the results. The C2 and C3 hydrocarbon isomers show very little separation on this column.

11.6.2 Detectors for sulphur gas analysis

Most universal detectors can be used for the detection of sulphur gases but preference is usually given to sulphur selective detectors. A detailed discussion of the detectors and their operating principles is given in Chapter 8. When using a TCD it should be kept in mind that it is used for relatively high concentrations and has some limitations. It uses hot metal filaments in a hot metal block that will react with the sulphur compounds and it is advisable to use corrosion resistant filaments since there is a possibility that

H_2S may react with water vapour to form sulphuric acid that could damage the filament. The lowest concentration of sulphur compounds that can be determined will be several hundred ppm.

Sulphur containing compounds in gases must usually be determined at very low levels making the correct choice of detector imperative. The selective detectors for sulphur compounds are the FPD, PFPD, SCD and AED but careful consideration should also be given to the matrix. Given the quenching that may occur with the FPD and PFPD a column should be used that will provide the needed selectivity so that the sulphur compounds are already separated from the hydrocarbons when they reach the detector. It is rather difficult to compare the specifications of the sulphur specific detectors given the potential loss of sulphur compounds in the analytical system. The analytical system, from sampling to injection and separation often contributes more to an inability to reach low detection limits than the sensitivity of the detector. Since these detectors are selective and respond to the number of sulphur atoms in a compound, one could theoretically calibrate using a single sulphur compound and use its response factor for all other sulphur compounds in the sample. This could however lead to errors as the assumption is that all the compounds will be affected the same by the analytical system. Since the sulphur containing compounds will have different boiling points and affinities towards the walls of the sample containers, transfer tubing, injector insert and column, the losses in the analytical system will probably not be the same for all the compounds.

11.6.3 System priming

For sulphur compounds, as well as all reactive gaseous compounds it is imperative to use suitably deactivated components for the construction of the analytical system. Even so, it is also necessary to ensure that the entire analytical system is properly passivated in order to get accurate quantitative results. One way to achieve this is by system priming which requires repetitive injection of a standard or sample until the ratios of the component peaks are stable as shown in Fig. 11.8. The priming injections will clearly show if any adsorption or absorption occurs in the system and might have to be performed regularly. For example, it may happen that results become repeatable after three injections and remain so for the rest of the day. However, when the system is left overnight with pure carrier gas purging the system, the active sites that were passivated during priming might be purged as well requiring re-priming the next morning.

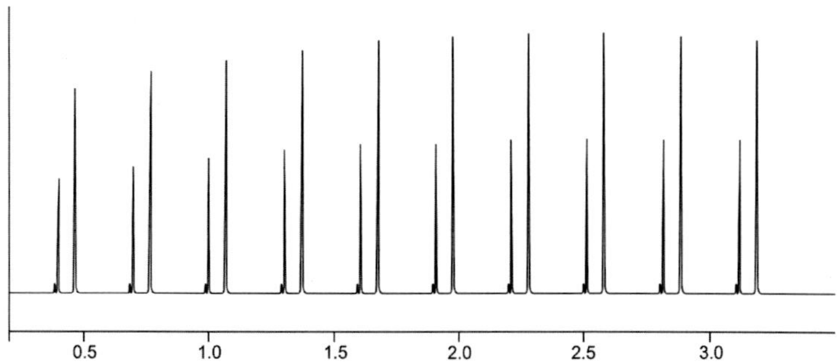

Fig. 11.8 This shows ten consecutive injections of a two component test mixture. The effect of priming is visible in the first injections with the loss of analyte occurring in the inlet, the column or detector.

11.6.4 Calibration

Commercial cylinder standards containing sulphur gases can usually not be certified below about 5 ppm (v/v) per component due to the adsorption of these compounds on the cylinder walls and the difficulty in preparation of these reactive species. Such standards are usually rather expensive and their shelf life limited. Preference is therefore given to point-of-use sample preparation using permeation devices or dynamic dilution of higher concentration cylinder standards; often in combination. One of the advantages of dynamic dilution is that instead of using pure helium or nitrogen as diluent, gases such as ethylene, propylene or mixtures can be used to match the sample matrix more closely than using pure gases as diluents. Of course one should always ensure that all lines are flushed properly when switching gas mixtures even if it is just different dilutions of the same standard. Cylinder standards, permeation systems and dynamic dilution are discussed in some detail in Chapter 4 but calibration of sulphur selective detectors at ppb levels can become quite challenging. It is advisable to always use multipoint calibrations with several replicates at each level as that would reduce the errors that may be introduced due to incomplete flushing of the lines and incomplete system priming.

11.6.5 Conclusions

Although the analysis of sulphur compounds is challenging it is possible with GC provided the system has been designed to allow low concentrations of sulphur gases to go through the system without modification. Columns that are guaranteed for the determination of sulphur compounds must be used, but whether to use a porous polymer or thick film non-polar phase will depend to a large extent on the compounds in the mixture and their relative concentrations. The choice of sulphur selective detector is less significant than the inertness of the system and special care should be taken with calibration, especially if dynamic dilution is not available since low concentration static standards of reactive gases are unlikely to last long. It is most important to realise that trace analysis of sulphur gases requires a dedicated GC system set up for optimal performance for these compounds. Attempting to use a GC set up for general purpose analysis of high concentrations of various different types of samples will severely compromise the ability to successfully analyse trace amounts of sulphur compounds.

11.7 Headspace analysis of transformer oil gases

The analysis of dissolved gases in transformer oils has become a very important means of diagnosing the condition of the transformer. ASTM Method D 3612 allows for three different methods of gas extraction, vacuum extraction, sparging the oil with the carrier gas on a stripper column and headspace. The interpretation of the results is rather complex and includes ratios of certain gas concentrations as well as calculating the total oxygenate content. The presence of acetylene is regarded as a warning but overall trends need to be followed by using regular sampling. The gases that must be quantified for diagnostic purposes include hydrogen, oxygen, nitrogen, methane, carbon monoxide, carbon dioxide and hydrocarbon gases. The C2's are always included although further information is available if C3's and even C4's are analysed. The analysis of C3's and C4's will lengthen the cycle time. Most of these systems use a gas chromatograph with argon carrier gas, a TCD, methaniser, FID and signal switching. Oxygen and nitrogen concentrations are high which could cause detector overload if the more sensitive PDHID were to be used. The ASTM method has been discussed as an example of series-bypass operation in Section 10.4.2 and a chromatogram of standards shown in Fig. 10.13.

Many transformers are equipped with on-line gas monitors which can add a helium artefact which must be clearly separated from the hydrogen. Using a porous polymer stationary phase in series with a Molecular sieve can give rise to a negative baseline dip in this early region of the chromatogram and can cause hydrogen to be reported incorrectly. Since the normal hydrogen concentrations that are measured are very close to the detection limit of the TCD, any small baseline variation could easily cause a measuring error of 50% or more.

Like vacuum extraction, headspace can be considered a partial extraction technique. However, by using constant incubation time and agitation of the sample bottle in the headspace sampler a highly repeatable extraction is achieved. Headspace is very popular because it eliminates a lot of tedious sample preparation steps and is relatively free from contamination and carryover problems that are common to other sample introduction systems because only volatiles are removed from the vial. It is also easily automated and batch sample preparation stations are available. The analysis time for the first sample will be rather long but the automatic sampler has a reduced cycle time for the second and subsequent samples by overlapping incubation time while the previous analysis is being run. Calibration is done using gas-in-oil standards and concentrations have to be corrected for altitude so that results are standardised for pressure. Inter laboratory calibration check samples are also available on a regular basis from institutions such as ASTM.

The range of concentrations in real TOGA analyses is difficult to predict and can lead to incorrect results. Concentrations of CO and CO_2 may be a few orders of magnitude higher than the concentration of acetylene or methane. While it is relatively easy to show good separations done on standards where most gases have similar concentrations, this is not as easy for real life samples.

In the system shown in Fig. 11.9 each column that is not used at the time for analysis is in the backflush mode and ensures that water vapour which is inevitably in the sample does not build up on the Molecular sieve column.

A typical analytical procedure requires about 10 mL of an oil sample, that is transported in a glass syringe, and which is carefully decanted into a 22 mL vial under an argon shield and immediately capped and crimped. Provided the correct septa are used, the sample can be kept at room temperature for a considerable period.

The sample vials are put into the auto sampler and a sequence is run. Typically 20 or more samples can be prepared in about 1 h. Since each vial is normally incubated for 30 min and although the GC run time may be shorter, allowing for cool-down of the oven, only two results per hour can be reported during unattended operation.

Chromatograms of a calibration standard and a sample of a healthy transformer oil is shown Fig. 11.10.

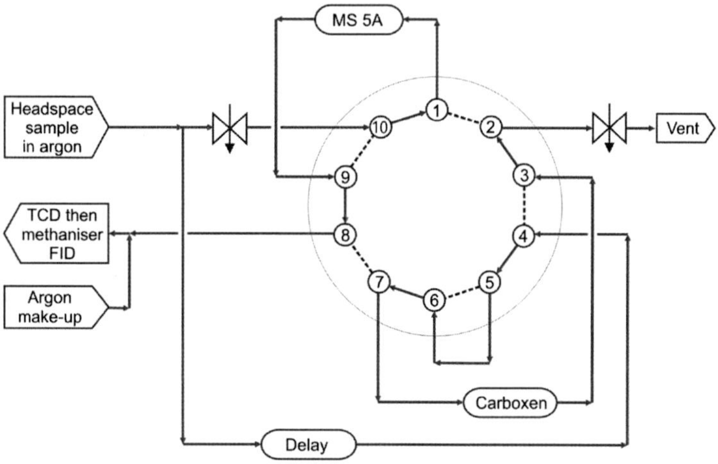

Fig. 11.9 Schematic plumbing diagram of a TOGA system. The 2 analytical columns are backflushed when not in use. The Molecular sieve 5A column is 8 m × 0.53 mm i.d. with a film thickness 50 μm. The Carboxen 1006 is a 30 m × 0.53 mm with a film thickness of 30 μm. The delay column is 60 m of any non-polar column such as a DB-1. The first detector is TCD then methaniser FID. Because the methane peak is measured on both channels the split ratio between the two columns can be set up. Usually 2/3 to the Molecular sieve and 1/3 to Carboxen set using the two needle valves. Since the hydrogen detection limit is vital reasonable values down to 2 ppm can be obtained using a short Molecular sieve column. Retention times for C3's and C4's would be very long as the ethane elutes close to the maximum temperature of the column. A different column system would be used if C3's and C4's were to be determined A temperature program of 70°C for 10 min and then 10°C/min to 220°C would typically be used since the FID is relatively flow insensitive.

11.8 Aldehydes

The low molecular weight aldehydes pose difficulties of their own. While formaldehyde is a gas at room temperature acetaldehyde is a liquid with a boiling point of 20.2 °C meaning that it is usually present in the vapour phase in many samples. Both show very poor responses on FID and it is therefore often missed in the compositional analysis of hydrocarbon type samples. Similar to water, these are also 'sticky' gases meaning that they are prone to adsorption on the surfaces of sample containers and tubing and gaseous standards are best prepared using permeation devices. Derivatisation of carbonyl compounds with 2,4-dinitrophenylhydrazine has been used successfully for many years; there are various scrubbing methods for collection of aldehydes in solutions and it can also be sampled using

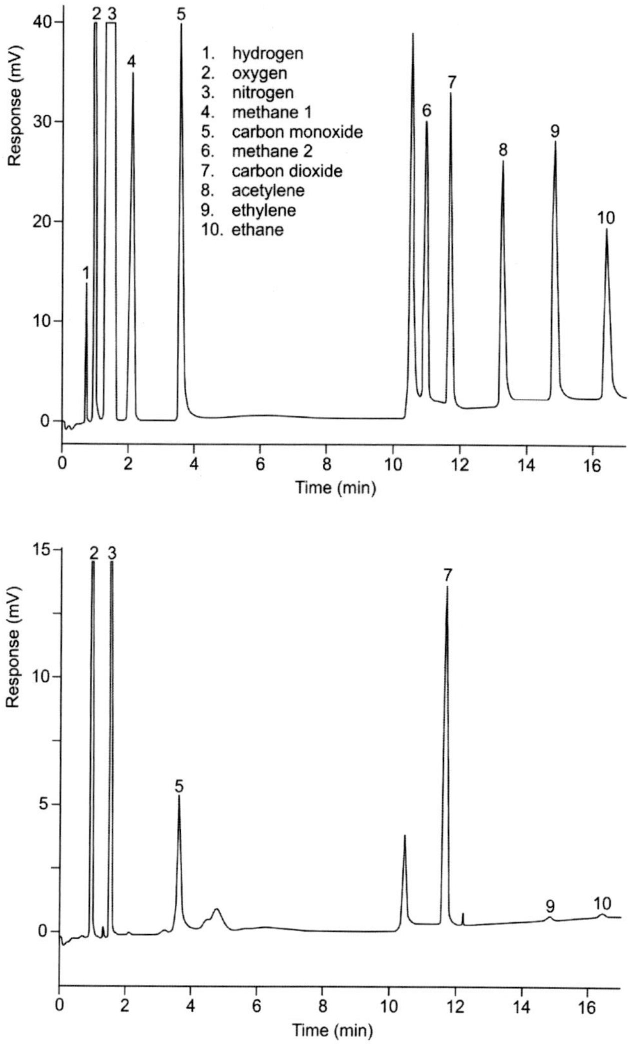

Fig. 11.10 Examples of typical TOGA analyses of a standard (top) and a healthy transformer sample (bottom).

thermal desorption tubes. There is also a need for direct methods, including headspace sampling of liquid and solid samples, without the need for cumbersome sample preparation. Spectroscopic methods are ideal for this but the low molecular weight aldehydes can also be successfully analysed by GC.

11.8.1 Formaldehyde

Formaldehyde is present in the atmosphere and is emitted by anthropogenic combustion and atmospheric oxidation of volatile organic compounds. It is also a by-product of many industrial processes and used in chemical intermediates such as urea-formaldehyde resins for adhesives and binders in wood, paper and pulp products. It is also a good disinfectant and can be found in many germicides, insecticides and fungicides while its application as an antibacterial agent is found in soaps, deodorants, cosmetics and other personal hygiene products. While its levels are low, the emissions from various sources such as decomposition of formaldehyde resins and vehicle as well as industrial exhausts, contribute to the poor quality of air especially in heavily polluted areas. Since inhalation is the main route of human exposure and it is a suspected carcinogen, the levels in air must be monitored especially in industries where exposure is significant as in mortuaries, pathology laboratories as well as wood, pulp and paper processing workers and furniture and fixture assemblers.

Since formaldehyde is soluble in water the column used should be capable of separating the formaldehyde from water. One's first reaction is that a polar column should be used since formaldehyde is water soluble and therefore polar. As it is in the gas phase, using a PLOT column seems equally obvious. However, it was reported that the PLOT columns only provides good peak shapes at temperature as high as 100 °C where separation from other components in the sample may be impossible. Using WCOT columns, the peak shape deteriorates with increasing polarity to the point where it becomes invisible on very polar columns [12]. Modern stable polar column are less prone to this effect and may well be suitable as will the columns specially designed for the analysis of aldehydes. However, a non-polar WCOT column with a thick phase of 5 μm works well provided that the concentrations are low, but the oven might have to be cooled to below ambient to effect adequate separation of the matrix components as in Fig. 11.11.

Since the FID is not suitable a PDHID is the detector of choice, especially at the low concentrations encountered in environmental monitoring samples. However, it was reported that a Ni catalyst can be used to convert formaldehyde and acetaldehyde to methane and ethane respectively, similar to what is done with a methaniser [12], but a PDHID is reportedly more effective for environmental samples [13].

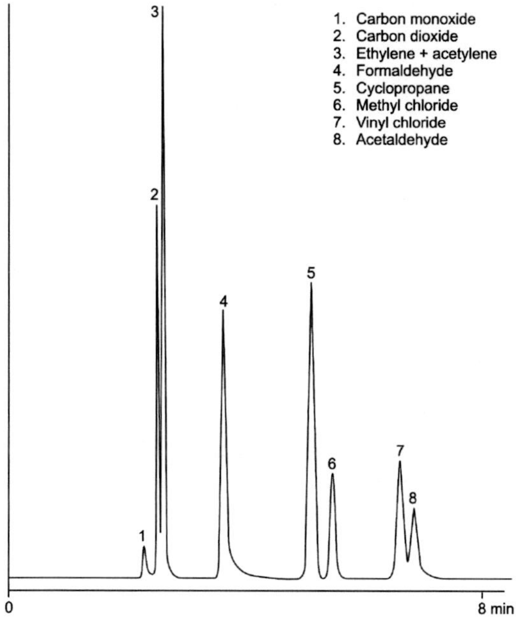

Fig. 11.11 A chromatogram of formaldehyde, acetaldehyde and other light gases using methaniser FID for detection. *(Reprinted from J. Luong, L. Sieben, M. Fairhurst, J. de Zeeuw, Determination of low levels of formaldehyde and acetaldehyde by gas chromatography flame ionization detection with a nickel catalyst, Journal of High Resolution Chromatography 19 (1996) 591–594. Copyright Wiley-VCH Verlag GmbH & Co. KGaA. Reproduced with permission.)*

For low concentration samples that contain very little or no moisture the sample loop can be cooled with liquid nitrogen to concentrate the formaldehyde and other volatile organics in the loop while passing a known volume of sample through it. Upon injection the sample loop must be heated rapidly and this can be done using resistive heating of the stainless steel loop [14].

11.8.2 Acetaldehyde

The determination of acetaldehyde is important in various industrial processes where it can take part in condensation and polymerisation reactions causing the formation of unwanted deposits in lines and valves and is also a poison for some industrial catalysts. It is an intermediate in alcoholic fermentation processes and may be present in small amounts in beer, wine and derived products. It can also be present in plant juice, essential oils and roasted coffee, to name but a few.

One of the major applications of acetaldehyde analysis is in the manufacture of beverage containers. Acetaldehyde is a degradation product that is generated during processing of polyethylene terephthalate (PET) polymers and high levels can impart off-flavours in some food and beverage products. This analysis is done with headspace sampling of the crushed polymers as in ASTM method F2013 [15].

In this method, the headspace incubation is done at elevated temperatures, typically 150 °C for 30 min despite the low boiling point of acetaldehyde. This ensures that as much of the acetaldehyde as possible escapes from the polymer and enters the headspace. The polymer must be finely crushed in a cryogenically cooled mill to prevent analyte loss or side reactions due to heat generated during crushing.

As with all trace analyses the quality of the solvent is of the utmost importance to eliminate interfering artefacts. Standards prepared in water require ultrapure water and HPLC grade is usually acceptable provided that the container has not been open for any length of time. Volatile organic compounds can be reduced by boiling but running solvent blanks is imperative. The authors have had success using commercial bottled water, still not sparkling, that is supplied in glass bottles. Clearly water stored in PET bottles will not be suitable. Since the boiling point of acetaldehyde is only 20.2 °C, the same as the laboratory temperature, the making of standards requires considerable skill. The acetaldehyde liquid and diluents must be cooled, transferred rapidly into pre-cooled containers and sealed immediately. The use of permeation standards in strongly advised for standards such as this.

Using a thick film non-polar liquid phase on a 0.53 mm i.d. column is recommended in the ASTM method, but it can also successfully be done on PLOT Q column which is preferred for applications where acetaldehyde and ethylene oxide must be separated. An example is shown in Fig. 11.12. Benzene may be present at low concentrations but will require an extended run time to ensure elution.

Although the FID is relatively insensitive to acetaldehyde, the PDHID can be used, for example for the determination of low acetaldehyde and formaldehyde concentrations in air in which case the air peak is cut to vent using a four port valve.

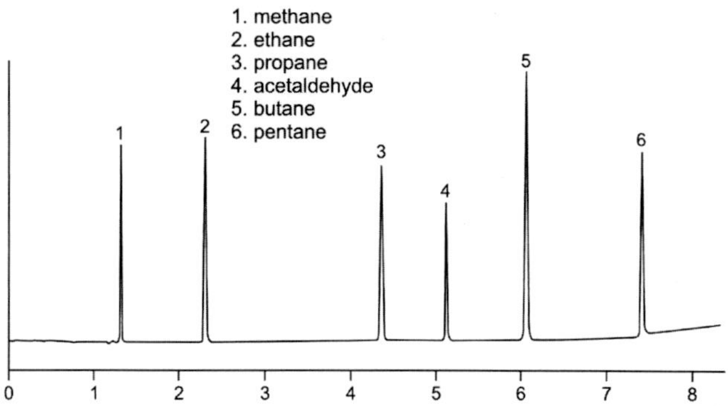

Column: Agilent PoraPLOT Q-HT, 25 m, 0.53 mm ID, 20 µm (Part no. CP7559). Split injection. Oven Temperature: 40°C (hold 1 min) to 250°C at 8°C/min (hold 3 min). Carrier Gas: He, constant pressure (8.7 psi, 60 kPa). Detector FID.

Fig. 11.12 Acetaldehyde in a hydrocarbon mixture separated on a PoraPLOT-Q column [16]. ©Agilent Technologies, Inc. 2011. *(Reproduced with permission, courtesy of Agilent Technologies, Inc.)*

11.9 Ethers

Similar to the aldehydes, the low molecular weight ethers are very volatile with dimethyl ether a gas at ambient conditions that will liquefy under modest pressure or cooling, similar to LPG. Diethyl ether is a liquid at ambient conditions but has a boiling point of 34.6 °C which means it can be found in significant concentrations in ambient air when a suitable source is available. Both are relatively non-polar and thermostable making their analysis by GC relatively easy with matrix interferences being the major stumbling block.

11.9.1 Dimethyl ether

Because dimethyl ether is flammable, easy to liquefy and transport, has a high oxygen content, lacks sulphur and other noxious compounds and provides clean combustion it is a promising renewable low carbon fuel for automobiles, power generation as well as domestic heating and cooking. It can be produced by gasification of coal or biomass and chemical conversion of natural gas which is very efficient. It has been used for many years as an aerosol propellant in the personal care industry. On the downside its decomposition may lead to polymer formation and is a particular problem in refineries producing methyl tert-butyl ether (MTBE).

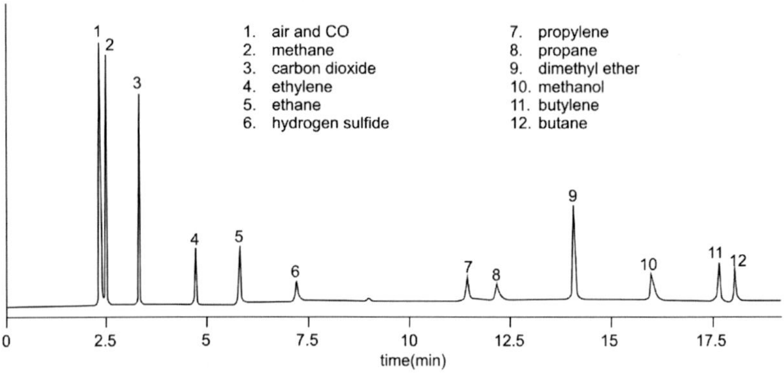

Column: Agilent J&W HP-PLOT Q PT, 30 m, 0.53 mm ID, 40 μm (Part no. 19095P-QO4PT). Injection 250 μL gas sampling loop, split 5:1, injector temperature 170°C.
Oven Temperature: 32°C (hold 5 min) to 70°C at 30°C/min (hold 5 min) to 160°C at 10°C/min.
Carrier Gas: Hydrogen, constant flow 40 cm/s (32°C). Detector TCD at 250°C.

Fig. 11.13 Chromatogram of a mixture of standards showing the separation of dimethyl ether from the hydrocarbon gases and methanol [17]. ©Agilent Technologies, Inc. 2013. *(Reproduced with permission, courtesy of Agilent Technologies, Inc.)*

It can be separated with relative ease from hydrocarbon matrix components using a PLOT-Q column and elutes between propane and 1-butene as shown in Fig. 11.13. Methanol, if present, will elute just after the dimethyl ether and if more separation is needed a PLOT-U column can be used [17]. In both cases heavier components should be back-flushed either to vent or if needed to another column suitable for separating the heavier components. On packed columns a mixed packing consisting of 80:20 Poropak N and Q was reported to provide good separation [18].

11.9.2 Diethyl ether

Diethyl ether has a high volatility and low flash point and has been used with petroleum distillates as a starting fluid for diesel engines. It was widely used as an anaesthetic agent but because of its flammability and post-anaesthetic side effects has been replaced with halogenated ethers. It is, however, still used as a recreational drug. The main application area is as a solvent in industrial and laboratory applications.

Diethyl ether can be separated from most common solvents using non-polar or moderately polar WCOT columns. In complex matrices such as gaseous hydrocarbons it is arguably best separated from most of the matrix components using the highly polar PLOT columns such as the GS-OxyPLOT column from Agilent. These columns will elute all the non-oxygenated

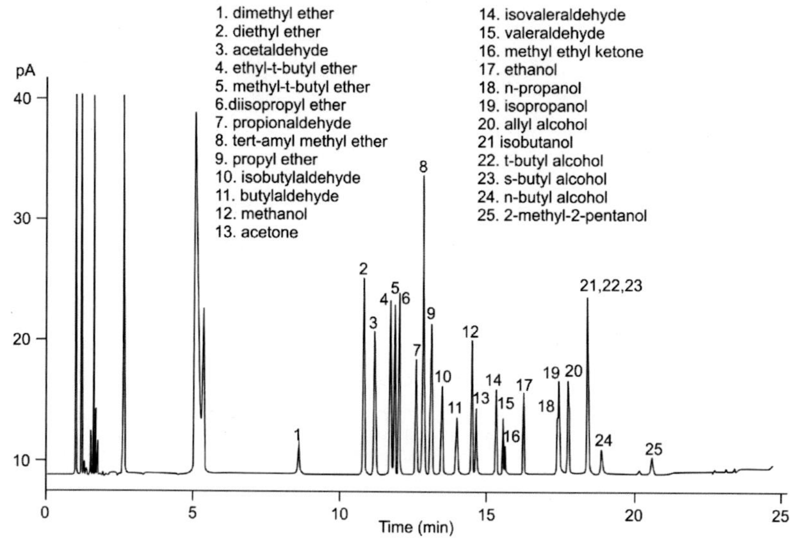

Column 1: Agilent J&W DB-1, 25 m, 0.53 mm ID, 1 μm (Part no. 125-102J). Column 2: Agilent GC-OxyPLOT, 10 m, 0.53 mm ID, 10 μm (Part no. 115-4912). Injection split. Oven temperature: 50°C (hold 5 min) to 240°C. Carrier Gas: Helium. Detector FID.

Fig. 11.14 Chromatogram showing the separation between the oxygenated and non-oxygenated hydrocarbons using a non-polar pre-column and highly polar PLOT analytical column [19]. ©Agilent Technologies, Inc. 2007. *(Reproduced with permission, courtesy of Agilent Technologies, Inc.)*

hydrocarbons up to octane first followed by the oxygenated compounds starting with dimethyl- and diethyl ether.

In an application as in Fig. 11.14, two columns are connected in series using a multiport valve that allows back-flushing of the pre-column to vent or to another column and detector. The sample is introduced into the non-polar pre-column through which the oxygenated compounds, being not well retained, pass quickly through onto the PLOT column together with the volatile hydrocarbons. On the PLOT column the hydrocarbons are not well retained and pass through to the detector first without significant further separation. The oxygenated compounds interact much more with the PLOT column and are separated and elute after the hydrocarbons [19].

11.10 Halogens and halogenated gases

The halogenated gases are rather unique in that they vary from being the most corrosive and poisonous gases known to some of the most inert gases.

While the latter are not too difficult to separate and detect, the corrosives require specialised equipment and creative configurations for sample handling as well as separation and detection. Before considering handling these gases the first requirement must be that the laboratory has the necessary infrastructure for example, adequate ventilation, the availability of breathing apparatus, first aid, secondary containment and everything else that will ensure the safety of the analysts that are working with it. The specific measures will depend on what must be analysed in what matrix and at what concentrations and a thorough review of the risks is imperative. Working with these samples, as with any gas samples, the standard personal protective equipment such as safety boots, laboratory overcoat, gloves and safety glasses will not provide the necessary protection against these gases.

11.10.1 The corrosive and reactive gases

Before looking at the possibilities for separating the corrosive gases it is important to take note of the requirements of the analytical system including the sample containers, tubing, valves and detectors. Sample containers and tubing should be nickel or the so-called super-alloys such as Monel, a nickel-copper alloy, Inconel, an austenite nickel-chromium-based alloy or similar types of materials. If trace contamination from ambient air is not a big concern, Teflon, PEEK or similar polymers may also be suitable. Deactivated glass sample containers cannot be used for fluorine or HF but could be used for chlorine and HCl, although the hazards associated with the glass breaking and releasing its contents into the laboratory would, in our opinion, preclude its use.

Packed columns should be in nickel tubing and fused silica capillary columns may be resistant to chlorine and HCl, but may not withstand the aggressive gases such as fluorine and ClF_3. There are few detectors that will be able to withstand the onslaught of these gases and using a conventional tungsten filament TCD to detect fluorine will work once after which the filaments will have disappeared in a small plume of WF_6 and the detector will be destroyed. Having said that, the TCD is probably the only suitable detector for these gases, but special TCD's that are nickel or gold plated and using nickel or gold filaments are required.

The first bit of advice is to avoid using GC if possible and to use alternative methods for these gases. In production environments, preference should be given to on-line techniques as it would remove the hazards associated with plant sampling, transport, storage and laboratory sub-sampling of the gases. For gases like HF, spectroscopic techniques may be much more amenable to the safe handling of the gas. If GC will be used, there are generally two approaches to analyse the corrosive gases depending on the analytical requirements. Obviously the requirements for determining trace amounts of these gases in air will differ substantially from having to determine the gaseous impurities in these corrosive gas matrices.

The first and also safest and easiest approach is to convert the corrosive gas into a compound that is more benign and that can be analysed more readily. This should always be considered before using the more difficult direct analysis approach. Upon opting for the direct analysis, care should be taken to ensure that the entire analytical system from sampling to detection is compatible with the matrix and components being analysed.

11.10.1.1 Chlorine and hydrochloric acid

Chlorine can be converted to chloro-acetone by passing the gas through acetone or adding acetone to the sample container. This is an exothermic reaction and can be hazardous so due care must be taken and the amounts should be kept low. The chloro-acetone will elute after acetone on a non-polar WCOT column with a retention index of 670–680 meaning that it will elute between hexane and heptane, but closer to the heptane peak. We used a packed nickel pre-column with 10% Kel-F coated on Chromosorb T to separate the permanent gases from chlorine, which is retained by the stationary phase and back-flushed while the permanent gases are further separated on a Molecular sieve column. Obviously this method does not allow the detection of chlorine and also does not need a special TCD as the matrix never reaches the detector. Since Kel-F coated columns have a very low T_{max}, oven temperatures should be kept below 175 °C or the pre-column may be kept outside the oven at ambient temperature. For low concentrations of Cl_2 in air, a PoraPLOT Q pre-column can be used to separate the Cl_2 from air and other permanent gases which can then be separated on a Molecular sieve column [20]. The effluent from the pre-column can be determined on a TCD and for trace amounts on an ECD.

Chlorine can also be converted to HCl in the sample loop by using hydrogen as carrier gas. Once again the reaction is exothermic and could be hazardous if amounts are not kept small. The heat generated by this reaction could also damage the injection valve so care must be taken to use a valve that can withstand the gas as well as the temperature. HCl is somewhat easier to separate and can be separated from other components using specialised stationary phases in a packed column such as 5% Carbowax 20 M on Fluoropak 80 [21] or on Hayesep R80-100 mesh as shown in Fig. 11.15.

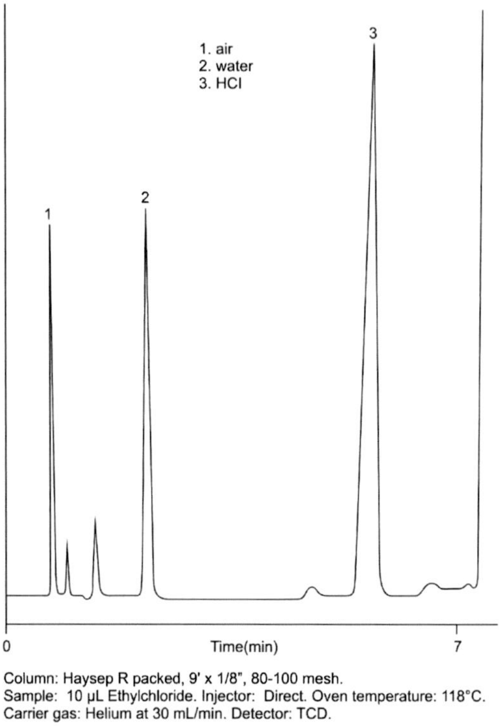

Fig. 11.15 HCL separated from water on a Hayesep R packed column [22]. ©VICI AG International. *(Reproduced with permission, courtesy of VICI AG International.)*

Using capillary columns HCl can be separated from air and water using a PoraPLOT U column on which the peak tails significantly and the best capillary column for this is probably the PoraPLOT Q which is highly inert and on which HCl elutes as a symmetrical peak.

11.10.1.2 Fluorine and hydrofluoric acid

The direct analysis of fluorine gas is, as far as we know, impossible by GC. It is simply too reactive and there are no stationary phases that are resistant to it, let alone being able to separate it. To analyse gaseous impurities in fluorine the sample can be passed onto an alumina trap with which the F_2 will react to form aluminium fluoride and release oxygen. The separation then becomes simply a matter of determining the impurities in oxygen. Similarly, conversion traps packed with alkali metal or alkaline earth metal hydroxide, oxide, carbonate, hydrogen carbonate, sulfate or nitrate can be used and the quantity of fluorine present in the sample determined by keeping the stoichiometry of the relevant reaction in mind [23]. If oxygen must also be determined the F_2 can be converted to Cl_2 using a conversion trap packed with a mixture of NaCl and NaF and held at 150 °C. After conversion the Cl_2 can be separated and back-flushed from a Kel-F column as described earlier while the permanent gases pass onto an analytical column for separation [24].

For very low concentrations of fluorine it can be reacted with hydrogen in the sample loop as described for chlorine but the reaction is one of the most exothermic reactions known and the hazards associated with this should be carefully considered. Fluorine will react with water to produce HF that can be determined by alternative techniques and will also react with most organic compounds to produce fluorinated analogues that are amenable to conventional GC analyses. Hydrofluoric acid is best determined using spectroscopic methods such as FTIR with an inert gas cell or if in aqueous solution, by ion chromatography.

11.10.1.3 Chlorine trifluoride and tungsten hexafluoride

These gases are used in the electronics industry for etching and vapour deposition respectively which places a high demand on their purity. Methods for analysing the impurities in these matrices were published by the authors [25,26]. To ensure an inert flow path, nickel tubing was used and was deactivated with ClF_3 that reacts with nickel oxide to form nickel fluoride and therefore forms an inert layer on the inner surfaces of the tubing. Both applications utilised nickel columns packed with 10% Kel-F on Chromosorb T as pre-columns to separate and back-flush the matrix. This allowed the separation of the permanent gas impurities using Molecular sieve columns and for WF_6, where CO_2 and SF_6 also needed to be determined, a column packed with Super Q was used on a second channel. Preference was given to packed columns to enable loading relatively large samples for the determination of low levels of the impurities. For the ClF_3 a nickel filament TCD was used and for the two channels used for WF_6, two PDHID's, necessitated by the impurity specification. Detection limits on the WF_6 system were below 10 ng/L with an RSD of less than 4% at sub μg/L levels and calibration remained stable for over a year at-line in a production facility (Fig. 11.16).

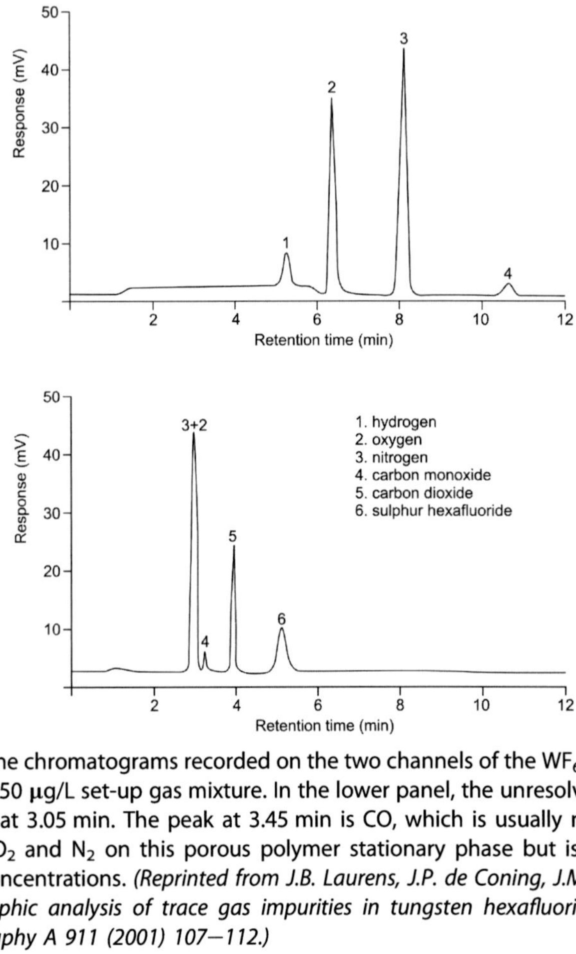

Fig. 11.16 The chromatograms recorded on the two channels of the WF_6 system using the nominal 50 μg/L set-up gas mixture. In the lower panel, the unresolved O_2 and N_2 peak eluted at 3.05 min. The peak at 3.45 min is CO, which is usually not fully separated from O_2 and N_2 on this porous polymer stationary phase but is separated at these low concentrations. *(Reprinted from J.B. Laurens, J.P. de Coning, J.M. Swinley, Gas chromatographic analysis of trace gas impurities in tungsten hexafluoride, Journal of Chromatography A 911 (2001) 107–112.)*

11.10.2 The inert halogenated gases

The first and major group of inert halogenated gases are the so-called chloro-fluoro-carbons or CFC's and there is a substantial amount of freely available literature on their analysis in air using direct and sorbent sampling, in water by headspace or purge-and-trap as well as other matrices. The vast amount of literature is due to the effect of the CFC's on the environment. While they have many trade names such as Freon and Arcton derived from their use as refrigerants, the use of the CFC number is most common as it gives an indication of the structure. The easiest way to obtain the structure from the CFC number is to add 90 to the number to obtain a three digit answer. The first digit refers to the number of carbon atoms, the

Table 11.4 Numbering convention for the CFC's.

CFC No	+90	C's	H's	F's	Cl's
11	101	1	0	1	3
12	102	1	0	2	2
22	112	1	1	2	1
23	113	1	1	3	0
32	122	1	2	2	0
113	203	2	0	3	3
142	232	2	3	2	1
143	233	2	3	3	0

second to the number of hydrogen atoms and the third to the number of fluorine atoms in the molecule. If the number of H's and F's does not add up to the valence of 4 of the carbon atoms, the remaining positions are occupied with chlorine atoms. A few examples are given in Table 11.4 [27]. In some cases as in CFC 113, two or more configurations are possible, for example CF_2Cl-$CFCl_2$ and CF_3-CCl_3; these are then numbered 113 and 113a respectively. Halons are fluorocarbons that contain bromine and no hydrogen and have a four digit number with the first to fourth digits referring to the number of C's, F's, Cl's and Br's respectively.

The choice of column would depend entirely on the analytical requirements and the matrix. Various columns have been used, such as Silica-PLOT, PLOT Q, Alumina-KCl, Alumina-Na_2SO_4 as well as non-polar and medium polarity WCOT columns with thick films, although these may require sub-ambient operation. Many of these compounds are also listed in the pollutants determined by EPA method TO-14 and others. Detection by TCD and FID is relatively common but for trace amounts the ECD would provide the lowest detection limits and highest sensitivity. The limited dynamic range of the ECD must be taken into account and can necessitate different sample sizes being used for different ranges of concentration. Using mass spectrometric detection provides additional confirmation, especially using the characteristic isotopic patterns of chlorine and bromine, while the use of MS-MS can improve greatly on the selectivity of the analysis.

Sulphur hexafluoride is often used as a tracer gas because of its detectability at low concentrations so the analysis of SF_6 is important and worth mentioning because of its interesting behaviour on a Molecular sieve column. On a Molecular sieve 5A column, SF_6 will elute very slowly with poor peak shape as it is held back very well by the pores in the stationary phase. On a Molecular sieve 4A, SF_6 elutes first as a very sharp peak and is not retained at all since the size of the SF_6 molecule is bigger than the pores in the stationary phase and it cannot therefore interact with the stationary phase. This makes it an ideal configuration for the determination of traces of SF_6 in most matrices which can easily be backflushed. However to determine impurities in SF_6, a Molecular sieve 4A column should not be used. For this application a Molecular sieve 5A column could be configured as a pre-column allowing back-flushing of the SF_6 matrix and other components such as CO_2, to a different column capable of achieving the required separation. For moderate concentrations a TCD would suffice; for trace analysis a PDHID can be used if other components must be determined as well and for trace, targeted analysis of SF_6, an ECD would be the detector of choice.

Nitrogen trifluoride is an oxidising gas used by several industries such as the manufacture of electronic components, high energy lasers and for the production of organic and inorganic fluorinated compounds. Once again, high purity is of the utmost importance and a detailed method using column sequence reversal was published by the authors [28]. This configuration substantially improved the peak shapes of compounds such as CO_2 and SF_6 that elute late when using heart-cutting of the matrix. Essentially column sequence reversal was used to change the elution of the matrix from the middle of the analysis to the last component eluted. To compensate for large concentration differences of some of the analytes, the range of the PDHID detector was switched to obtain peaks that do not saturate the detector.

There are many more of these rather exotic halogenated gases, such as trifluoromethyl sulfur pentafluoride (SF_5CF_3) and it is not possible to elaborate on the analysis of all of them, especially if the fluorinated hydrocarbon gases such as perfluorobutane (C_4F_{10}), perfluorocyclobutane (cC_4F_8) and the highly poisonous perfluoroisobutene (C_4F_8) are added to the list.

Clearly there are hundreds more applications in gas analysis but it is just not practical to try and add everything to this chapter. Some of the applications such as NOx's have purposely been omitted because they, with the exception of N_2O, are difficult to analyse by GC. Many further applications will be found on the Internet.

References

[1] Restek Corporation, Restek Searchable Chromatogram Library, Restek Corporation, Bellefonte, PA, USA, 2017 [Online].

[2] A. Brown, G. Varhga, M. Downey, N. Hart, G. Ferrier, K. Hall, Methods for the Analysis of Trace-Level Impurities in Hydrogen for Fuel Cell Applications. NPL Report AS 64, Queen's Printer and Controller of HMSO, Teddington, Middlesex, UK, 2011.

[3] G. Kaye, T. Laby, Tables of Physical and Chemical Constants, fifteenth ed., Longman, New York, 1986, p. 219.
[4] M. Trubyanov, G. Mochalov, S. Suvorov, E. Puzanov, A. Petukhov, I. Vorotyntsev, V. Vorotyntsev, Towards the interaction between calcium carbide and water during gas-chromatographic determination of trace moisture in ultra-high purity ammonia, Journal of Chromatography A 1560 (2018) 71–77.
[5] W. Thomson, J. Joule, On the thermal effects of fluids in motion, Philosophical Transactions of the Royal Society of London (Downloaded from: http://rstl.royalsocietypublishing.org/ on February 26, 2017) 1853 (143) (1853) 357–365.
[6] L. Felix, W. Farthing, J. Irvin, T. Snyder, Apparatus and Method for Maintaining Multi-Component Sample Gas Constituents in Vapor Phase during Sample Extraction and Cooling, May 11, 2010. United States Patent US 7,712,384 B2.
[7] International Standards Organisation, Natural Gas - Calculation of Calorific Values, Density, Relative Density and Wobbe Indices from Composition, vol. ISO 6976:2016, 2016, pp. 1–57.
[8] Bureau of Indian Standards, Natural Gas - Calculation of Calorific Values, Density, Relative Density and Wobbe Index from Composition, vol. IS 14504 (1998), 1998, pp. 1–49.
[9] Ametek Process Instruments, Heating Value Measurement of Natural Gas Using a Gas Chromatograph, 2014 [Online]. Available: www.chandlerengineering.com.
[10] A. Attari, D. Klass, Natural Gas Energy Measurement, Elsevier Applied Science Publishers Ltd, London, 2005.
[11] R. Morehead, J. de Zeeuw, J. Bromps, G. Stidsen, Column Selection for Speciation of Sulfur Compounds in Natural Gas, Restek Corporation, Bellefonte, PA 16823, not dated.
[12] J. Luong, L. Sieben, M. Fairhurst, J. de Zeeuw, Determination of low levels of formaldehyde and acetaldehyde by gas chromatography flame ionization detection with a nickel catalyst, Journal of High Resolution Chromatography 19 (1996) 591–594.
[13] M. Hunter, K. Bartle, A. Lewis, J. McQuaid, P. Myers, P. Seakins, C. Van Tilburg, The use of the helium ionization detector for gas chromatographic monitoring of trace atmospheric components, Journal of Separation Science 21 (2) (1998) 75–80.
[14] M. Hunter, K. Bartle, P. Seakins, A. Lewis, Direct measurement of atmospheric formaldehyde using gas chromatography-pulsed discharge ionisation detection, Analytical Communications 36 (1999) 101–104.
[15] ASTM F2013 - 10, Standard Test Method for Determination of Residual Acetaldehyde in Polyethylene Terephthalate Bottle Polymer Using an Automated Static Head-Space Sampling Device and a Capillary GC With a Flame Ionization Detector.
[16] Agilent Technologies, Inc, Application Note. Hydrocarbons, C1–C6. Trace Analysis of Acetaldehyde in Hydrocarbon: 'Symmetry of Acetaldehyde Peak', Agilent Technologies, Inc., USA, 2011.
[17] Y. Zou, Application Note: Coal-To-Chemical Process Gas Analysis Using Agilent J&W HP-PLOT Q PT and HP-PLOT U PT, Agilent Technologies, Inc., USA, 2013.
[18] J. Lin, R. Hendrickson, M. Ladisch, Gas chromatographic determination of dimethyl ether in a mixture containing propane, C4 hydrocarbons, n-pentane, methyl tert-butyl ether and methanol, Journal of Chromatography 367 (1986) 195–200.
[19] A. Vickers, Application Note: A "Solid" Alternative for Analyzing Oxygenated Hydrocarbons - Agilent's New Capillary GC PLOT Column, Agilent Technologies, USA, 2007.
[20] A. Hunt, J. Alder, Quantitative analysis of chlorine in air by gas chromatography, Analytical Communications 33 (1996) 61–64.

[21] E. Obermiller, G. Charlier, Gas chromatographic separation of hydrogen chloride, hydrogen sulfide, and water, Analytical Chemistry 39 (3) (1967) 396–397.
[22] Chrompack Application Note. Application 207 - GC, "Water in Ethylchloride" Varian Scanview 8 CD, Middelburg, NL.
[23] K. Hakuta, M. Aramaki, T. Suenaga, M. Kodama, H. Nakano, S. Nakagawa, Method for Analyzing Fluorine Containing Gases, May 21, 1991. USA Patent US5017499 A.
[24] R. DeGrazio, R. Auge, Gas Chromatographic Investigations for the Determination of Fluorine and Oxygen in Mixtures, Dow Chemical Company, Rocky Flats Division, Colorado, USA, 1967 (RFP-880).
[25] J. Laurens, J. Swinley, J. de Coning, Gas chromatographic analysis of trace impurities in chlorine trifluoride, Journal of Chromatography A 873 (2000) 229–235.
[26] J.B. Laurens, J.P. de Coning, J.M. Swinley, Gas chromatographic analysis of trace gas impurities in tungsten hexafluoride, Journal of Chromatography A 911 (2001) 107–112.
[27] N. Reuter, Nomenclature of CFC's/Freons/Halons/Coolants, Agilent Community, May 15, 2018 [Online]. Available: https://community.agilent.com/docs/DOC-2232-nomenclature-of-cfcsfreonshalonscoolants.
[28] J. de Coning, J. Swinley, Optimisation of a gas chromatographic method for trace gaseous impurities in nitrogen trifluoride by column sequence reversal, Journal of Chromatography A 1180 (2008) 151–158.

CHAPTER 12

Troubleshooting

Contents

We cannot stress enough that prevention is better than cure. Consistent preventive maintenance must be practiced such as exchanging spent in-line filter cartridges before contaminants migrate into the analytical system. For troubleshooting specific detector problems please refer to Chapter 8.

It is rather difficult to generalise when it comes to finding faults but the rules that must be followed to ensure success are well established. First and foremost never attempt to diagnose any problem before it can be repeated. Troubleshooting involves methodical detective work arising from some clues that are not always obvious. Like any good detective story, false accusations must be avoided as they will be totally misleading. A lot of time

A practical guide to gas analysis by gas chromatography
ISBN 978-0-12-818888-0
https://doi.org/10.1016/B978-0-12-818888-0.00012-7

can be wasted hunting problems within the hardware, when the problem is actually due to the sample that is being analysed. The biggest danger in jumping to conclusions or making the wrong diagnosis, is that one stops looking for the real problem and starts treating symptoms rather than the root cause. There will be some occasions when it is blatantly obvious what is wrong such as an empty gas cylinder but this is the exception. If the system fails to switch on there are only two possible reasons; there is no power at the socket or the instrument has an electrical problem. This may be as simple as the replacement of a fuse but there is a good chance that this is an early warning sign for a more serious problem.

The first thing that should be eliminated is operator errors by referring to standard operating procedures and the instrument log book. Check and measure all operating parameters from flows to temperatures to ensure that correct procedures are adhered to. Even if the digital readout shows 100 mL/min check this flow with your own flow meter. Eliminate sampling changes by running standards and comparing these runs to those on file in the log book.

To efficiently identify problems the two most important requirements are good documentation coupled with a strictly systematic approach. We have advocated customising dedicated systems for specific applications and the only one who knows how that system has been assembled is the builder of that system. A plumbing diagram must be available before anyone else can even attempt to troubleshoot such a system. Clearly the system functions as a whole although it is made up of many interrelated units. Systematic troubleshooting allows these constituent parts to be isolated as far as practical and tested individually where possible. Any troubleshooting exercise also requires a large serving of common sense but the biggest problem with common sense is that it is rather uncommon. Today's sophisticated computer controlled systems can be a lot more difficult to troubleshoot than older analogue systems because the software is attempting to correct some portion of the problem and could be masking the real problem. However advanced the system control, certain parameters must be correctly set for optimum operation. For example configuring the software with carrier gas as nitrogen but plumbing the system with helium could allow the system to go to the ready state but all settings will be inaccurate. This will appear to manifest itself as though the electronic flow control is faulty whereas it is purely an operator error. It is vitally important that we must search for causes and not be misled by symptoms.

Another basic necessity is to change only one thing at a time and to identify the real problem and not create a lot of unnecessary and irrelevant repair work. In the past, it was expected that discrete components would be replaced on circuit boards by an electronics technician equipped with a soldering iron. In the 21st century, multilayer circuit boards and the high level of computer control means that no components other than software chips can be exchanged. Only board replacement can be done and this by a technician equipped with earthing straps on their wrists to avoid static charges on their hands. Since this is an expensive exchange, it is really important that the diagnosis is accurate. Never work on any electrical or electronic part with the power on; not only can you cause irreparable damage to circuitry by creating short circuits but you may also cause similar damage to yourself from burns or electric shock. It is also vital that one understands what can or cannot be disassembled or repaired to avoid damage to electronic or mechanical parts. As an example, it would be foolhardy to open the radioactive source of a detector.

The first sign that something has changed will be in the quality of the results. This would be confirmed by running a standard sample repeatedly and confirming the observed change. Some very small and seemingly trivial changes may affect results. Do not limit your search to obvious changes in the laboratory, but look for changes outside the laboratory as well. Electrical interference from air conditioners or elevator motors can influence the laboratory noise levels. Changing laboratory temperature between the day and night will create a bias in results. Could this be changed by the fact that at a certain time of the year sun shines directly on the equipment in the afternoon? Check out all the things that may have changed since the last set of good results. Has anything changed in the way the sample is taken? Have there been any changes of staff? Has that staff member been absent and just returned to the laboratory? Have new gas cylinders been installed?

As stated previously, on any system that is equipped with any valving more complex than a sampling valve, a plumbing diagram is imperative before any troubleshooting can be done. It is important to remember that valve systems have multiple pathways depending on which valve is switched in which position.

The most effective way to identify the source of the problem is to eliminate as much of the system as possible. The 'divide and conquer' technique will help to focus attention on the part of the system causing the problem. Ideally one would split every system and sub-system into two equal parts and eliminate 50% each time. As an analogy consider driving along and a noise starts in the vehicle. Is it from the front or the rear? Is it on the left side or on the right side? Quickly the problem has been isolated away from 75% of the whole.

On complex systems this may not always be practical but there are quick checks that help. For example, when no peaks appear on the data system we can divide the system by heating the column oven and monitoring the background signal of the detector. If this can be observed on the detector read-out but not the data system then the problem lies in the data handling side and not with the detector or column or injection. Conversely, if no change is shown on the detector read-out then there is a problem with the detector or the flow to the detector.

Another aid in identifying the cause of a problem is asking the 'if … then …' question. For example, if we increase a flow do we observe the correct increase in pressure? If we increase column oven temperature on a flow controlled system does the column head pressure increase accordingly?

12.1 System pressure monitoring

A lot of valuable information can be gleaned from observing pressure changes. As discussed in Chapter 2, one of the initial leak checks done is to pressurise the system, then close the cylinder valve and monitor the pressure drop of the gauge measuring cylinder pressure. Unless extreme flows are being used, the small volume of gas in the regulator should deliver gas to the system for 10 min or more. This test can be repeated after each valve has been switched. Although time consuming, this is very valuable information and quickly highlights one leaking pathway. Care must be taken to ensure that the sample loop is not filled with sample while doing these tests.

Depending on the plumbing of the system, further diagnostic information may be available on screen. For example, in a column sequence reversal system only one flow controller is used. In a leak tight system the pressure and flow must be identical in either valve position provided the temperature is kept constant.

If you suspect that a valve is switching slowly or erratically it is also possible to monitor the recovery rate of the pressure after the valve switch.

Detailed information regarding leak checking is provided in Section 2.3.6.

12.2 Baseline effects

The value of running blanks, that is, injecting nothing but pure carrier gas, cannot be overestimated. Not only does it provide confirmation that the valves are switching when they should, that the system is leak free, the flow path is not contaminated and that the detector is working, but it proves useful diagnostic information if everything turns out to be less than perfect.

12.2.1 Ghost peaks (blanks)

It is always a good start in any analysis to start with a blank. One of the toughest challenges in chromatography is to inject nothing and see nothing. In gas analysis this is particularly difficult as can be seen from Fig. 12.1. Even when we evacuate air from the sample loop to a good vacuum, there is still at least a few hundred ppm of residual nitrogen and oxygen in the loop. However there are some ways to try to simulate this situation. For example, the helium vented from the PDHID is very pure when no analyte is present so it can be used to purge the sample loop. When injected there should only be small peaks in the blank chromatogram.

Another important test for sampling valves is to set up a run that injects immediately after the valve has been switched to the load position. There must be no sample flow. Since the sample loop is full of carrier gas when switched to the load position and this carrier is again injected, a blank run should be obtained. If the valve is allowed to stand in the load position the carrier gas in the loop will gradually be replaced by air. This is also a good check against carry-over from one sample to the next.

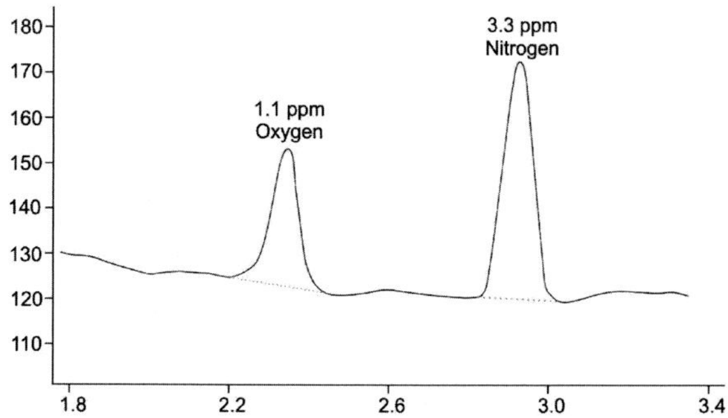

Fig. 12.1 Chromatogram of commercial helium 5.0 analysed on a custom built gas chromatograph. The same helium is used as carrier and ionisation gas after purification.

It is also very important to check the complete sampling system for carry-over. Samples are collected in all manner of containers and these must be purged between samples. Some form of valve is involved in containing the sample and many valves use elastomer seals which can have memory effects. Even gas tight syringes that use Teflon seals are prone to memory effects.

Another potential source of ghost peaks is decomposition of one or more of the compounds in the sample. Although more prevalent in GC analyses where high temperatures are used, it can be observed in gas analyses, especially for gases that can be converted from one form into another such as some of the NOx's.

With method development it is not uncommon to see a peak that appears to elute too early, in other words faster than the unretained peak. From Chapter 2 this should not be possible and this peak is also broader than a peak should be without retention. The reason for this is that it is a late eluting peak from the previous injection.

If a split injector is used remember that there is an in-line filter in the split vent line and the majority of sample exits through this. When saturated there will be backstreaming of compounds from this filter and the peak size will most probably be dependent on the time the system has idled between injections. To confirm this contamination the filter needs to be heated a few degrees and then the ghost peaks will increase significantly. These filters can be baked out as they are essentially just a small packed column but if this facility is not readily available they should be replaced.

12.2.2 Noise

All measuring systems must have a certain level of noise. A perfectly flat and straight baseline on any GC detector is a fault condition for example an incorrect zero setting. Background and noise measurements should be taken at regular intervals and records maintained. When the system signal is amplified there will be signs of short term noise as well as drift and wander. Even with the detector fuel gases switched off there will still be electronic noise visible and this must be lower noise than when the detector is operational. Different types of noise are shown in Fig. 9.2. It is always useful to know when these levels have changed. Has this been a gradual change or a step change? Did it occur after a change of gas cylinder? Are there any other systems sharing the gas supply and have they similar symptoms? The frequency of noise is rather important. Short term noise is typically a problem at the detector end of the system since any disturbance on the injector side of the column will be attenuated by the delay through the column.

Air leaks will affect some detectors such as TCD or ECD but will not have a noticeable influence on a FID or FPD. A dirty injector, splitter trap or contaminated column may contribute to the noise and background as would a contaminated or defective detector. The incorrect ratio of combustion gases and inserting the column the incorrect distance into the detector could also increase the noise.

12.2.3 Spikes

A fairly common problem is the appearance of spikes in the chromatogram in random positions especially if the system has not been used for some time. Spikes appear as a peak without width, in other words as a line that may be positive or negative and can be easily distinguished from peaks as shown in Fig. 9.3. They are usually caused by particulate matter passing through the detector either from dirty gas, dirty detector or column material. It is not uncommon for spikes to be present after an instrument has been shut down for a period of time or on a new system. The spike frequency will reduce with time and eventually disappear. External influence from other equipment sharing the same electrical supply as the GC is quite common and rather difficult to isolate. It may be difficult to operate at certain times of day when certain other users are loading the electrical supply. Spikes can also show on flow sensitive detectors due to the opening or closing of the laboratory door and hence disturbing the ambient pressure in the laboratory. Electrical spikes can also be generated by any poor connection especially on the detector and data system side of the GC. After eliminating all of the above causes, it might actually be due to an electronic problem on the detector controller.

12.2.4 Baseline shifts

Ideally the chromatographic trace should not show any steps when valves are switched and well-designed systems are constructed to try and minimise this effect. As with noise, a perfectly smooth baseline is an indication that things are not working properly.

Many of the detectors used are flow sensitive and even the most rapid electronic flow controller is unable to correct for valve switches. For some short time during the valve switch the flow is actually blocked as the rotor is turning. Large baseline steps are a good indicator of flow changes. For example, if a TCD system shows a baseline step after injection, there is almost certainly a leak on the sample loop or on the valve itself. If a column switch shows a large baseline change the quality of carrier gas through one of the columns is contaminated. This could be due to a leak, column bleed or difference in the carrier and auxiliary gas quality. Such a step is shown in Fig. 12.2.

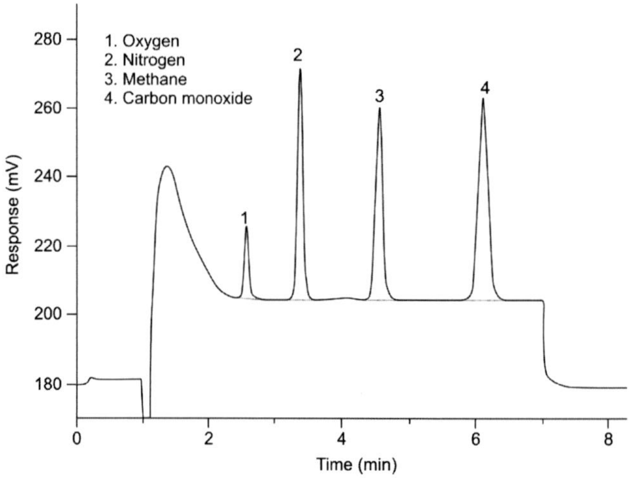

Fig. 12.2 Example of the flow disturbance due to a valve switch and the increased baseline due to differences in the carrier gas purity or flow before and after the valve switch. *(Reprinted from J.P. de Coning, J.M. Swinley, Optimisation of a gas chromatographic method for trace gaseous impurities in nitrogen trifluoride by column sequence reversal, Journal of Chromatography A 1180 (2008) 151–158.)*

With multidimensional systems one is sometimes obliged to accept a certain amount of baseline shift after a valve switch. For example, consider a system that is plumbed for backflush-to-vent using a 10-port valve for injection. As the 200 μL sample is injected the whole 2 mL of carrier gas in the pre-column is also injected. Since this is ten times larger than the actual sample, this will cause some baseline shift if this gas is not very clean or that leg of the system is not totally gas tight.

12.2.5 High background signals

Assuming that a logbook has been maintained, the normal value of the detector background will be on file over a period of time. This information can easily be lost if only chromatograms or method print-outs are retained for record purposes as the system will automatically zero the background and the actual background signal will not be known. It is important to know whether the increased background signals have been gradual or sudden. Gradual changes may be rather common in that normal wear and tear is taking place, for example the electrodes of the PDHID require periodic exchange. Please refer to Chapter 8 for detector information.

A sudden increase of background signal is something that requires immediate attention. Something must have changed recently such as a cylinder of gas, gas line, the composition of the sample or an extended power outage. This could well be caused by an in-line gas filter reaching saturation and there is a breakthrough of contaminants. Some detectors require a long warm-up period and patience is required. If no such sudden change can be identified, good detective work is required. For example, if a small column oven temperature change makes a large change in signal the column is contaminated and may be cleaned with a bake-out. Alternatively, if a 5% flow change in one of the fuel gases causes a major change in background signal, that gas or gas line will probably be contaminated. A heat gun may well allow one to pinpoint this problem.

12.3 Peak shape

The shapes of the peak in a chromatogram are probably the single most important diagnostic tool for the experienced chromatographer. Interpreting the defects in an analysis as shown in the chromatogram is a skill that becomes easier with practice. A fundamental understanding of chromatography theory and of the system configuration is essential. Being observant, curious, critical and able to explain what happens and why, helps a lot.

12.3.1 Tailing

While we know that the ideal peak shape will be Gaussian, a certain amount of peak shape distortion will be regarded as normal due to column overload or activity of some analytes. Depending on the types of columns used, front tailing or back tailing must be expected when there is some overload. Front tailing shows as a slowly increasing signal which returns rapidly to baseline. Obviously the asymmetry of such as a peak will be considerable less than 1. Front tailing predominantly occurs with sample overload on WCOT columns coated with liquid phases. The quickest way to confirm this is to inject less sample or a lower concentration of that compound. When practical, this can be as simple as diluting the sample in a gas-tight syringe using air or carrier gas as diluent or increasing the split if the injector is part of the flow path. In practice few peaks have an asymmetry of 1 as in perfect Gaussian peaks but an asymmetry of 0.8—1.2, determined as in Fig. 1.12, must be aimed at.

Back tailing can be caused by small dead volumes in the plumbing of the system which will be lowest on the earliest eluting peaks. However, they are also characteristic of sample overload on adsorption columns such as PLOT and micro-packed columns. These peaks will have an asymmetry >1. The difference between front and back tailing is illustrated in Fig. 12.3. When using alumina columns tailing will show somewhat more frequently than on other columns.

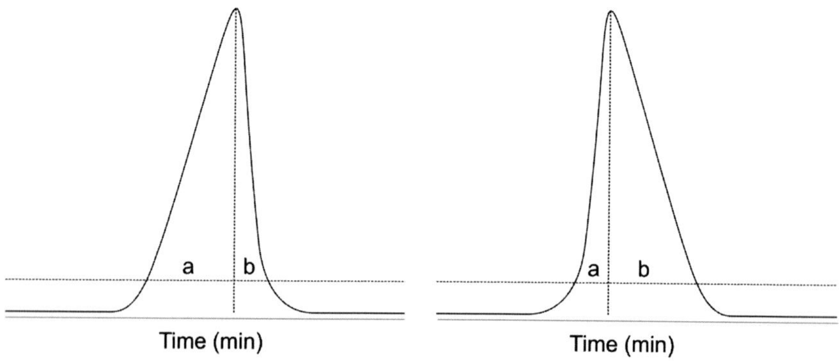

Fig. 12.3 Front and back tailing as seen when overloading WCOT and PLOT columns respectively.

Asymmetry is often observed when analysing polar compounds on a non-polar column where even very small peaks of polar compounds will still show symptoms of overloading due to the limited capacity of the non-polar phase to dissolve the polar compounds. When looking at asymmetric peaks bear in mind that detector sensitivity is not equal for all compounds. A small peak can show severe tailing due to overload because that compound is relatively insensitive on the detector.

12.3.2 Flat topped and rounded peaks

A common desire in gas analysis is injecting more sample to improve sensitivity. This is done by increasing the size of the sample loop or by increasing the sample pressure in the loop. This often leads to flat topped peaks that do not increase in height according to the change in sample amount as was demonstrated in Fig. 6.15. Some detectors are especially prone to overload because of their limited dynamic range.

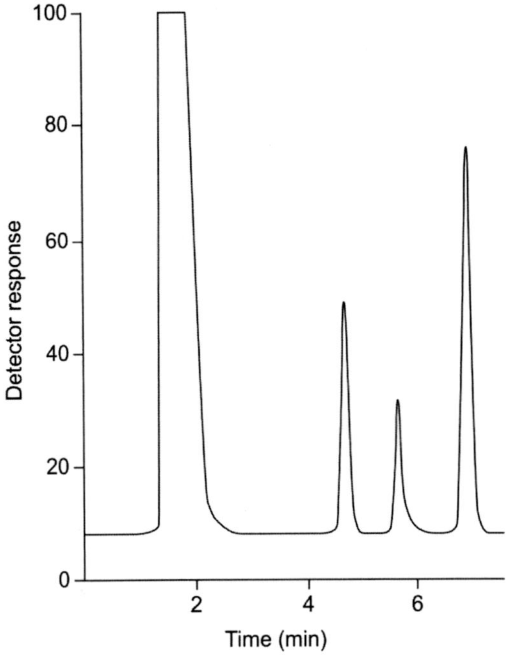

Fig. 12.4 The first peak has a flat top meaning that its concentration was too high and saturated the detector or data system.

Data systems also have a limited dynamic range which is determined by the number of bits in the analogue to digital converter (ADC). Many systems have a maximum analogue input voltage of 1000 mV which means that all peak heights must be less than 1000 mV. Protective circuitry is employed to prevent damage to the ADC so any signal greater than 1000 mV is cut off at this value. The only conclusion one can draw from such a flat topped peak, as in Fig. 12.4, is that the true area is greater than the area reported. Most detectors have range settings which must be used in conjunction with the amount of sample to ensure that the ADC does not go out of range while some modern systems have an auto ranging feature that can prevent this.

12.3.3 Negative peaks

Negative peaks can be expected in certain cases especially when valve switching is used and therefore can often be regarded as normal. Certain detectors can give erroneous negative peaks such as a PDHID giving negative peaks where the impurity in the carrier gas is higher than the level

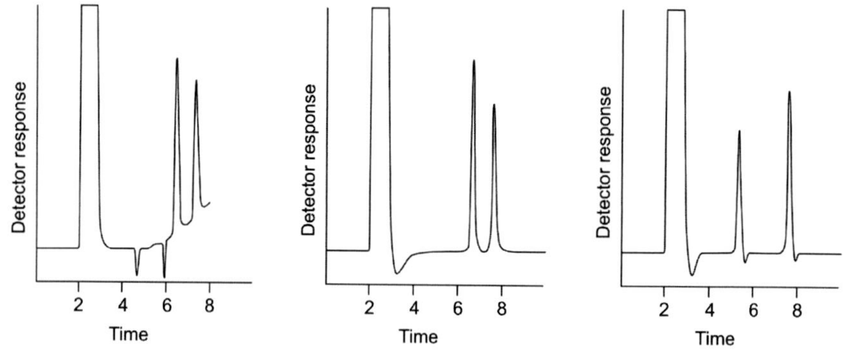

Fig. 12.5 Examples of negative peaks caused by impurities (left) and the negative dip after an overloaded peak (middle) and negative dips at all peaks as would be observed with a dirty ECD (right).

in the sample. Other detectors may give a negative dip before or after a peak if not working correctly. For example a dirty ECD will show a negative dip after each peak as shown in Fig. 12.5. The above negative peaks are error conditions that should be rectified but need to be taken cognisance of when integrating. Very rarely it may be necessary to integrate negative peaks as it is easy to time program polarity switching, do baseline zeroing and integrating the positive response. Remember that the ADC typically has a range of only −5 mV but +1000 mV.

12.3.4 Split peaks

In gas analysis it is rather rare to produce split peaks unless the actuator is erratic because valve injection is usually highly repeatable. However it easily happens when trying to inject a large volume directly with the gas tight syringe. The syringe plunger is pressed down upon injection but the carrier gas pressure causes the plunger to shoot back up and when pressed down again two injections are made and all peaks will appear as doublets as shown in Fig. 12.6. With valve injections, the only time split peaks are visible is with serious detector overload especially on very sensitive detectors.

12.3.5 Loss of all peaks

First look for the obvious. Is the column connected to the correct detector? Is the data system monitoring the correct channel? Is the detector switched on and, if applicable, has it been ignited?

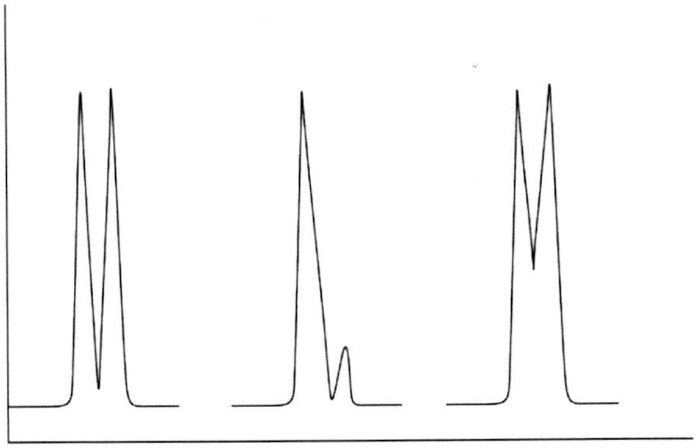

Fig. 12.6 Examples of split peaks.

As already stated, one quick check is to heat the column oven and watch the detector background signal change. If this can be seen on the detector read-out but not on the data system, the data system is not processing the correct signal. This will most likely be only a software change to correct this. If the detector and data system respond to this temperature change then it is probable that the injection valve is not injecting sample. If a full analytical cycle has been interrupted for any reason there is a good chance that the sampling valve was not returned to the load position.

It is not always possible to see if the valve actuator of the sample valve has switched fully. Many of the valve actuators can be manually overridden to force immediate switching of the valve. Verify that the correct sound is made when the actuator should switch and refer to Chapter 5 on valves for realignment procedures. It can occur that the internal mechanical coupling between the valve and actuator can break and this can only be found by disconnecting the valve and actuator.

A very frustrating but fairly common cause of total peak loss is a broken column or a column connection, such as a Quickfit coupling, that came loose. This is one of those problems that is so obvious that it is often overlooked. Fused silica columns can break when the polyamide coating is rubbed off by the column vibrating against the side wall of the oven due to the turbulence of the oven fan. With time and high temperature usage, fused silica does tend to become a little brittle and can break. Monitoring the pressure during a run is therefore important as it would provide a clue that the column is broken although it may not indicate if a particle trap that was connected to the end of a column, just before the detector, came loose.

12.3.6 Loss of resolution

Resolution loss is nearly always related to a change in retention time and a change in the properties of the column. Some columns such as Molecular sieve are known to gradually change polarity with the absorption of water vapour which adversely changes their separating capability. The typical symptom observed is poor separation between oxygen and nitrogen or reduced methane and carbon monoxide separation. This can be corrected by appropriate bake-out. If the column oven temperature is limited due to the presence of valves or other columns, it will be necessary to remove the column and bake in another oven capable of the required temperature. A similar change in polarity is well known for alumina columns, Fig. 12.7, and the water vapour can be baked out without damage to the column [2]. Alumina PLOT columns also have limited sample capacity and overloading can be confused with loss of resolution.

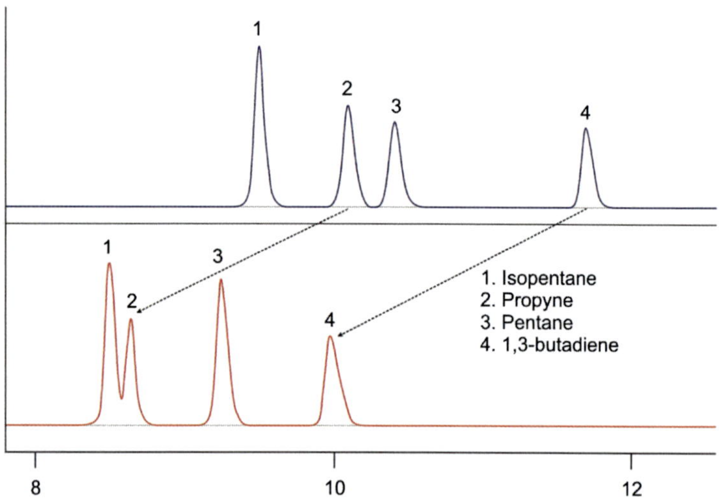

Fig. 12.7 Partial chromatograms showing the effect of water on the peak position on an Alumina KCl column; it clearly becomes less polar when water is adsorbed. *(Figure reproduced with permission of American Laboratory. J. de Zeeuw, R. Morehead, T. Vezza, B. Bromps, Chromatographic Behavior of Activated Alumina Adsorbents for the Analysis of Hydrocarbons, October 01, 2011 [Online]. Available: http://www.americanlaboratory.com/913-Technical-Articles/35226-Chromatographic-Behavior-of-Activated-Alumina-Adsorbents-for-the-Analysis-of-Hydrocarbons/.)*

12.4 Quantitation problems

Irreproducibility problems are among the most difficult to pinpoint since any one or several parts of the overall system are malfunctioning. We must initially turn to blank runs for some clues. Again we must emphasise the systematic approach.

One of the first diagnostics is to analyse the results and decide whether the problems relate to variations in retention times or in peak areas or heights or in both simultaneously. This will give some indication of where to start. Differing concentrations in sample composition and matrix will cause some changes in run to run reproducibility but the differences discussed here are larger and more random.

12.4.1 Peak area or height changes

If only the peak area is changing but not retention times, the system probably has a problem on the sample loop side of the sampling valve. This may be a valve leak but this would most probably cause a retention time change as well. More likely would be a problem in the sampling system. All inlets will be fitted with a particulate filter of some type and partial blockage can cause such problems. In practice it is best to replace a filter than to attempt cleaning although some fittings have replaceable filter elements.

Another possible source of error could be a leak at the detector or column interface or minor changes in fuel gas or make-up gas flows. On modern instruments these can be monitored on screen while a flow meter may be needed on older systems.

It is also important to confirm whether peak areas and peak heights vary in unison. If one is better than the other it may indicate that the integration parameters have not been optimised.

12.4.2 Retention time fluctuations

If only retention time is affected the chances are good that either the column oven temperature is not repeatable or the column is aging. If the relative reproducibility of retention times is worse on early eluting peaks this will be an indication that the column initial temperature is not stable. Columns are generally made from poor thermal conductors and after cool down at the end of a temperature program, equilibrium time must be allowed for the inside column material to reach oven temperature. Many GC systems have very rapid oven cool down and therefore have default of

2–3 min for equilibrating these temperatures. However, this can be changed in the method to a fraction of a minute to speed up the cycle time. Although the readout of oven temperature looks totally correct, the actual temperature of the stationary phase inside the column can vary significantly from run to run. This problem can also be closely related to small changes in the ambient temperature in the laboratory.

Another factor that will affect retention time is the amount and composition of the matrix gas. If a large sample of a very dense matrix gas such as sulphur hexafluoride or nitrogen trifluoride is injected, this matrix will drastically change the density of the carrier gas causing all peaks to elute earlier. This could be as much as a 5% change between a standard in helium and the actual sample.

A slight but continuous drift in retention time in either direction is rather common and can often be corrected by baking of the column. However, rapid drift or irreversible drift is likely to be a warning sign that there are compatibility problems between the sample and column material.

12.4.3 Simultaneous area and retention time changes

If the problem affects the peak area and retention time, the problem will most likely be on the column side of the injection valve. This could cover the valve or any connections between the valve and columns. If the system involves column switching, then there could be any number of problems relating to the overall system. Again it will be necessary to try and isolate the problem to a smaller part of the system. For example, find a sample that does not require back-flushing and inject the sample several times on the two columns in series.

To investigate the influence of data acquisition on the results, the questions that must be asked include: are peak heights more repeatable than peak areas? Are retention times repeatable? Are peaks fully separated? Are baseline corrections as calculated by the data system the same for standard and sample? If the standard has a tangential baseline correction but the sample has a vertical split correction, results will vary. Likewise, if higher concentrations of sample cause different baseline allocations the results will be suspect. Another thing to check is to make sure that the time windows for peak identification do not overlap as overlapping windows force the data system to make decisions for peak identification. A good understanding of Chapter 9 on data handling is essential.

12.4.4 Evaluating calibration results

Once the system has been calibrated and the calibration data reprocessed as though they are samples it is important that one does a critical inspection of the results. This is typically what is needed in a forensic evaluation of these values. In Chapter 8 we discussed how detectors respond differently to certain compounds and using the regression results one must confirm that the calculated relative responses meet the criteria introduced. For example using argon or nitrogen as carrier gas on a TCD, the response of hydrogen and helium must be similar and at least ten times larger than for any other gas. On a standard FID the responses must be in proportion to the number of carbons and must decrease as the level of oxidation increases. For example, formaldehyde response must be much less than the response for methane although both compounds only have one carbon. The methaniser FID must give very similar responses for methane, carbon monoxide and carbon dioxide. As shown in example Section 11.1.5 when the calibration covers more than one range the responses for each range must be similar. The PDHID should give similar responses for all gases except neon although there will be some discrimination between hydrogen and oxygen or nitrogen. All Flame Photometric Detectors must give a quadratic response for sulphur compounds. For most other detectors there are similar 'common sense' rules that apply. If there are any anomalies in these values then it is necessary that peak identification is confirmed and that the sampling and inlet system be checked for leaks and carry-over. If no problems are found it could also be an indication that there may be problems with the standard used for calibration.

12.5 Low column lifetime

The expected life of any column is difficult to specify since it will depend on the type of column and particularly on the constituents of the sample. Poisoning of stationary phases by some sample components can reduce column life significantly. In backflush or heart-cutting systems this will happen if switching times have not been correctly set whereby it will allow the wrong sample components to migrate onto a particular column.

Carrier gas purity will also play a significant role in determining column lifetime. Nearly all stationary phases can be affected by the presence of oxygen at elevated temperature. Since the carrier gas flows continuously, even small impurities add up to large amounts of impurity over a period of time. This is often more significant than the same impurity injected with the sample on a discrete basis. Guidelines for gas purification were discussed in Chapter 2.

In a laboratory where columns are changed often to accommodate different analysis requests, it is good practice to have a column log for each column. This log should indicate when it was used and for what type of sample, what T_{max} was used, how it was baked out and how many samples were analysed. This can prove invaluable if the column does not perform as it should when next used. These columns must be flushed with an inert gas and tightly capped when not in use.

12.6 What performance is acceptable?

For clarity we define repeatability as the precision obtained by one operator using one analytical system whereas reproducibility is the precision obtained by multiple operators using multiple systems. Repeatability will always be better than reproducibility but the latter provides a better estimation of the long term performance of a method.

One would typically connect a medium concentration standard or known sample and allow multiple injections. Since this is a best case scenario, a lot of effort must be expended to get these values absolutely repeatable. Depending on the sample and detector used one should aim at an area repeatability of less than 0.5% relative. On a modern GC, retention time repeatability should be better than 0.1% relative. Once satisfied at this level other concentrations of the standard can be run to ensure that the repeatability is acceptable at all concentrations. Clearly the lower concentrations which result in smaller signals will always be slightly less repeatable.

12.7 Tools of the trade

As previously indicated this is a list of equipment that is recommended for the competent analyst to troubleshoot their analytical system but not enough to be able to construct a system from scratch. Also refer to the tools mentioned in Chapter 2 on the gas analysis laboratory.

1. A set of imperial spanners is imperative. 1/4″ fittings require 9/16″ spanners and not 14 mm, 1/8″ fittings require 7/16″ spanners and not 11 mm and 1/16″ fittings require 1/4″ spanners. Also 1/2 ″ not 13 mm spanners are needed on some fittings. The use of the correct sizes certainly prolongs the lifetime of any nut without rounding of the

hexagon. Ideally it is useful to have two of each as one needs a counter wrench when opening or closing fittings. Adjustable wrenches are not recommended as the play in the jaws will cause some damage to the nuts.

2. Actuators and clamping rings use imperial sized hexagon headed screws that require a set of imperial Allen keys. It is not possible to find a metric key that will work on a 7/64″ hexagon headed screw without damaging it. A small set is all that is required.
3. Some of the modern instruments have their covers fastened with screws having Torx heads. Generally these GC's are supplied with one or two Torx spanners of the correct size, usually T20. A full set of Torx wrenches can be obtained in most hardware shops.
4. Leak detector. An electronic leak detector is preferable because it has the advantage that hot connections can be tested in situ. For cool connections and for air or nitrogen leak checking use a 50:50 solution of isopropanol in water and not any soap based solution.
5. Flow meter. Again a digital meter is preferable as it allows direct measurement of flows of all the gases used. It must be used to confirm that electronic flow controllers are operating correctly. A bubble flow meter can also be used for flow measurements. Although this is more tedious, it is probably more accurate than the digital meter that should be calibrated annually. For highly diffusive gases such as hydrogen the soap bubble will be a little permeable resulting in a slightly low flow measured value. Flow meters are often used to check fuel gas flow ratios where absolute accuracy is of limited concern
6. Heat gun. A normal hair dryer could be used but it is preferable to have a heat gun with a power rating of at least 1500 W for heating gas lines and observing changes in signal levels.
7. A syringe reserved for troubleshooting preferably with a small test mixture for this purpose. This could be as simple as air or lighter gas depending on the application being done.
8. Instrument manuals. These may well take the 'help' file format as part of the software. This is very general information concerning the make and model of your GC.
9. Full documentation with updated plumbing diagrams. This covers all specific adaptations of the basic GC to perform your application and must be readily available.
10. A log book per instrument. This records all historic data concerning the performance of each of these systems in your environment. This

allows checking for trends and also to record preventive maintenance history. Many GC's have settable counters that will warn the operator when this maintenance is due. This is also a good place to record any changes that are made to operating parameters. As recommended previously it may be necessary to also keep logs for each column.

11. A tubing cutter capable of cutting $^1/_8''$ and $^1/_{16}''$ tubing can be of value when it is necessary to replace ferrules that no longer seal. Micropacked columns cannot be trimmed in this way as metal retainers are used.
12. Trace gas analysis systems use crimped tubing as flow controllers and users of these systems require a crimping tool to be able to adjust flows.

12.8 Concluding remarks

Finally, to reiterate and summarise, keep good records of routine maintenance, operating parameters and samples analysed. Identify and isolate the problem and do not overlook what is obvious including any recent changes to the system or sampling procedures.

Read the manual and help files before everything else fails and discuss problems with knowledgeable colleagues as well as the instrument manufacturers. When getting technical support from the suppliers, stay close to the service engineer, ask questions and understand what went wrong, why and how it is being fixed and if it could have been prevented.

While most suppliers offer free technical advice there are a multitude of useful sources on the Internet varying from application notes, blogs, user forums and other publications. Keep track of what you found – it may come in useful again later, especially when you have to hand over to, or train another analytical scientist, prepare a report, a presentation or a publication of your own.

References

[1] J.P. de Coning, J.M. Swinley, Optimisation of a gas chromatographic method for trace gaseous impurities in nitrogen trifluoride by column sequence reversal, Journal of Chromatography A 1180 (2008) 151–158.

[2] J. de Zeeuw, R. Morehead, T. Vezza, B. Bromps, Chromatographic Behavior of Activated Alumina Adsorbents for the Analysis of Hydrocarbons, October 01, 2011 [Online]. Available: http://www.americanlaboratory.com/913-Technical-Articles/35226-Chromatographic-Behavior-of-Activated-Alumina-Adsorbents-for-the-Analysis-of-Hydrocarbons/.

Index

Note: 'Page numbers followed by "f" indicate figures, "t" indicate tables.'

D

I

K

L

M

Made in the USA
Las Vegas, NV
13 May 2026

46925064R00264